X. Yu, J. Kacprzyk (Eds.)

Applied Decision Support with Soft Computing

Springer

Berlin
Heidelberg
New York
Hong Kong
London
Milano
Paris
Tokyo

Studies in Fuzziness and Soft Computing, Volume 124
http://www.springer.de/cgi-bin/search_book.pl?series=2941

Editor-in-chief
Prof. Janusz Kacprzyk
Systems Research Institute
Polish Academy of Sciences
ul. Newelska 6
01-447 Warsaw
Poland
E-mail: kacprzyk@ibspan.waw.pl

Further volumes of this series can be found on our homepage

Vol. 108. L.M. Sztandera and C. Pastore (Eds.)
Soft Computing in Textile Sciences, 2003
ISBN 3-7908-1512-8

Vol. 109. R.J. Duro, J. Santos and M. Graña (Eds.)
Biologically Inspired Robot Behavior Engineering, 2003
ISBN 3-7908-1513-6

Vol. 110. E. Fink l. 112. Y. Jin
Advanced Fuzzy Systems Design and Applications, 2003
ISBN 3-7908-1523-3

Vol. 111. P.S. Szcepaniak, J. Segovia, J. Kacprzyk and L.A. Zadeh (Eds.)
Intelligent Exploration of the Web, 2003
ISBN 3-7908-1529-2

Vol. 112. Y. Jin
Advanced Fuzzy Systems Design and Applications, 2003
ISBN 3-7908-1537-3

Vol. 113. A. Abraham, L.C. Jain and J. Kacprzyk (Eds.)
Recent Advances in Intelligent Paradigms and Applications", 2003
ISBN 3-7908-1538-1

Vol. 114. M. Fitting and E. Orowska (Eds.)
Beyond Two: Theory and Applications of Multiple Valued Logic, 2003
ISBN 3-7908-1541-1

Vol. 115. J.J. Buckley
Fuzzy Probabilities, 2003
ISBN 3-7908-1542-X

Vol. 116. C. Zhou, D. Maravall and D. Ruan (Eds.)
Autonomous Robotic Systems, 2003
ISBN 3-7908-1546-2

Vol 117. O. Castillo, P. Melin
Soft Computing and Fractal Theory for Intelligent Manufacturing, 2003
ISBN 3-7908-1547-0

Vol. 118. M. Wygralak
Cardinalities of Fuzzy Sets, 2003
ISBN 3-540-00337-1

Vol. 119. Karmeshu (Ed.)
Entropy Measures, Maximum Entropy Principle and Emerging Applications, 2003
ISBN 3-540-00242-1

Vol. 120. H.M. Cartwright, L.M. Sztandera (Eds.)
Soft Computing Approaches in Chemistry, 2003
ISBN 3-540-00245-6

Vol. 121. J. Lee (Ed.)
Software Engineering with Computational Intelligence, 2003
ISBN 3-540-00472-6

Vol. 122. M. Nachtegael, D. Van der Weken, D. Van de Ville and E.E. Kerre (Eds.)
Fuzzy Filters for Image Processing, 2003
ISBN 3-540-00465-3

Vol. 123. V. Torra (Ed.)
Information Fusion in Data Mining, 2003
ISBN 3-540-00676-1

Vol. 125. M. Inuiguchi, S. Hirano and S. Tsumoto (Eds.)
Rough Set Theory and Granular Computing, 2003
ISBN 3-540-00574-9

Vol. 126. J.-L. Verdegay (Ed.)
Fuzzy Sets Based Heuristics for Optimization, 2003
ISBN 3-540-00551-X

Vol 127. L. Reznik, V. Kreinovich (Eds.)
Soft Computing in Measurement and Information Acquisition, 2003
ISBN 3-540-00246-4

X. Yu
J. Kacprzyk (Eds.)

Applied Decision Support with Soft Computing

Springer

Professor Xinghuo Yu
Faculty of Engineering
Royal Melbourne Institute of Technology
GPO Box 2476V
Melbourne VIC 3001
Australia
E-mail: x.yu@rmit.edu.au

Professor Janusz Kacprzyk
Systems Research Institute
Polish Academy of Sciences
ul. Newelska 6
01-447 Warsaw
Poland
E-mail: kacprzyk@ibspan.waw.pl

ISSN 1434-9922
ISBN 3-540-02491-3 Springer-Verlag Berlin Heidelberg New York

Library of Congress Cataloging-in-Publication-Data applied for

A catalog record for this book is available from the Library of Congress.

Bibliographic information published by Die Deutsche Bibliothek
Die Deutsche Bibliothek lists this publication in the Deutsche Nationalbibliographie; detailed bibliographic data is available in the internet at <http://dnb.ddb.de>.

Springer-Verlag Berlin Heidelberg New York
a member of BertelsmannSpringer Science+Business Media GmbH
http://www.springer.de

Printed in Germany

Typesetting: camera-ready by editors
Cover design: E. Kirchner, Springer-Verlag, Heidelberg
Printed on acid free paper 62/3020/M - 5 4 3 2 1 0

Preface

Today few organizations can survive without the support of automated assistance. The complexity of modern decisions has caused organizations to become increasingly dependent on advanced information technologies to quickly process large quantities of data. Computerized decision support has been and will always be the key for the success of any organization.

Soft computing has provided sophisticated methodologies for the development of intelligent decision support systems. Fast advances in soft computing technologies, such as fuzzy logic and systems, artificial neural networks and evolutionary computation, have made available powerful problem representation and modelling paradigms, and learning and optimisation mechanisms for addressing modern decision making issues.

There have been substantial developments in the soft computing technologies for decision support over the last few years, which have been extensively reported in the literature through numerous technical journals and books.

Different from most existing books, this volume will predominately focus on "applied" decision support with soft computing, that is, the real applications of soft computing in decision support. Areas covered in this book include planning, management, finance and administration in both the private and public sectors.

The first part of the volume is concerned with more general paradigms, solutions and analyses related to a broadly perceived decision support using some soft (and also hard) computing tools and techniques. In line with an application oriented character of the volume, discussions are always illustrated on examples and applications.

We begin with the chapter "Modeling Knowledge: Model-Based Decision Support and Soft Computations" by M. Makowski and A.P. Wierzbicki which provides an overview of foundations, rationale, tools and techniques, architectures, etc. of a model-based support for modern decision making, with remarks and comments on the role of soft computing.

W. Cheetham, K. Goebel and P. Bonissone in the paper "benefits of Decision Support Using Soft Computing" show, on the basis of successful applications, advantages of using soft computing tools that have allowed developers at General Electric to maximize the benefits and minimize the costs of decision support systems they have created, reducing costs of design, automation, maintainance, and maximizing the value of the decision process. These benefits are shown on two applications.

In the chapter "Evolving Connectionist-Based Decision Support Systems", N. Kasabov gives, first, some background information on decision support systems and then introduces a framework for building dynamic, adaptive decision support systems that evolve their structure and improve their knowledge-base through a continuous learning from data and through interaction with the environment; they use evolving connectionist systems (ECOS), adaptive learning neural network models, thus their name is evolving connectionist based decision systems (ECBDS).

In the paper "An Agent-Based Soft Computing Society with Applications in Financial Investment Planning", Chengqi Zhang and Zili Zhang, propose an agent based soft computing society framework that can be of use for the construction of decision support systems in various domains. For illustration, an application in financial investment planning tasks is presented.

In her paper "A Rough Sets/Neural Networks Approach to Knowledge Discovery for the Development of Decision Support Systems", I. Jagielska proposes a conceptual and a technological framework for adaptive development of decision support systems (DSS) through knowledge discovery from historical data, advocating the use of hybrid rough sets/neural networks framework. For illustration, an example related to industrial and business decision problems where historical data exists is shown.

The next paper concern applications and implementations of decision support systems applied to various domains exemplified by socio-economic, technological, environmental, agricultural, etc. problems. This emphasis on applications is what makes the volume special in the existing literature on the subject.

In "Decision Support Systems in Healthcare: Emerging Trends and Success Factors," written by T.-L. Leong, a comprehensive survey of developments of decision support systems in healthcare, including applications of soft computing techniques, is given.

The chapter "Relational Structures for the Analysis of Decision Information in an Electronic Market" by Bartel van der Walle is concerned with electronic markets , i.e. web-based information systems that provide communication and decision support for online business activities of buyers and sellers, and focuses on a relatively small but active electronic job market for students (the buyers) and companies (the sellers) who had registered for a forthcoming university career fair.

In their paper "A Fuzzy Evaluation Model: A Case for Intermodal Terminals in Europe", R.A. Ribeiro, A.M.D. Moreira and E. Declercq are concerned with an analysis of development of a friendly transport concept, an intermodal terminal selection, to minimize possible negative effects of an expected increase of the commercial freight transport due to a continuously increasing demand for speed, reliability, reduced transportation costs and value added services.

H. Bury and D. Wagner discuss, in their paper "Application of Kemeny's Median for Group Decision Support", general problems concerned with the choice of a method of group judgement for expert opinions given in the form of preference orders, with a special attention Kemeny's median and its modification proposed by Litvak. Heuristic computational algorithms are proposed, and a real-life example for ranking university professors is described.

The chapter "An Internet-based Group Decision and Consensus Reaching Support System" by Sławomir Zadrożny and Janusz Kacprzyk is concerned with a group decision making and consensus reaching decision support system and its implementation within the framework of the Internet (WWW), with emphasis on consensus reaching, under linguistic preferences and majority.

J.I. Peláez , M.T. Lamata and F.J. Fuentes present, in their paper "Limpio: A DSS for the Urban Waste Collection Problem", a decision support system for a company involved in urban waste collection, taking into account problems such as the initial volume of information needed to run the systems, the direction of traffic in the city, or the measures needed to establish how valid a solution is.

In their paper, "A Decision Support System for Air Quality Control Based on Soft Computing Methods", P. Holnicki and A. Kałuszko show how to support an optimal allocation of financial means for emission reduction in a given set of power and heating plants in a region, with emphasis on sulfur. A case study reported concerns a set of major power plants in the Silesia Region (Poland).

R. Alcalá, J.M. Benítez, J. Castillas, J.L. Castro, O. Cordón, A. Gonzàles, F. Herrera and R. Pérez discuss, in their chapter "Multicriteria Genetic Tuning for the Optimization and Control of *HVAC Systems*", the use of genetic algorithms for the optimization and control of heating, ventilation and air conditioning (HVAC) systems employing fuzzy controllers to increase efficiency and peformance..

K.P.Lam's chapter "Intelligent Information Systems for Crime Analysis" discusses a recent development of in the use of intelligent techniques for crime analysis, in particular related to crime map and facial identification. An integrated man-machine approach is described. An application in Hong Kong is presented.

The chapter "Fuzzy Decision Trees Application to Reservoir Recognition" by X. Z. Wang, D. S. Yeung, E. C. C. Tsang, J. W. T. Lee, outlines an application of fuzzy decision trees in reservoir recognition for an oilfield in Northern China.

In their paper "A Decision Support System for the Construction of Cattle Diets", D.A. Pelta, J.L. Verdegay and J.M. Cadenas discuss the problem of determining a best diet for cattle, a important problem plagued by uncertainty, imprecisely known parameters and lack of data. Fuzzy logic is employed, and an application for an Argentinian farms, where the cattle is not confined, is shown.

In the paper "Prediction of Parthenium Weed Dispersal using Fuzzy Logic on GIS Spatial Image," written by A. Chiou and X. Yu, the application of fuzzy logic in the prediction of parthenium week dispersal based on GIS spatial image for Australia's Central Queensland region is discussed, a project funded by the Australia's Commonwealth Government.

Overall, this book provides a comprehensive coverage of up to date conceptual frameworks in broadly perceived decision support systems, and successful applications.

We are sincerely grateful to the many contributors who have given their valuable time and expertise to this project, without whom, this collection of contributions would not have been possible. We want to particularly thank them for bearing with us a lengthy period of editing process for the book. Finally special thanks to our families for their support and devotion.

Melbourne, Australia
Warsaw, Poland
January, 2002

Xinghuo Yu and Janusz Kacprzyk

Contents

Part I:

General Issues

Modeling Knowledge: Model-based Decision Support and Soft Computations

Marek Makowski[1] and Andrzej P. Wierzbicki[2]

[1] International Institute for Applied Systems Analysis (IIASA)
Schlossplatz 1
A–2361 Laxenburg, Austria
E-mail: marek@iiasa.ac.at

[2] National Institute of Telecommunication
ul. Szachowa 1
04–894 Warsaw, Poland
E-mail: a.wierzbicki@itl.waw.pl

Abstract. This chapter provides an overview of model-based support for modern decision making. It starts with discussing basic elements of decision making process, including characteristics of complex decision problems, concepts of rationality, and various requirements for model-based support at different stages of decision making process. Then the characteristics of models, and of modeling processes aimed at decision-making support for complex problems are presented. In this part guidelines for model specification and instantiation are illustrated by an actual complex model. This is followed by a discussion of modern methods of model analysis, which include a more detailed discussion of reference point optimization methods, and an outline of methods for sensitivity analysis, and of softly constrained inverse simulation. Finally, an overview of architecture of model-based decision support system is presented.

Keywords: decision making, decision support systems, fuzzy sets, goal programming, intuition, knowledge, mathematical modeling, model analysis, model specification, multi-objective optimization, Pareto-optimality, robustness, sensitivity analysis, simulation, soft computing.

1 Introduction

Rational decision making is becoming increasingly difficult despite fast development of methodology for decision support and even faster development of computing hardware, networks and software. At least three commonly known observations support this statement:

- First, complexity of problems for which decisions are made grows even faster.
- Second, the amount of knowledge potentially applicable to rational decision making is also growing very fast. However, this knowledge is available in diversified forms (encoded in analytical models, stored in large

and heterogeneous data sets, or as tacit knowledge of practitioners and scientists).

- Third, knowledge and experiences related to rational decision making develop rapidly but in diverse directions, therefore integration of various methodologies and tools is practically impossible.

Most of decision problems are no longer well structured problems that are easy to be solved by intuition or experience supported by relatively simple calculations. Even such types of problems that used to be easy to be defined and solved, are now much more complex because of the globalization of the economy, and a much greater awareness of its linkages with various environmental, social and political issues. Therefore, modern Decision Makers (DM) typically need to integrate quickly and reliably knowledge from these various areas of science and practice in order to use it as one element of the Decision Making Process (DMP). Unfortunately, the culture, language and tools developed for knowledge representations in key application areas (such as economy, engineering, finance, environment management, social and political sciences) are very diversified. This observation is known to everybody who has ever participated in a team-work with researchers and practitioners having background in different areas. Given the great diversity of knowledge representations in various disciplines, and the fast growing amount of knowledge in most areas, the need for efficient knowledge integration for decision support becomes a challenge that deserves to be addressed.

Good DMs make decisions by combining own knowledge, experience, and intuition with that available from other sources. Experienced DMs have a fairly stable set of ways for thinking, evaluating, judging and making decisions; this is often called a Habitual Domain (HD), defined by Yu [85]. HD is specific for each person, and it differs often substantially, even for persons with similar cultural and professional background, experience and duties. Moreover, whenever a complex problem is considered, then no one individual system of management (represented by one person) is sufficient. Therefore, typically a DMP involves more than one person, its participants often have very different HDs, and it is the integration of their ideas that creates a powerful force for change and improvement, which typically drive DMPs for complex problems.

For computerized decision support, the key question is how to represent knowledge about the decision situation – including external, objective knowledge about the underlying physical, biological, technical, environmental etc. aspects, and the subjective, individual knowledge and preferences of decision makers. The experiences in applications of decision support have shown that these two types of knowledge should be expressed separately, and then used in a combined way.

There are various approaches even to the question how to express the external knowledge. One approach is to deny the existence of any external knowledge except that contained in relevant data sets and in logic; this perspective of knowledge engineering leads to so-called *data-based decision support.* Another approach is to accept relevant knowledge coming from selected disciplines of science and express it in the form of computerized mathematical models; this perspective leads to so-called *model-based decision support.* The latter developed historically from the tradition of various disciplines, including engineering design and operations research (OR).

The OR methods were developed and first applied to well structured decision problems, where optimization focused approaches provided solutions corresponding to optimization of a precisely defined goal in a precisely formulated mathematical programming problem that can adequately represent only a well structured problem. Due to the successes of early applications and limitations of computing power, OR for relatively long time was focused on single-criterion optimization approaches, which has limited application to actual decision-making support for complex problems. However, the assumption made for application of the classical OR methods are not adequate for most of actual support for decision-making related to complex problems, because of at least the following reasons:

- Actual complex problems can hardly be represented by only one precisely defined mathematical model.
- Already about 50 years ago Simon [62] has pointed out that actual decisions, particularly in large organizations, are optimized only until a satisfying (in technical terms called *satisficing*) level of key criteria is reached; at this point instead of a further optimization other attributes of decisions are considered.
- DMs clearly prefer to be sure that their sovereignty in making decisions (for which he is responsible) is not jeopardized by developers of the model used for decision support, nor by a computer used for its analysis.

The limitations of the traditional OR approaches have been discussed e.g. by Ackoff [1,2], and Chapman [11]. Motivated by the needs for model-based decision-making support for complex problems, various scientific communities have over last several decades developed diversified methodologies and tools (often referred to as *modeling paradigms*) aimed at providing adequate model-based support to DMPs for such problems.

A scientific paradigm, as defined by Kuhn [30] and by Hloyningen - Huene [24] embodies the consensus of a scientific community on an approach to a problem (or to a class of problems), and consists of the theories, laws, rules, models, concepts, and definitions that are generally accepted in science and practice. Due to the unquestionable success of modeling in problem solving, various modeling paradigms have been intensively developing over last few decades. The following list of publications provides a relatively small sample of diverse successful attempts to develop modeling methods and apply to various Decision Support Systems (DSSs) which over the years have resulted in many recognized modeling paradigms: [87,12,27,70,90,72,34,58,68,84,65,55,66,77,57,53,8,54,88,81,79,38,56,9,18,32].

All these developments have been rational from the point of view of providing more efficient solutions for specific types of models or elements of modeling process. However, efficiency requirements in developing specific methodologies and tools for each modeling paradigm resulted in paradigm-specific problem representations, and corresponding modeling tools for model specification, analysis, documentation, and maintenance. As a consequence of this long-term development, it has become increasingly difficult to apply all pertinent paradigms to a problem at hand, because these resources are fragmented and using more than one paradigm for a problem at hand is expensive and time consuming in practice. Models have become complex and/or large, therefore the development and use of a model with even one specific paradigm is a costly and time-consuming process. Moreover, incompatible model representations used by different paradigms imply that resources used for modeling with one paradigm can hardly be reused when another paradigm is applied to the same problem. Therefore multi-paradigm modeling support is currently one of the most challenging problems faced not only by the OR community, but by a broader audience of computerized decision and design support or even generally scientific community. Various modeling paradigms could be used much more broadly, if scientific communities will get a better understanding of methodologies developed by other communities, and if modeling resources (model specifications, data, modeling tools) will be available in a uniform representation.

This chapter gives an overview of model-based decision making support methods, which are applicable not only to soft computing, but also to other modeling paradigms. Therefore it provides a context in which advanced soft computing methods can be used within multi-paradigm modeling, which in turn will make them available to a much larger scope of applications than that presented in this book.

The chapter is organized as follows. Section 2 discusses the issues of modern decision making, including the characteristics of complex problems requiring decision support, concepts of rationality, and the role of intuition in decision making, as well as the dynamics of decision processes. Section 3 presents an overview of key modeling paradigms relevant to the development of models aimed at decision-making support. Section 4 concentrates on basic concepts related to the architecture and implementation of model-based decision support systems.

2 Modern Decision Making

By a *decision maker* we could understand anybody who makes decisions or takes a part in a decision making process – but everyone makes daily many decisions. Thus, we limit the concept of a decision maker to those persons who are aware of importance of decisions made by them or at least reflect how these decisions are made. Nevertheless, this concept is very rich in meaning. We can distinguish many types of decision makers, such as a *top-level*, economic or political *decision maker*, or an *analyst* preparing the analysis of a given decision situation, or generally a *user* of a decision support system. We can also distinguish various types of decisions: operational, tactical, strategic, creative, *etc.*

But what is a *modern decision maker?* Is a person modern because of using computerized help, such as decision support systems or computer network communications? Is a person modern because of understanding computerized modeling and using its various means - logical, graph-theoretical, operational research type? Certainly, these are some dimensions of being modern. However, there is also a richer meaning of being modern: *a person is modern because of using basic concepts that are adequate for understanding the contemporary, complex world.* This meaning is actually related to creativity and thus adaptability – two features that are especially important now, because the contemporary world is not only complex, but also changing fast.

2.1 Changing Civilization Eras

We live in a period of fundamental technological and social change. The reasons of this change are variously described, but one of the most widely accepted is the hypothesis of **the turn of cultural eras** – from the era of *industrial civilization* to the *information civilization era.* The information civilization society is related to many more specific terms, such as *service society, information society, networked society, knowledge based economy,* etc. These concepts are variously though similarly defined; for example, information society and knowledge based economy can be jointly defined as society and economy in which information and knowledge become essential productive factors. Other definitions put more stress on technological developments, such as the emergence of the *global information infrastructure.* However, we maintain that the most important is the change of basic concepts.

A cultural era is indeed defined by a system of basic concepts which are used for perceiving and understanding the world. Such a system must be widely accepted, embedded in the culture and education of the era; it constitutes a *cultural platform* of the era. The basic concepts that constitute the cultural platform of industrial civilization were developed much earlier than the beginnings of this era, usually dated at the 18-th century and coinciding with the Enlightenment era. These concepts were actually based on the discoveries of Gutenberg, Copernicus, Galileo, Newton, Kepler and others that helped to perceive the universe as a giant machine, to construct smaller machines and industrialize the world. Thus, the cultural platform of industrial civilization is *mechanistic* and *reductionist*: the analysis of the universe or a society is reduced to the analysis of their parts and relations between these parts. Most of our thinking about the world is still based on this platform.

The cultural platform of informational civilization is being formed now, though it uses concepts developed during over hundred years. It started with the theories of Maxwell and Einstein, challenging the mechanistic perception of the universe, proposing fundamentally different, relativistic interpretations. Later, many developments of telecommunication and control science contributed to the understanding of the role of information in complex systems, including the fundamental concept of feedback which changed the classical understanding of the relation of cause and effect. Finally, the development of computers and computer science resulted in diverse understanding of information processing.

The new cultural platform is not fully formed yet; some essential concepts – such as the deterministic theory of chaos – are fully known only to a handful of specialists. The full formation of the new cultural platform, its acceptance and spread by our educational systems might need yet several generations. There is no doubt today, however, that the mechanistic vision of the world of industrial civilization era will be replaced by a new vision. Some computer scientists propose to perceive the world as a giant computer; however, this is a simplistic vision which does not take into account relativism, feedback, chaos. More probable is that the new vision will have a systemic character and will include a deeper understanding of chaos.

Thus we cannot say that the information civilization has already arrived, though it has started. The date of its beginning might be counted at the discovery of the TCP-IP protocol in computer networks just before 1980; but full information civilization has not arrived with this date, similarly as industrial civilization did not arrive around 1760 with the improvement of steam engine, it only started then. There exist today information technologies that will fundamentally change the society even as we know it today. These are, for example, microprocessors with computing power comparable to that of supercomputers a few years ago, which will become cheap and generally used in the next few years; teleconferencing programs that make it possible to construct sophisticated videophones, available in every home; the explosive development of computer networking; computer integrated design, manufacturing and management; widespread use of computer intelligence, called today the *vision of ambient intelligence,* implying also widespread use of computerized decision support systems; and many others. But in order to use such systems effectively, we should be aware of basic concepts concerning decision making, used when developing such systems.

2.2 Concepts and Creative Processes

Language, logic, mathematics are just instruments of description, discussion, formalization of our deeper thoughts that might occur on a nonverbal level, in terms of images and basic concepts. It is obviously better for communication that basic concepts are well understood and have a well-defined meaning.

On the other hand, *a concept is the more valuable the richer it is*, in a synergetic sense; and the number and value of associated images related to a concept depend on the level of expertise. We shall discuss here mainly such concepts that are essential for modern computerized decision support.

For modeling external knowledge about decision situations, especially important is the concept of *cause and effect*; but it is understood still by most people, even philosophers, somewhat mechanistic and in terms of *binary logic*, while we know today that there are situations in which *fuzzy, rough or other multi-valued logic* is more adequate, and that this concept is even more fundamentally changed if we include also *feedback* mechanisms. The concept of feedback is essential for a modern, systemic understanding of the world – but it is differently perceived by an economist who has often been taught only the basic definition of this concept, differently by a control theory specialist who spent a life investigating various feedback effects, and differently by a specialist in the *theory of chaos*, who would say *feedback with nonlinear dynamics might produce chaos in a deterministic system, which will change the way we should think about the indeterminism of the universe.*

Indeterminism is deeply related with the concept of *uncertainty*, essential for decision theory. If we could be sure about consequences of every planned action, decision making would be much simpler. In fact, decision theory developed together with statistics, with probabilistic descriptions of uncertain results of decisions. However, today we know that probabilistic description of uncertainty is only one of such possible representations and is not always acceptable. It is a very good tool of describing uncertainty if we have access to large amounts of data from which probability distributions can be empirically derived or at least the parameters of some assumed (e.g. normal) distributions experimentally tested. If the data are more scarce, other ways of expressing uncertainty might be more adequate. One is suggested by the *rough set theory* of Pawlak [47]: classify existing data into three logical categories – *true, false,* and *may be*, then assess how large is the last, uncertain category. Another, actually older, is to use not trinary, but infinite-valued logic as suggested by *fuzzy set theory* of Zadeh [87]; this is particularly appropriate if the uncertainty is expressed by people, by linguistic descriptions instead of numerical data.

Another basic concept is that of *complexity*. It has a rather precise meaning in computer science. It gives then an approximation of the number of basic computations needed to solve a finite computational problem, in the worst possible case and in the sense of the character of the dependence of this number on the dimensions (number of variables) of the model of the problem. For very simple problems (which are rather rare) this dependence is linear; for simple problems, polynomial. Truly complex problems have exponential or combinatorial dependence of computations on the dimension of the model. This means that while we can solve, say in several hours of computing time, a problem of dimension n, solving the same type of problem with dimension $n+1$ can take several years. Complexity theory is actually much more complex than shortly described here: sometimes we cannot estimate the worst case complexity, sometimes we can only prove that the complexity is no less than some estimation, etc. However, these are technical details; the essential issue is that most truly difficult computational problems have at least non-polynomial (exponential or more) complexity. And this concerns only finite problems i.e. problems solvable in finite time. There are many problems – e.g. nonlinear programming problems – that can be solved only approximately, their exact solution would require infinite time. But again, complexity of solving them with prescribed accuracy is very often non-polynomial.

This describes only the issue of computational complexity. The issue of complexity of real life problems would have to take into account the mapping of the real life into computerized models. We shall show below that there is no precise way to evaluate such mapping; hence complexity of real life problems must have only an approximative, heuristic meaning. It must take into account that computerized models are very often composed from many submodels; that each of the submodels might have parameters determined with statistical procedures using available data; that a composition of submodels does not behave as a simple sum of them; that computational errors in complex models not only accumulate, but also might explode; that feedbacks between submodels not only increase computational complexity, but also might result in a chaotic behavior of the entire model; etc. To deal with all these problems, we must follow Einstein advice: *keep things as simple as possible, but not too simple.*

We mentioned that computerized models are used in decision support to represent external knowledge about a decision situation. We also distinguished knowledge from information. But what is *knowledge?* Typical definitions distinguish between at least two types of knowledge. One is *personal knowledge* – what a person actually knows and can apply usefully – which might be learned by education or derived from experience; the latter is often difficult to be described and communicated by words and is called in such case *tacit knowledge.* Another is *social knowledge* – entire body of knowledge established by various sciences including hard sciences, but also social sciences and humanities, recorded in books and by other means, transmitted by educational institutions. Selected part of social knowledge, pertinent for a given decision situation, we describe here as external knowledge – in opposition to the personal and tacit knowledge of the decision maker. But there are also various ways of representing knowledge in a computer. Classical definitions of knowledge engineering in computer science define knowledge as patterns that can be learned from data. However, this definition assumes in a sense that the only source of knowledge are pertinent data, and such an assumption is certainly too narrow for applications in decision support. Much better is to define knowledge as any model, either deduced from data or from social knowledge; patterns are only a type of logical models and we use very often other, analytical types of models to represent computerized knowledge.

We see that a precise agreement on a single meaning of a concept is often not possible. Moreover, such an agreement would not necessarily guard us against hidden assumptions in its accepted definition. The philosophy of science (see *e.g.* [29,69]) stresses that our *mesocosmic experience* results in the danger of such hidden assumptions; special techniques are used – *e.g.* that of thought experiments – to clarify such presuppositions. While the clarity and a unique meaning of a concept are needed for communication, there might be other aspects of cognitive processes for which the *richness of meaning is more valuable*; one of such aspects is *creativity*.

Any scientist with some experience in creative processes – such that require not only mere search and logical or experimental testing, but also result in new cognitive insights – would acknowledge the basic role of the richness of concepts related to her/his field of study. We often perceive uneasiness about an aspect of a scientific problem; then we try to look at it from various angles, through diverse images related to our basic concepts; after some time, *a moment of cognitive enlightenment, a heureka effect* comes. The mechanism of this effect relates to subconsciousness and intuition.

Similar creative processes occur also in other fields – in art, politics, even managerial decisions. It obviously helps to study your subject thoroughly and to discuss it with others – we are forced then to at least partly formulate our thoughts, and even casual remarks of others might trigger associations; it also helps to relax after some time of study, forget the problem for a while – an intuitive answer might need such a gestation period. It might also be helpful to extend your interests beyond the strict boundaries of your discipline or even embark on some interdisciplinary research; it pays to travel and be exposed to cross-cultural influences. How soon you perceive a heureka effect and how good your answer is, depends on many aspects: on the level of expertise, on the thoroughness of study, on cross-fertilization through discussions, interdisciplinary and cross-cultural contacts – and on personal creativity or intuitive capabilities, admittedly hard to define.

Thus, *creative and cognitive processes are relying on the richness of images and concepts* – or also on the richness, even redundancy, of information and knowledge; *they often occur* on a deep level of nonverbal associations, often *subconsciously or intuitively*. Intuition is usually (probably for historical reasons) considered as an opposite concept to rationality. We shall discuss first the concept of rational decisions, then show that it is possible to rationally define and analyze intuition.

2.3 Concepts of rationality

During the last century, joint developments in statistics and economics resulted in an increasing understanding how economic markets function. This was also related to the study how economic decisions are made and resulted in the development of *decision theory*. Actually, decision theory includes various branches considered today as separate disciplines: mathematical models of analyzing conflicts and cooperation in game theory, general models of decision preferences in value theory, graph theory and theory of relations as applied to decision making, etc. However, all these branches rely on the same concept of rationality, resulting from the study of economic decisions.

In decision theory, *a decision maker is considered rational if she/he chooses the most preferred decision*, defined usually as *maximizing her/his utility or value function*. Actually, such a definition is slightly oversimplified; but the assumption of choice based on preference, on the maximization of a utility or value function has been very productive in the developments of economic theory and mathematical game and decision theory. It has a long research tradition and has led to such an expansion of mathematical economics, game and decision theory that no single book today would even list nor summarize the existing literature of the subject. A *paradigm* in the sense of Kuhn [30] has been formed around this assumption, with all its features – including also a psychological commitment to this assumption, reinforced by the sociological aspects of a field of science.

There were also many schools of criticism of this central assumption of decision theory, usually dependent on a certain disciplinary or cultural perspective; we shall call such a school a *framework* of perceiving rationality of decisions.

The most notable is the framework coming from managerial and microeconomic perspective (consistent also with an engineering type of behavior). This is the theory of *satisficing behavior* developed originally by Simon [60,61]. According to Simon, individual decisions are not optimized because it is too difficult (particularly in situations with uncertainty), because the access to information is limited, and because an individual optimization might lead to unnecessary conflicts in non-market situations with a limited number of actors (as in management). Instead of optimizing, an individual decision maker forms her/his own *aspirations*. They are formed adaptively, through learning processes that play an important role in this theory. The central point of the theory of satisficing behavior is that the decision maker stops searching and is satisfied with a given decision, if the aspiration levels are attained.

The satisficing framework of rationality was further developed through intensive studies, including empirical tests; finally, it was assimilated also in the utility maximization paradigm under the name of *bounded rationality* (this illustrates a self-reinforcing aspect of the paradigm: while believing that a perfectly rational individual should optimize utility, it can be admitted that most individuals are less than perfect). Related to the satisficing framework but distinct by stressing a big organization perspective (in corporate or military planning, also in centrally planned economies) is the framework of *goal-oriented behavior*, discussed *e.g.* by Glushkov [19], Pospelov and Irikov [52] independently made operational by Charnes and Cooper [13] and many other researchers as the *goal-programming method* of multi-objective decision support.

From the perspective of computerized decision support in multi-objective situations, a methodology of *reference point optimization* was later developed – see *e.g.* [71,74,31,77]. This methodology uses both optimization and adaptive formation of aspirations, but rather as instruments of interaction between the decision maker and the computer than as assumptions concerning human behavior. While this methodology generalizes the goal-programming method and can support decision makers which adhere to various rationality frameworks (*e.g.* such that believe either in satisficing or goal-oriented or utility-maximizing behavior), it also stresses the dynamics of a decision process and adaptive learning.

There were also many other schools of criticism of the assumption of utility maximization, *e.g.* coming from the psychologist's perspective, such as the *theory of regret* by Tversky [26]. Another important avenue of criticism of individual utility maximization resulted from *the evolutionary perspective*, with the introduction of the concept of *evolution of cooperation*, see [5,55]. This approach shows that evolutionary interests of a population – and also individual interests of members of this population – might be better served by adopting certain individual decision rules that take into account cooperation principles than by pursuing the maximization of purely individual interests (no matter whether short- or long-term). The concept of evolution of cooperation does contradict more extreme interpretations of the paradigm of individual utility maximization (that it is better if everyone pursues her/his own individual interests) and contributes to a better understanding of emerging issues that require common action, such as the importance of preserving clean environment or, more generally, preserving our planet for the survival of the human race.

These were the schools of criticism of the simplified meaning of the term *rationality* related to decision analysis; however, this term is interpreted yet differently in various scientific disciplines. A possibly broadest sense of rationality or validity of scientific statements is implied by the philosophy of science, see *e.g.* [49–51]. Even in this field, there are diverse interpretations of this concept: *falsificationism* by Popper, *historicism of changing paradigms* by Kuhn, *competition of scientific programs* by Lakatos, *etc.*, see [23] – versus *evolutionary epistemology* as initiated by Lorentz [33], see [82]. From the more normative, Popperian perspective, rationality of a statement concerning the real world implies its empirical testability and falsifiability (at least in principle); if a statement relates rather to a method of communication than to the properties of the real world itself (as, for example, a mathematical theorem) then it must be logical and consistent with the assumptions of the method.

When seen from this perspective, the term *rationality* in decision theory has been used rather unfortunately. Interpreted as a mathematical statement, the assumption of utility maximization is logical and consistent with many other developments of decision theory. However, interpreted as a description that real *rational* decisions are always made in the best perceived interests of the decision maker, this assumption is itself not rational in the normative, Popperian sense of the theory of science. Popper uses precisely this as an example of an unrefutable theory. If any other concerns – honor, altruism, *etc.* – can be subsumed into the individual utility, then the assumption of best perceived individual interests cannot be falsified in principle and expresses rather an ideology than science – at least, according to Popper, while Kuhn [30] argues that certain components of ideology are always present in a paradigm of normal science. Thus, we will use sometimes the concept of rationality of a decision maker in its paradigmatic sense, but with an adjective; we shall call it *economic rationality* to distinguish it from other possible perspectives.

We shall use the general term *rational decision* in a possibly broadest sense: while stipulating that the specific meaning of this term might depend on professional and cultural background (see *e.g.* [22] or [85]), we shall require that a rational explanation of the way decisions are made should itself be – at least in principle – empirically refutable. In this sense, we can speak also about *procedural rationality* in the sense that we can empirically investigate the process of decision making and analyze whether it is reasonable. But before examining decision making processes, we must outline the role of intuition in decision making.

2.4 A rational definition of intuition

The experiences in political decision making result in the following pattern. *The higher the level and experience of a decision maker, the less inclined is she/he to trust in various tools and methods of decision analysis and support.* When asked to explain such an attitude, they point to tacit knowledge, personal experience and intuition. A clear conclusion is that decision scientists must make more efforts to better understand the dichotomy of rational versus intuitive decisions and the role of intuition in decision making processes.

In everyday language, we tend to use the word *intuitive* with some connotation of *irrational.* This is probably due to Bergson [6], who attached a great importance to intuition but interpreted it as a somehow mystic force which by definition could not be an object of research by rational means of inquiry. However, almost a hundred years of research in various fields of science motivates today a reversal of this interpretation. We shall show that *intuition*, when appropriately defined, *might be a subject of rational inquiry.*

We first turn to an avenue of criticism of the mainstream decision theory which claimed that individual decisions are too complicated to be explained by *analytical* theory, that decisions are rather made in a *deliberative* way. This criticism relates to *general systems theory* [7] or so-called *soft systems approach*, stresses the role of *synergy* – that the whole is bigger than the sum of its parts – as well as the role of *heureka* (after Archimedes) or *aha effect* – an enlightenment moment in a cognitive process. Thus, soft or deliberative decision making consists in trying to perceive the whole picture, to observe it from various angles, and finding the right decision by expert intuition.

This avenue of criticism is also presented in the book *Mind over Machine* by Dreyfus [15] that presents a well documented thesis as follows: *the way decisions are made changes with the increasing level of expertise* of the decision maker. While a beginner, novice or apprentice needs analysis to support her/his decisions, specialists, experts and master experts either make decisions instantly, intuitively, or rely increasingly more on deliberation.

Another reason to analyze the concept of intuition comes from the research on hemispheric asymmetry of the brain, see e.g. [64]. All this research indicates that the typical left hemisphere is somehow related to *verbal, sequential, temporal, digital, logical, analytic, and rational* thought activity, while the right one is rather related to *nonverbal, visuo-spatial, simultaneous, analog and intuitive* thought activity. Best substantiated by experimental results is the distinction between verbal and visuo-spatial processing in distinct hemispheres; further dichotomies are more speculative. Note that in the dichotomies presented above after [64] rationality is still rather traditionally opposed to intuition.

Nevertheless, we need a definition of intuition that is rational in the sense of rationality of scientific theories: it should be testable experimentally. Such a definition can be obtained from two premises. One is the knowledge derived from modern telecommunications and complexity theory: transmitting images (video signals, typically 2 MHz) requires 10^2 times more volume (broadband) than transmitting speech (audio signals, typically 20 kHz). Since the complexity of processing such signals is certainly not linear, at least quadratic, processing images requires circa 10^4 times more capacity than processing words.

Another is a thought experiment: what happened to human mind at the moment of the discovery of language? *The development of language was a powerful evolutionary shortcut in the history of humanity*: it was easier to think in words and the communication with other individuals not only helped evolutionary in dealing with the environment, but also opened all development of cultural achievements of the race.

However, this shortcut left powerful parts of our brain utilized only subconsciously. If we acknowledge the essential difference of difficulty between processing images and processing words, we can postulate that *our mind works also on a deeper layer of nonverbal image processing; this task employs a large part of the mind's processing potential, is sometimes conscious but often unconscious, and uses rather multi-valued than binary logic.* Our language can only try to approximate this deeper level by words and by a simpler, more decisive binary logic; but words will always have multiple meanings.

Based on such premises, we can *define intuitive decisions as quasi-conscious and subconscious*[1] *information processing, leading to an action*, utilizing aggregated experience and training and performed (most probably) by a specialized part of the human mind.

Note that this definition includes intuitive decisions such that have conscious components but take advantage of the parallel processing by the parts of our mind specialized in subconscious or quasi-conscious operations. Thus, *every day our mind makes many intuitive operational decisions.* The role of training in making these decisions well is fully understood and institutionalized in human society through our education systems. All training in repetitive activities is actually aimed at delegating them to the quasi-conscious level, automating them – by establishing shortened or more easily used connections between synapses, some specialized *roads in the brain.*

However, this definition includes also strategic and creatively intuitive decision making and makes it possible to develop empirical studies of intuitive decisions. While an empirical study of repetitive, operational intuitive decisions might be simpler (we can ask subjects of experiments not to use conscious parts of their minds, we can make sure that they do not use them by turning their attention to other matters or, as done by Dreyfuses, by saturating the conscious part of the mind), such a study of strategic and creatively intuitive decisions will be certainly more valuable. Unfortunately, not many such studies have been concluded yet, see [78,80].

2.5 Dynamics and phases of decision processes

The above discussions lead us to the following questions: *if most decisions are made intuitively, how can we support them using computerized information and models?* In particular, how can we better understand and support creative or strategic intuitive decision making?

The complexity and the speed of change of information in the modern world make the use of computerized decision support self-evident. However, the reflection on the role of intuition changes the focus of decision support. We can use here the analogy to training in repetitive decision activities and automating them: decision support is aimed at automation of decisions, the issue is only to what extent, at which stages of a decision process and which parts of such a process will be, after all decision support, reserved for human decision makers.

The classical decision theory concentrated almost exclusively on automating the actual choice. In his criticism of the mainstream theory, Simon [61] defined the following *essential phases of an analytical decision process*:

---

[1] Our mind can work quasi-consciously when we want to do something but leave the details to subconsciousness, or subconsciously when we do not know that we want to do something and are not aware of doing this, but we can become aware a posteriori of the results of subconscious work of our mind.

- **Intelligence** - information and data gathering;
- **Design** - selecting or constructing a model of the decision situation;
- **Choice** - examining decision options and choosing between them.

Later, see *e.g.* [31], another essential phase of

- **Implementation**

was added. These phases could be subdivided into more detailed sub-phases. However, such a subdivision of a decision process puts the emphasis on analytical decision making. For the case of creative or strategic, intuitive decision processes, the following subdivision of their phases was proposed in [76]:

- **Recognition**;
- **Deliberation or analysis**;
- **Gestation and enlightenment**;[2]
- **Rationalization**;[3]
- **Implementation**.

An essential question is: how to support various phases of an intuitive decision process, as in an analytical decision support? Or, if we cannot support them directly, could we support them somehow indirectly during an analytical decision process?

It is clear that *we must concentrate more on the phases of intelligence and of design and analysis, preceding the phase of choice.* During a decision support process, there must be enough time and opportunities for learning and thus enriching intuition, helping in gestation and possible enlightenment – all this well before the phase of choice or rationalization.

How should these conclusions influence the way of constructing or using decision support systems? Any decision process is actually multi-objective – particularly if it includes elements of intuition and should be creative: it should aim at novel solutions. If the process is supported analytically by computerized information processing and models, the information not only should be presented to the decision maker in graphic terms, but it should also be rich, multi-dimensional; we do not know a priori on which piece of information her/his subconsciousness will focus. In the organization of decision support, we should avoid an early aggregation of objectives into a utility or value function, avoid even pair-wise comparison and other known techniques

[2] This is an extremely important phase – we must have time for forgetting the problem in order to let our subconsciousness work on it. The expected heureka effect might come but not be consciously noticed; for example, after a night sleep it is simply easier to generate new ideas (which is one of the reasons why group decision sessions are more effective if they last at least two days).

[3] For example, when writing a paper, we obviously rationalize our deeper thoughts – and sometimes also advertise them. Note the similarity of this phase to the classical phase of choice.

related to the paradigm of utility maximization, nor should we require consistency from the decision maker – since if she/he knew from the beginning what she/he precisely wants, the solution could not be creative. The support should be rather concentrated on helping to generate new options, even to formulate new problems, on organizing the dynamics of the decision process in such a way that it leaves enough time for gestation.

However, if we concentrate on multi-objective decision theory and optimization, we can use the tools of this field in order to support the intuition of an analyst rather than to replace his intuitive power of choice by a computerized system. This is achieved by applying such tools to the earlier phases of problem formulation and analysis and by *treating optimization as a flexible tool of analysis, not as the goal.* In particular, optimization can be used as a tool of *multi-objective model simulation* which takes into account hard and soft constraints, includes inverse simulation problems (in which model output, not model input is specified by the analyst) and various aspects of multi-objective model analysis. This provides a sufficiently flexible toolbox to help in treating computerized models as a kind of virtual reality, in which the attainability of various aspirations and the consequences of various approaches to the decision problem are analyzed.

2.6 The decision maker of the future

We observed that decision support systems are used today mostly by analysts or modelers. Although this might change in generations to come, we should not expect this change too soon. Note that a high-level political decision maker is elected because of her/his ability and quality of intuitive generalizations in decision making – thus, she/he will not use computerized decision support unless she/he grows up in a culture where such tools are used everyday and by everybody.

On the other hand, the volume and the speed of change of information processing will grow together with the maturing of the information society. The education systems take these facts into account already today. With the development of information society, the role of education will also substantially change. As the information processing becomes one of the essential productive factors, the access to information and the ability to process it with modern means will become one of the main social issues. Thus, the access to good education, adequate for the challenges of information society, will be essential for a success in life. However, education will at the same time concentrate less on preparing for a designated profession or position in life, and more on providing general abilities and teaching a system of concepts adequate for understanding the new civilization era of information society. All this indicates that computerized decision support will be used much broader in the future than today.

What characteristics, then, will or should a future decision maker have? Evidently, she/he must be able to use various computerized information and model processing tools. However, even more important is being more informed in the sense of basic concepts that allow to understand the coming cultural era. Finally, in decision making, she/he should know in what proportions to blend computerized, analytical decision support and her/his intuition – or, in other words, how to use decision support in order to enlarge her/his *habitual domain*, see [86].

3 Modeling for Decision-making Support

In previous sections we have discussed characteristics of modern decision makers, of decision problems, and of decision making processes, as well as the role of mathematical models in representing knowledge relevant also to decision making processes and thus supporting making better decisions. In this section we outline the characteristics of models, and of modeling processes that pertain to modeling activities aimed at decision-making support for complex problems.

One needs to stress that such characteristics are different from those of models used in introductory text books that focus on providing basic knowledge on mathematical modeling. Small and simple models are typically used for educational purposes to illustrate basic concepts, methods, and techniques. However, adequate model-based support for complex problems typically requires also complex and/or large models, which in turn require also more complex modeling methodology and tools than those adequate for the toy-type models. Therefore, in this section we focus on modeling issues that are relevant for model-based support of decision-making for complex problems.

3.1 Context

Generally speaking, mathematical modeling is the process of creating a model, which is an abstract representation of a system (or a phenomenon, or a problem) developed for gaining a better understanding of the modeled object. A purpose must be defined for any modeling project because modeling even a small part of reality and using even a part of accumulated knowledge would inevitably result in a model far too complex to be implemented using justifiable resources. Consider, e.g., modeling a cup of coffee. Very different models are suitable for studying various aspects, e.g., how something (sugar, cream) is dissolved in a content of the cup, or in what conditions the cup might break from thermal stresses, or what shape of the cup is most suitable for use in aircraft, or how a cup of coffee enhances productivity of an individual or of a group. An attempt to develop a model that covers all these aspects, and represents all accumulated knowledge on even such a simple topic like a cup of coffee would hardly be rational.

A mathematical model describes the modeled object by means of variables, which are abstract representations of these elements of the object that the users of the model want to study. Many commonly used models can be classified as Algebraic Models (AM).[4] AMs are widely used for supporting decision-making[5] in many areas of science and industry for predicting the behavior of a system under particular circumstances, when it is undesirable or impossible to experiment with the system itself. The understanding of the system gained through a comprehensive examination of its model can greatly help in finding decisions (controls) whose implementations will result is a desired behavior of the system. Thus AMs support finding better solutions to real problems than those that could be found without model-based problem analysis.

AMs have many common analytical features and are used whenever decisions require analyses of a large amount of data and logical relations, in a wide range of application domains including (but not restricted to) planning problems in environmental systems analysis, telecommunication, logistics, transportation, finance, marketing, production, distribution, as well as in science, research and education. The systems that are modeled have very different characteristics (including the nature of the underlying physical and/or economical processes, their complexity, size, types of relations between variables). There is also a great variation in the use of models, which depends on various factors (like the decision making process, the background and experience of model users, available data, resources assigned for modeling activities). However, modeling activities have many similarities also when the modeled systems are very different.

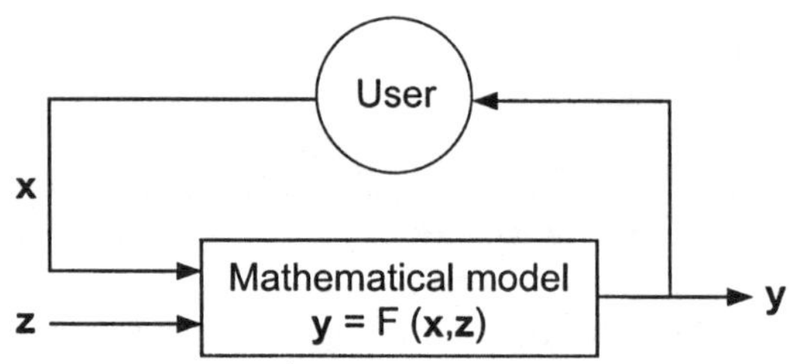

Fig. 1. A mathematical model represents relations between decisions (inputs) x, external decisions (inputs not controlled by the user) z, and consequences (outcomes) y

[4] Following the Oxford Dictionary and a common understanding, we use the term *Algebraic Model (AM)* for a set of relations (such as equations or inequalities) between quantitative inputs (decisions) and outputs (performance indices) that measure consequences of implementation of decisions.

[5] The term *decision support* is typically associated with management and policy-making but in practice similar model-based activities are being performed also in almost all fields of industry and research.

The concepts essential for explanation of model-based decision support are illustrated in Figure 1. The basic function of a DSS is to support the user in finding values of his/her decision variables, which will result in a behavior of the system that fits best to the preferences of the user. A model-based DSS is typically developed using the following concepts:

- decisions (inputs) $\boldsymbol{x}$, which are controlled by the user,
- external decisions (inputs) $\boldsymbol{z}$, which are not controlled by the user,
- outcomes (outputs) $\boldsymbol{y}$, used for measuring the consequences of implementation of inputs,
- mathematical model F, which represents the relations between decisions $\boldsymbol{x}$ and $\boldsymbol{z}$, and outcomes $\boldsymbol{y}$, which can be symbolically represented by: $\boldsymbol{y} = F(\boldsymbol{x}, \boldsymbol{z})$.

Before discussing these basic concepts in the following subsections we will introduce an example of a complex model, which will be used to illustrate not only these concepts but also several other issues.

3.2 The RAINS model

In order to be able to later present various modeling problems we outline now an illustrative example of the RAINS model used for supporting international negotiations aimed at improving the European air quality, and described in more detail in [3].

RAINS provides a consistent framework for the analysis of emission reduction strategies, focusing on acidification, eutrophication, and tropospheric ozone. RAINS comprises modules for emission generation (with databases on current and future economic activities, energy consumption levels, fuel characteristics, etc.), for emission control options and costs, for atmospheric dispersion of pollutants, and for environmental sensitivities (i.e., databases on critical loads). In order to create a consistent and comprehensive picture of the options for simultaneously addressing the three environmental problems (acidification, eutrophication, and tropospheric ozone), the model considers emissions of SO_2, NO_x, ammonia (NH_3), and VOC (volatile organic compounds). The quality of air is assessed by several indicators computed at a few hundred grids, and their values depend on locations and amounts of emissions of various pollutants. Hence the decision (input) variables are emissions (represented in this simplified presentation by a vector $\boldsymbol{x}$). The vector of output variables $\boldsymbol{y}$ includes subvectors of various air quality indicators, each composed of vectors representing values of such indicators at each of the grids into which Europe is divided for the purpose of air quality assessment.

The simplest model of air quality can be defined by $\boldsymbol{y} = F(\boldsymbol{x})$, where an operator $F(\cdot)$ represents relations between emissions $\boldsymbol{x}$ and air quality indicators $\boldsymbol{y}$. In the actual implementation these relations are defined by a collection of several thousands equations. Depending on the type of pollution, components of $F(\cdot)$ have different forms (e.g. linear for sulphur oxides and non-linear for trophosperic ozone). Such a model can be used for computing air quality corresponding to given emissions. However, it is not very useful for finding cost-effective strategies to improve air quality. For this purpose one needs to develop another model, which relates emission levels to costs of reducing them, and then to combine/couple both models. Such a combined model can be used for finding minimum-cost strategies for achieving air quality that conforms to a given standard, or for maximizing an overall index of air quality for given cost limits, or for examining various trade-offs between costs and air quality. The combined model contains more outcome variables, namely those representing costs of using various technologies that result in corresponding emission levels. Separate models need to be developed for different types of emissions, with corresponding air quality indices, and it is rational to re-combine them into one model, because some measures for decreasing emissions impact on more than one type of emission. Moreover, air quality should be assessed for several types of indices jointly, hence all relevant types of pollution should also be considered jointly.

Why not start with the "full" model instead of gradually developing one? The gradual approach has evolved in practice. First, a relatively simple (linear, with few hundred variables) model was developed, and used in practice. Then various needed submodels were developed, validated separately, and combined in larger models, until a multi-pollutant model was obtained, a complex (non-linear, non-convex with about 30,000 variables) model equipped with several methods and tools for its analysis. These submodels have been developed by several collaborating institutions, which in turn use other models to compute the data needed for specification and validation of the developed models. Such a bottom-up approach is typical for development of complex models, which are usually composed of submodels developed and validated separately.

3.3 Multi-paradigm modeling

The RAINS model is an example of modeling a real-life complex problem, but it is by no means unique. Similar approaches are applied in all fields of research and policy-making, when models are used for solving complex problems. Over time the complexity of the problems being solved has grown, and therefore the requirements of users have also grown substantially. In order to meet such requirements, the modeling technology faces new challenges: models are not only becoming more and more complex, their development typically requires results from other models, many of them using huge amount of data; in many cases a model is composed of submodels developed by various teams of specialists in different disciplines. Therefore, it is practically impossible to rationally select one modeling paradigm for a best model representation of a complex problem. On the contrary, one needs to experiment with different modeling paradigms to finally find the most appropriate one. Moreover, best paradigms are often different for different submodels.

The development of diversified modeling paradigms has been to a great extent driven by the needs of supporting decision-making for problems for which existing methods have not provided adequate support. In order to address these needs diverse types of models were developed to possibly best represent various problems by a selected type of model. These types of models are characterized by diverse types of variables and relations between them, e.g. *static, dynamic, continuous, discrete, deterministic, stochastic, set-membership, fuzzy, softly constrained, etc.* Moreover, various methods of model analysis (e.g. simulation, optimization, softly constrained inverse simulation, multicriteria model analysis) have been developed to best support various types of modeling tasks for different purposes and/or users. Finally, due to the growing complexity of various computational tasks, so-called *solvers* (dedicated software for performing such tasks) have become more and more specialized, even for what was originally the same type of mathematical programming problem. Each modeling paradigm embodies a lot of accumulated knowledge, expertise, methodology, and modeling tools specialized for solving many of the problems belonging to each modeling paradigm.

Multiparadigm modeling, defined as an efficient application of all pertinent modeling paradigms, is one of the key issues of modeling complex problems. In some situations it is possible to use a more general paradigm, which "includes" a simpler paradigm, e.g. building a non-linear model that contains a linear part.[6] Typically, instead of using a more general paradigm (one can formally treat a linear model as a non-linear one), it is rational to use a *unifying paradigm*, e.g. supporting non-linear models with a (possibly large) linear part. However, in other situations it is not practicable to unify different paradigms. In such situations one needs to *switch paradigms*. The substantial difference between switching (within a properly organized modeling process) paradigms, and applying different paradigms "independently" consists of appropriate handling of these elements of modeling process, which are common for different paradigms, and to support comparative analyses of results obtained with the help of applied paradigms.

While various modeling paradigms may differ substantially, most of them are based on a number of common concepts and elements, which are briefly discussed below.

Decision variables In model-based decision support it is assumed that decisions have quantitative characters and therefore can be represented by a set of the model variables, hereafter referred to as decisions $\boldsymbol{x} \in E_x$, where E_x denotes a space of decisions. In a trivial case $x \in R$, which denotes that a decision is represented by a real number. However, in most cases $\boldsymbol{x}$ is a vector

[6] However, even in such a case it is typically rational to use non-linear and linear paradigms to respective parts of such a model.

composed of various types of variables. For larger problems the components of $\boldsymbol{x}$ are aggregated into various types of algebraic structures.[7]

In the RAINS model the vector of emission levels $\boldsymbol{x}$ is composed of subvectors $\boldsymbol{x}_i$, where the index i corresponds to countries. Each of $\boldsymbol{x}_i$ is in turn a vector composed of vectors $\boldsymbol{x}_{i,j}$, where the index j corresponds to types of pollution. Each of $\boldsymbol{x}_{i,j}$ is also a vector composed of vectors $\boldsymbol{x}_{i,j,s}$, where the index s corresponds to industrial and consumption sectors composed of activities that emit air pollution. Each of vectors $\boldsymbol{x}_{i,j,s}$ is composed of variables representing emission levels associated with various technologies that belong to s-th sector of i-th country, and emits pollution of type j. Hence, $\boldsymbol{x}$ is actually a four-dimensional matrix, whose dimensions correspond to countries, pollution type, sectors, and technologies, respectively. Additionally, optional decision variables are considered for scenarios that allow limited violations of air quality targets. For such scenarios variables corresponding to each type of considered air quality targets are defined for each receptor. Such variables represent violations of given environmental targets. Optionally, violations of targets can be balanced with surpluses (understood as a part of the difference between a target and its corresponding actual value).

External decisions Figure 1 illustrates two types of inputs to the core model: decision variables $\boldsymbol{x}$ controlled by a user, and external decisions denoted by $\boldsymbol{z}$. In practice inputs $\boldsymbol{z}$ may include representations of various quantities that substantially influence the values of outcomes $\boldsymbol{y}$ but are not controlled by the user, for example:

- Regulations or commitments on environmental standards for air or water quality management models.
- Meteorological conditions assumed for modeling physical relations in environmental models, e.g. *an average*, or *a wet*, or *a dry*, or *a worst* year data for a water model.
- Forecasts of changes in demand for services, e.g. in telecommunication or transportation models.

In the RAINS model the external decisions $\boldsymbol{z}$ are represented by:

- Values representing the environmental standards that define constraints for various indices (such as maximum concentrations of various air quality indicators).
- Set of meteorological data used for calibration of the air pollution transport submodel.

[7] Actually, components of $\boldsymbol{x}$ are often implemented as multidimensional (typically sparse) matrices. However, for a simplification of the presentation we use the term *vector* for containers of various types of variables used in this chapter. Also for the sake of brevity we call decision variables simply decisions, and we use the term *decision* for vectors of decisions $\boldsymbol{x}$.

While the external decisions are beyond control of the user of a DSS, he/she typically wants to examine a range of scenarios with various representations of external decisions in order to find out not only a solution which will best respond to a most likely representation of external inputs z but also a solution that will be *robust*, i.e. will be good also for various other compositions of z that should be considered. From the mathematical programming point of view, the external decisions are represented as given fixed values of variables, and therefore are treated as parameters of the model.

Outcome variables The consequences of the implementation of various decisions x are evaluated by values of outcome variables $y \in E_y$. In various fields of applications outcome variables are named differently, e.g. outcomes, metrics, goals, objectives, performance indices, attributes.

In the RAINS model one outcome variable represents the sum of costs of reductions of emissions, and four sets of other outcome variables correspond to various indices of air quality. While the definition of the cost is conceptually rather simple[8], an appropriate definition of air quality indices is rather complex. Environmental effects caused by acid deposition, excess nitrogen deposition (described by a two-element linear critical loads function), and by eutrophication are evaluated at each receptor by PWL (piece-wise linear) functions that represent accumulated excesses over the thresholds of the environmental long-term targets. If optional violations of environment standards are allowed, then a maximum (over a set of receptors) violation of each type of air quality indicator is also considered as an output variable. Moreover, other outcome variables are based on measures of a distance between the emission levels, and the so-called reference levels, which correspond to various commitments.

In other models a great variety of variables is used as outcome variables. For an illustration we mention: various types of costs (e.g., investment, operating and maintenance, and total annual cost), various probabilistic estimations (e.g. expected income, probability of insolence), attributes of elements of a solution (often represented by variables of the linguistic type, e.g., having values such as excellent, good, average, bad).

Types of variables The above discussed taxonomy of variables (decision, external decisions, outcome)[9] is of a primary interest for users of a model. There is however another taxonomy of variables, namely according to the type of variable in the sense of mathematical representation, e.g. continuous,

[8] However its implementation is complex, and it requires advanced methodological and software approaches described in [3] and [39], respectively.

[9] In fact in a typical complex model a majority of variables are so-called *auxiliary variables*, which are introduced for various technical reasons, and are therefore not interested for users.

integer, binary, fuzzy, stochastic, dynamic, linguistic.[10] In many situations a choice of the variable type is determined by its nature, e.g. a continuous variable for a concentration, binary variable for a logical (yes or no) choice. However, in many other situations a choice of variable's type implies a type of simplification made for a model specification, e.g. a continuous variable for a number of cars produced by a factory.[11] A choice of variables' types is often a result of a compromise between adequacy of a problem representation and resources needed for the development of a corresponding model. Such a choice is done together with a choice of type of relations, which is discussed below. Both choices require a deep understanding of their consequences, which can hardly be formally assessed. Therefore they are based on a mixture of modeler's knowledge, experience and intuition. This is why modeling complex problems remains to be a kind of art (or craft), see e.g. [46] for a more detailed discussion of related topics.

Core (substantive) model A mathematical model used for model-based decision support is often referred to as a *core model*, or a *substantive model*, to stress its role, which is to provide an evaluation of consequences that will result from an implementation of given decisions $\boldsymbol{x}$ and $\boldsymbol{z}$ without imposing any constraints on the user's preferences.[12] As already illustrated in Figure 1, a mathematical model is used for predicting the consequences of decisions $\boldsymbol{x}$, which can be either proposed by a DM or computed by a DSS. The consequences are measured by values of outcome variables $\boldsymbol{y}$. Therefore, a model can be represented by mapping $\boldsymbol{y} = F(\boldsymbol{x}, \boldsymbol{z})$, where $\boldsymbol{x} \in E_x$, $\boldsymbol{z} \in E_z$, and $\boldsymbol{y} \in E_y$ are vectors of values of decisions, external decisions, and outcomes, respectively. For the sake of brevity we will assume further on that the external decisions z are given and represented as parameters of the mapping F.

In other words, a core model is composed of decision, output and auxiliary variables, and of the relations (inequalities, equations, etc.) between these variables that indirectly determine the set of admissible (feasible) decisions and corresponding solutions.

As already mentioned, a model specification (definitions of variables and relations between them) is based on both understanding of the modeled problem, and on modeling knowledge, experience, and intuition. It is therefore

[10] A linguistic variable is an enumeration (using a set of descriptive labels, such as high, average, low) of a finite (typically small) number of values of integer variables (or of ranges of continuous variables).

[11] Such a choice is often done consciously – because rounding large real values to a nearest integer has a very small relative error, and solving an integer programming problem is much more difficult than a relaxed (when integer variables are replaced by continuous) one.

[12] The adjective *substantive* has been used for core models in order to stress their distinction from *preferential* models that express the preferences of decision makers.

impossible to provide a complete prescription for model specification. Hence we provide only several general guidelines.

A model is a representation of a part of knowledge done for a specific purpose. As illustrated by the example of a cup of coffee in Section 3.1, such representations even of the same problem may be very different. The following practical examples illustrate some of the related issues:

- Not always much more detailed models provide better representations. A good illustration of this statement is given in [16]: the prediction of the temperature rise by doubling CO_2 in the atmosphere made in 1903 by Arrhenius is similar to predictions by modern climate models run on supercomputers.
- Application of integer variables and/or non-linear relations typically increases dramatically the demand of computing resources. In many situations a relaxed (instead of integer) problem and/or piece-wise linear (or quadratic) approximations of non-linear relations provide a satisfactory simplification of the more accurate model.
- Even for a given mathematical problem an alternative formulation (which is equivalent from the optimization point of view) can be dramatically easier to solve, see e.g. [44] for a practical example.
- Models described in text books (or used as tutorial examples in modeling software) are typically rather simple, and the corresponding modeling technology (the way of model specification and analysis) is adequate to such models. However, for complex and/or large models (typically composed of sub-models) a very different modeling technology needs to be used, see e.g. [39] for more detail.

Models are often classified into white-box or black-box models, according to how much knowledge about the types of relations between variables is used. Typically, a priori knowledge about the problem is used directly for a model specification. While this is a natural and recommended approach, sometimes the resulting model would be too complex to be actually used. In such situations it is often rational to apply a specific type of analysis to a more detailed model, which provides a basis for a specification of a simplified model, to use the simplified model for more computing resource demanding types of analysis (e.g. optimization-based), and then to check selected results with the detailed model. Such an approach has been successfully used e.g. for the RAINS model described above.

However, for many problems there is no a priori information about the relations between variables that can be effectively used for model specification. In such situations black-box models are often built based on analysis of available data. There are many successful implementations of this type of models. However, much less is published about pitfalls and risks related to using models built without a priori analytical knowledge. Such models typically rely on statistical analysis of available data, which sometimes provide also experienced modelers with very wrong results, see e.g. [28] for a recent example. However, even a correct statistical analysis may result in wrong conclusions. Consider for example analysis of a large sample of data from observations of two variables: pressure p and volume v of a gas contained in a cylinder with a variable volume. A statistical analysis of such laboratory data will likely result in a conclusion that $p \times v = const$ unless a person uses a priori knowledge, and includes the temperature into consideration. Such a trivial modeling mistake is unlikely but it illustrates the importance of using the relevant knowledge in model specification. A more elaborated discussion of simple examples can be found e.g. in [10].

Together with defining the mapping F, the core model defines a set of feasible decisions $X_0 \subseteq E_x$. In other words, $\boldsymbol{x}$ is feasible, if and only if $\boldsymbol{x} \in X_0$. The set X_0 is usually defined only implicitly by a specification of a set of constraints that correspond to logical and physical relations among all the variables used in the model.

The substantive model should include only logical and physical relations that are necessary to adequately represent relations between inputs $\boldsymbol{x}$ and outputs $\boldsymbol{y}$. In addition to inputs and outputs, a model contains various intermediate and parametric variables (balance and/or state variables, resources, external decisions), conventionally called auxiliary variables. In a typical complex model decision and output variables are a small fraction of all variables. However, auxiliary variables are introduced for easing the model specification and handling, and are typically not interesting for an end-user of the model.

Therefore, the specification of a substantive model that defines F and X_0 *should include neither any relations that reflect conditions for acceptability of a solution by a user nor a preferential structure of a DM*. In other words, a *core model* accounts only for logical and physical relations between all the variables that define the mapping F and the set X_0 of feasible solutions. All other constraints and conditions that implicitly define acceptability of a solution by a user and those that represent a preferential structure of a DM should be included into an interactive procedure of the model analysis.

Finally, we should point out that the value of a mathematical model as a decision aid comes from its ability to adequately represent reality. Therefore, there is always a trade-off between the requested accuracy (realism) of the model and the costs (also time) of its development and providing the model with data. Hence the requested accuracy should be consistent with the accuracy really needed for the model and with the quality of the available data.

Substantive model versus mathematical programming models A reader familiar with mathematical programming may be surprised, that a core model does not contain any goal function. This is a very different approach from the traditional formulation of mathematical programming problems in the form:

$$\hat{\boldsymbol{x}} = \arg\min_{\boldsymbol{x} \in X_0} \mathcal{F}(\boldsymbol{x}) \tag{1}$$

In order to provide a solution that corresponds well to the preferences of the user, the problem (1) requires:

- A unique specification of one criterion $\mathcal{F}(\boldsymbol{x})$ that adequately represents a preferential structure of a DM, which is very difficult, if at all possible, for most real-life situations.
- A definition of the set of feasible solutions X_0, which is typically not fixed, because it depends on definition on constraints corresponding to those objectives of the user, which can not be included in $\mathcal{F}(\boldsymbol{x})$.

This distinction between mathematical programming models and core (substantive) models is a very important one. Using substantive models is the recommended way of implementation of any model-based DSS.

We shall explain now *why* the core model should not contain any representation of a preferential structure of a DM.

It is usually not possible to uniquely specify a model that can yield a unique solution reflecting the preferences of a DM. For example, very often it is practically impossible (even for a good analyst or an experienced DM) to specify e.g. values for a group of constraints that would cause a feasible solution that corresponds well to preferences of a DM. In order to illustrate this point let us consider the regional water quality model described in [43]. A DM typically considers different waste water treatment technologies and the related costs, as well as standards for water quality. However he/she knows that specification of constraints for a group of (either ambient or effluent) water standards may lead to solutions that are too expensive. On the other hand, assuming constraints for costs (with water quality standards being goals) could result in an unacceptable water quality. Values of constraints are in such cases formally parameters in a corresponding optimization problem. But those values are in fact decisions that reflect the preference structure of a user. Setting constraints' value too tight would result in restricting the analysis of the problem to a (possibly small) part of feasible solutions (often resulting in analysis of only very expensive solutions, or even in making the set X_0 empty). A typical advice in such situations is to specify two types of constraints, so called hard and soft constraints which correspond to *must* and *should* types of conditions, respectively. But, in fact, dealing with soft constraints can easily be done within multiobjective model analysis, which is discussed in Section 3.6.

3.4 Modeling process

Modeling is a network of activities, often referred to as a *modeling cycle*, or a *modeling process*, composed of:

- problem formulation,
- model (symbolic) specification,
- collection, cleansing, and verification of data,

- model implementation, verification and validation,
- model analysis,
- model management, including documentation of the whole modeling process.

Typically, a modeling process starts with an analysis of the problem, including a description of objectives and questions to be answered. Subsequently, a conceptual (qualitative) version of a model is set up to support further discussions between modeler and user. In this phase selections have to be made of types of variables and the mathematical relations between them, to be used for calculating answers with the model. In the next step the modeler translates this conceptual model into a *model specification.* The latter is of a generic nature. It is composed of mathematical (symbolic) relations, and implemented using either a general purpose modeling tool or by developing a problem specific model generator. Different types of variables and relations are used depending not only on the kind of the modeled problem, but also on the choice of model type that is relevant to its future use, available data, and resources for model development, analysis and maintenance. For any non-trivial problem, model specification is an iterative process which involves a series of discussions between developers (typically OR specialists) and users until a common understanding of the problem and its model representation is agreed. Substantial changes of model specification are usually made during this process.

A *model instance* is defined by the model specification and a selection of data that define parameters of its relations. During the model implementation several model instances are created and tested in order to verify that the symbolic model specification is properly implemented. Model instances differ by various selections of data used for *instantiations* of the model specification, which typically corresponds to various assumptions about the modeled problem. Typically many instances of a model are used for different sets of data corresponding to various assumptions that the user wants to examine in order to check to what extent the model adequately represents the problem. The data typically come from different sources (often also as results of analysis of other models); therefore, assembling and making data complete and consistent (e.g. defined in units consistent with specification of model relations) is a resource consuming process. An instance of the model is also called a *substantive model,* and is composed of relations between variables defined by a selected set of data, which is used for the definition of the model parameters.

The next phase of the modeling process is *model analysis.* A typical decision problem has an infinite number of solutions, and users are interested in those that correspond to their preferences (assumptions, trade-offs), which is often called *preferential structure* of the user. A preferential structure takes different forms for different ways of model analysis, e.g. for the classical simulation it is composed of values of input variables, for the single criterion optimization it is defined by a selected goal function, for multicriteria model analysis it is defined by an achievement scalarizing function. A preferential structure typically induces partial ordering of solutions (characterized by output variables) obtained for different combinations of values of inputs. Preferential structure in a well-organized modeling process is not included in the core model, but it is defined during the model analysis phase, when users typically modify their preferences.

A properly organized analysis of a model is the essence of any model-based problem support. Properly organized means that the user is supported in using all relevant methods of analysis, comparing the results, documenting the modeling process, and also in moving back to the first stage, whenever he/she wants to change the type of the model (e.g. for handling uncertainty, or imprecision of model parameters using a different type of variables or relations). During the model analysis different *computational tasks* are generated and solved by *solvers*, which are software specialized for specific types of mathematical programming problems.

While the scheme of the modeling process has not changed substantially over time, the complexity of each of its components has substantially increased. Therefore for complex models one needs to use a new modeling technology that adequately supports development of such models, see e.g. [39] for a review.

3.5 Model analysis

While the users need to be involved in the whole modeling process, they do not need to understand all methodological and technical details of its elements. However, some more detailed understanding of the methods for model analysis is helpful for using them efficiently. Therefore further on in this chapter we concentrate on one of the most successful approaches to analysis of complex models, namely on the Reference Point Optimization method. Here we outline the links of this method with the whole modeling process.

The first step is a selection by a user, out of the set of outcome variables $\boldsymbol{y} \in E_y$ of a subset of objectives $\boldsymbol{q} \in E_q$, where E_q is a space of objectives. Quite often objectives are referred to as criteria, and in this chapter these two terms will be used interchangeably. Usually E_q is a subspace of E_y, that is, the DM select some criteria q_i from the set of outcomes E_y. Sometimes also some of the decision variables $\boldsymbol{x}$ are used as criteria, but for the sake of consistency we assume that such a variable is simply represented by one of the outcomes $\boldsymbol{y}$. Such a set of objectives is typically modified during model analysis.

A partial preordering in E_q is usually implied by the decision problem and has obvious interpretations, such as the minimization of costs competing with the minimization of pollution. However, a complete preordering in E_q cannot usually be given within the context of a mathematical programming model. In other words, it is easy to determine for each objective separately, which solution (represented by vectors $\boldsymbol{x}$ and $\boldsymbol{q}$) is the best one. However, for conflicting objectives there are two sets of solutions:

- Pareto-optimal (often called efficient), i.e. a solution, for which there is no other solution with at least one criterion that has a better value while values of the remaining criteria are not worse.
- Dominated, i.e. solutions which are not Pareto-optimal.

Obviously, a Pareto-optimal solution is preferred over any dominated solution (assuming that the selected criteria represent well the preferential structure of a DM). However, a set of Pareto-optimal solutions (often called Pareto-set, or Pareto frontier) is typically composed of a infinite number of solutions, many of which are very different. Pareto-optimal solutions are not comparable in a mathematical programming sense, i.e. one can not formally decide which is better than another one.

However, DMs are able to express own preferences for various efficient solutions. One of the basic functions of multiobjective decision support is to provide various ways in which a DM may specify her/his preferences, analyze the resulting solutions, and use the results of comparative analysis of various solutions for modifications of preferences. There is no reliable formal way for separating a specification of preferences from a process of learning from the model analysis. It is a commonly known fact that decision making is not a point event, even in situations where it is realistic to assume that the problem perception does not change during the DMP. Therefore, the possibility of using a DSS in a learning and adaptive mode is a critical feature, which has been a key motivation for the development of the reference point method discussed below.

3.6 Reference point optimization

The reference point optimization, often also called the reference point methodology, or the aspiration-based methodology of multi-objective optimization, has been presented in detail elsewhere, see, e.g., [31,81]. Here we shall recall some basic concepts related to this methodology starting, however, with some concepts basic for multi-objective optimization.

Pareto-optimality and efficiency We mentioned in Section 3.5 the concept of Pareto-optimality also called vector-optimality or multi-objective optimality. Its essence is the evaluation of a decision or its outcome by several criteria (selected between decision outcomes): the decision is Pareto-optimal if we cannot improve (find another decision that improves) one of the criteria outcomes without deteriorating other criteria outcomes. Strictly speaking, the term Pareto-optimal is usually reserved for when we either maximize or minimize all criteria. In more general cases, when we minimize some criteria, maximize others, or even do different things with other criteria (e.g., keep them close to prescribed values), we speak about efficient decisions and outcomes.

The concept of efficiency is essential for model-based decision support, because it helps to select such decisions and outcomes that might be interesting for the decision maker or decision support system (DSS) user. If we define only one criterion, represented by a so-called objective function in a mathematical programming model, then its optimization usually defines only one decision and its outcomes.

Because there are usually many efficient decisions, their evaluation and overview must be organized somehow. Suppose that the general form of a substantive model for decision support is:

$$\boldsymbol{y} = \boldsymbol{f}(\boldsymbol{x}); \quad \boldsymbol{x} \in X_0. \tag{2}$$

In this model $\boldsymbol{x} \in \mathbf{R}^n$ denotes a vector of decision variables, X_0 is a set of admissible decisions that is usually defined by a set of additional inequalities or equations called constraints, and $\boldsymbol{y} \in \mathbf{R}^m$ is a vector of model outputs or decision outcomes.

$Y_0 = \boldsymbol{f}(X_0)$ is called the set of attainable outcomes. The modeler, when analyzing the substantive model, might specify several model outputs as especially interesting; we call these objectives or criteria, and denote them by $q_i = y_j$, forming an objective vector $\boldsymbol{q} \in \mathbf{R}^k$, a vector in the objective space. While this vector and space might change during the decision process (according to specific tasks and changes of preferences specified by the modeler), we denote the relation between decisions and their outcomes by $\boldsymbol{q} = \boldsymbol{F}(\boldsymbol{x})$. $Q_0 = \boldsymbol{F}(X_0)$ is called the set of attainable objectives.

Let us assume that all objectives are, e.g., maximized. A Pareto-optimal decision and its outcome are such that there are no other admissible decisions and thus attainable objectives that would improve any objective component without deteriorating other components. A decision and its outcome that are not Pareto-optimal are called Pareto-dominated. Equivalently, Pareto-optimal decisions and outcomes can also be called Pareto-nondominated.

The sets of Pareto-optimal decisions and outcomes typically contain many elements, not just a single decision and its outcome. Thus, Pareto-optimality is an essentially weaker concept than single-criterion optimality. Pareto-optimality does not tell us which decisions to choose, it tells us only which decisions to avoid. This nonuniqueness of Pareto-optimal decisions has been considered a drawback in classical decision analysis; thus, in addition to a substantive model, a complete preferential model was usually assumed that specified a definite utility or value function whose maximum defined – hopefully, uniquely – "the optimal" decision and outcome.

Such an approach, however, is not necessarily the best for interactive decision support, where the user of the DSS (or the modeler in our case) is supposed to exchange information with the DSS, experiment with the model and modify possible solutions. The nonuniqueness of Pareto-optimal decisions is then an advantage, not a drawback. To use this advantage, we need only an additional way of controlling the selection of Pareto-optimal decisions by parameters specified by the user.

There are essentially two main methods of parameterizing Pareto-optimal decisions:

- By using weighting coefficients, i.e., specifying how much relative importance we assign to various objectives. Mathematically, the method corresponds to, e.g., maximizing the weighted sum of all objective functions over the set of admissible decisions. When the weighting coefficients are all positive, the maximization of the weighted sum results in Pareto-optimal decisions. However, more important is the issue of whether we could produce all Pareto-optimal decisions (which is called a complete parametric characterization of the Pareto frontier). When using the maximization of a weighted sum, we can sometimes produce all Pareto-optimal decisions and outcomes by changing weighting coefficients, but only under restrictive assumptions – e.g., the set of attainable objectives must be convex (or even strictly convex in order to get unique answers).
- By using goals or reference objectives in decision space, i.e., specifying what objective outcomes we would like to achieve. This method might work in a much more general setting than the method of using weighting coefficients, but it is more complicated mathematically. At first glance, an appealing mathematical method would be to minimize a distance measure or simply a norm of the difference between the goal and the attainable objective vector. Such techniques of norm minimization were first used historically, either in the displaced ideal method of Zeleny [89] or in the broad family of goal programming techniques starting with the work of Charnes and Cooper [13]. However, simple examples show that norm minimization might produce decisions that are not Pareto-optimal, thus additional assumptions are necessary. They amount, generally, to limiting the use of goals to objective values that are highly unrealistic. This motivated the development of a different approach – the reference point approach – that uses reference objectives that can be realistic, but avoids norm minimization and instead uses more complicated functions to be optimized (usually, maximized), called order-consistent achievement functions.

Thus, reference point methodology could be considered as a generalization of goal programming, aiming at using arbitrary (not only unrealistic) goals or reference objectives and obtaining only efficient outcomes, at the cost of avoiding norm minimization and replacing it by optimization of a more complicated function. We shall discuss now the relations between these methods in more detail.

The main advantages of goal programming are related to the psychologically appealing idea that we can set a goal in objective space and try to come close to it. Coming close to a goal suggests minimizing a distance measure (usually a norm of the difference) between an attainable objective vector (decision outcome) and the goal vector.

The basic disadvantage relates to the fact that this idea is mathematically inconsistent with the concept of Pareto-optimality or efficiency. One of the basic requirements – a general sufficient condition for efficiency – for a function to produce a Pareto-optimal or vector-optimal outcome (when minimized or maximized) is an appropriate monotonicity of this function. However, any distance measure is obviously not monotone when its argument crosses zero. Therefore, distance minimization cannot, without additional assumptions, result in Pareto-optimal solutions.

Consider, for example, the simplest case when the goal vector is in itself an attainable decision outcome but not an efficient objective vector; norm minimization then leads to the obvious solution with objectives equal to the goals. Even for convex outcome sets, either special tricks or restrictive assumptions are needed in goal programming to provide for efficiency of obtained decision outcomes. If, however, the set of attainable objectives is not convex (e.g., discrete, as in Figure 2), then norm minimization generally cannot result in efficient outcomes. Both objectives q_1 and q_2 in this figure are to be maximized and the Pareto-optimal outcomes, denoted by circles, are to the "North–East" of the set of attainable objectives. There are many intuitively reasonable vectors of goals, such as $\bar{\boldsymbol{q}}^1$, which would produce inefficient outcomes, such as $\boldsymbol{q}^1$, if a norm as a measure of the distance is minimized.

The problem with setting a goal and trying to come close to it is how to provide for efficiency of resulting outcomes. There are two ways to do this: either to limit the goals or to change the sense of coming close to the goal.

Trying to limit the set of goals is the essence of the displaced ideal method of Zeleny [89]: if we select goals that are sufficiently distant from the set of attainable objectives, then we can prove that norm minimization will result only in efficient outcomes, regardless of what norm we use or what properties the set of attainable objectives have. This is illustrated in Figure 2(b), where the goal $\bar{\boldsymbol{q}}$ is in the displaced ideal area and the outcomes resulting from norm minimization are efficient. However, such limitation of admissible goals means losing the intuitive appeal of the goal programming approach: if unrealistic goals have to be set, the approach loses its basic advantages.

Trying to change the sense of coming close to the goal changes the nature of the goal. Reference points are goals interpreted consistently with basic concepts of vector optimality; the sense of "coming close" to it is special and certainly does not mean norm minimization. If we accept the logic of Pareto-optimality, then coming close to a given reference point should mean:

- Objective outcomes are in some sense uniformly close to the given reference point, if the latter is not attainable (while the precise sense of uniform closeness might be modified by demanding that the resulting decisions and their outcomes remain Pareto-optimal).

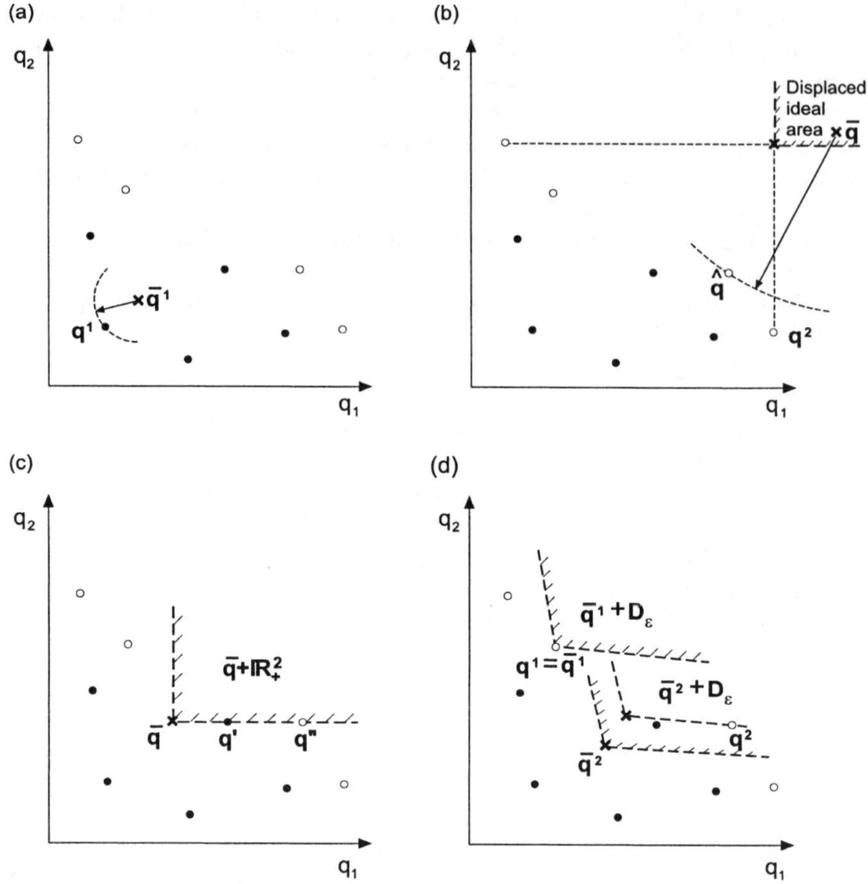

Fig. 2. Examples of discrete outcomes using various methods: (a) goal programming or norm minimization; (b) displaced ideal; (c) maxmin method; and (d) reference point method. Circles indicate nondominated points, crosses indicate goals or reference points

- Objective outcomes are precisely equal to the given reference point, if the latter is Pareto-optimal, which, somewhat simplified, means attainable without any surplus.
- Objective outcomes are in some sense uniformly better than the given reference point, if the latter is attainable with some surplus. Such reference point is inefficient, not Pareto-optimal (where the sense of its uniform improvement can be variously interpreted).

The first two cases (almost) coincide with goal programming; the third case is, however, essentially different: it means not coming close in any traditional sense, but "coming close or better".

This change of the sense of coming close is deeply related to how people make decisions in reality and how computers should support decision-making. In turn, this is related to the concept of satisficing decisions of Simon [61], which was used as a description of how people make actual decisions (particularly in large organizations) and the concept of quasi-satisficing decisions of Wierzbicki [72], which describes how a computerized DSS should help a human decision maker.

According to Simon [61], real decision makers do not optimize their utility when making decisions, for many reasons. Simon postulated that actual decision makers, through learning, adaptively develop aspiration levels for various important outcomes of their decisions. They then seek decisions that would result in one of two outcomes, that is:

- Objective outcomes as close as possible to the aspiration levels, if the latter are not attainable (which corresponds to an optimization of decisions, but in the sense of the distance from aspiration levels).
- Objective outcomes equal to aspiration levels, if the latter are attainable (which corresponds to stopping improvements).

Satisficing decision making can be mathematically represented by goal programming. In the case of attainable aspiration levels, the decision maker might learn to increase them, but usually not for current decisions, only for future ones. Many studies have shown that such satisficing behavior of a decision maker, though it might seem peculiar, is often observed in practice. In particular, the use of various reference levels by decision makers (such as aspiration levels, but also including reservation levels, of which the latter are very important in the theory of negotiations) has been repeatedly confirmed in practice.

Independent of whether a real, human decision maker would (or could, or should) optimize in all cases, we can require that a good computer program supporting decisions through model analysis should behave like a hypothetical, perfectly rational decision maker. An important exception is that the program should not "outguess" its user, the real decision maker, by trying to construct a model of her/his preferences or utility function, but should instead accept simple instructions that characterize such preferences.

Thus, the methodology of reference point approaches assumes two things: first, that the instructions from a user to the computerized DSS have the convenient (for users) form of reference points, including aspiration levels and possibly reservation levels, and second, that the user is not asked how she/he determines the reference points. An essential departure from Simon's assumptions and from goal programming techniques, however, is the following feature: the methodology of reference point approaches assumes that the computerized DSS tries to improve a given reference point, if this point is attainable.

Therefore, the behavior of the DSS – not that of its user – is similar to perfect rationality. It does not minimize a norm, but optimizes a special achievement function which is a kind of a proxy utility or value function (of the DSS) such that the decisions proposed by the DSS satisfy the three cases of "coming close or better" described previously. Because of the difference – in the last case of "coming better" – with the satisficing behavior, we call such behavior quasi-satisficing. The latter can be compared to the behavior of a perfect staff (either a single staff member or a team of them) that supports a manager or boss, who gives instructions to the staff in the form of reference levels. The staff works out detailed decisions that are guided by the given reference point.

However, being perfect, the staff does not correct attainability estimates (real, human staff might behave otherwise) and does not report to the boss that the reference point is attainable when it really is not. Instead, the staff proposes decisions that result in outcomes as close as possible to the desired reference point and reports these decisions together with their not quite satisfactory outcomes to the boss. If the reference point is attainable without any surplus, the perfect staff works out the decisions needed to reach this point and does not argue that a different point and different decisions might be better. If the reference point is attainable with surplus, the perfect staff does not stop working and start gossiping over drinks – as Simon's model of satisficing behavior would suggest. Instead, the staff works out decisions that would result in a uniform improvement of outcomes as compared to reference levels, and proposes decisions together with improved outcomes to the boss. Obviously, only a computer program could behave in this perfect, quasi-satisficing manner at all times.

However, goal programming corresponds precisely to satisficing behavior: if the aspiration levels are attainable, then there exist outcomes precisely equal to them, thus the corresponding distance is zero. Since we cannot get a distance less than zero, the optimization is stopped (the staff prepares drinks for relaxation).

Thus, reference point optimization is a generalization of the goal programming approach to such cases when we can and want to improve (minimize or maximize) certain outcomes beyond their reference points. For this purpose, a special class of order-consistent achievement functions, sometimes similar but always different from distance functions, was developed, investigated in detail, and applied in many examples and DSSs. Here we shall present only some basic concepts related to them.

Objective ranges and objective aggregation While the complexity of various types of multi-objective models has been already discussed, here we stress the characteristics of some models, already described in the simple form of equation (2). Such a model describes a mapping from the set of admissible decisions X_0 into the set of attainable objectives Q_0. The latter set is not given explicitly – we usually only know how to compute its elements – and the issue is how to analyze the elements and properties of the Pareto ("North–

East") frontier of this set. This is usually called a multi-objective analysis of a model.

After specifying any set of variables of a model as objectives q_i, the modeler should first know – at least approximately – the ranges in which these variables might vary. This is also important because we often aggregate objectives, i.e., combine them into one function (not necessarily by summation). Many objectives might also have various units of measurement and must be rescaled to dimension-free units before aggregation. Thus, any system supporting multi-objective optimization and model analysis must include a function for estimating these ranges.

The usual way of such an estimation is to compute a so-called ideal or utopia point by optimizing separately each objective, and to estimate its counterpart – the nadir point, which is composed of worst values of the objectives in the efficient set. A simple way of approximating nadir components (although certainly not the best) is to take the worst values of objective components that occur when we are computing the best values of other components during the calculations of the utopia point. The exact computations of the nadir point might be much more difficult than the computations of the ideal point, see e.g. [25].

In any case, we can assume that the modeler has defined (either arbitrarily or by computing the utopia point and estimating the nadir point) some estimates of ranges of each objective value:

$$q_{i,lo} \leq q_i \leq q_{i,up} \quad i = 1, \ldots, k \tag{3}$$

where $q_{i,up}$ (for maximized objectives, and $q_{i,lo}$ for minimized ones) is at least as high or low as the corresponding utopia point component. The range $q_{i,up} - q_{i,lo}$ is approximately as large as the range utopia–nadir. After specifying such ranges, we can reduce objectives to dimension-free scales (e.g., percentage) and then discuss the relative importance of criteria, their weights, and interpret further concepts such as the trade-off coefficients. Note that without such a reduction (therefore without estimating objective ranges) any specification of the relative importance of objectives is pointless.

For dimension-free objectives or criteria, a classical way to compute an efficient point (e.g., a Pareto point for all objectives maximized) is to maximize a weighted linear scalarizing function:

$$s_1(\boldsymbol{q}, \alpha) = \sum_{i=1}^{k} \alpha_i \frac{q_i - q_{i,lo}}{q_{i,up} - q_{i,lo}}, \quad \alpha_i \geq 0, \ \sum_{i=1}^{k} \alpha_i = 1 \tag{4}$$

where α_i are weighting coefficients that might be interpreted as weights of importance of objectives. If these coefficients are positive, then each maximal point of the above function is Pareto-optimal.

Conversely, if the set Q_0 is convex, for each Pareto-optimal decision $\hat{\boldsymbol{x}}$ there exist weighting coefficients $\hat{\alpha}_i \geq 0, \forall i = 1, \ldots k$, such that the maximum of the function $s_1(\boldsymbol{F}(\boldsymbol{x}), \hat{\alpha})$ with respect to $\boldsymbol{x} \in X_0$ is attained at $\hat{\boldsymbol{x}}$.

Thus, it might seem that the weighted linear scalarizing function covers all that is needed for multi-objective optimization. However, there are at least two important reasons to consider such functions as highly inadequate. First, the necessary condition of Pareto-optimality, stated above with the help of the linear scalarizing function, does not account for nonconvex (particularly discrete) sets Q_0. The second reason for the inadequacy of a weighted sum for multi-objective analysis is the fact that maximal points of linear scalarizing functions might depend (in important cases) discontinuously on the weighting coefficients.

For these reasons, it is better to use other functions rather than the weighted sum in order to aggregate various objectives. Historically, and particularly in goal programming approaches (see [13]), much attention has been given to the use of a norm that characterizes a distance of an attainable outcome $\boldsymbol{q} \in Q_0$ from a reference point $\bar{\boldsymbol{q}} \in \mathbf{R}^k$:

$$s_2(\boldsymbol{q}, \bar{\boldsymbol{q}}) = \parallel \boldsymbol{q} - \bar{\boldsymbol{q}} \parallel . \tag{5}$$

We assume that all objectives are maximized, but the distance function should be minimized. Unfortunately, such a distance function is inadequate if we would like to use the reference point as the main parameter to change or control the selection of Pareto-optimal solutions. We would like to admit the use of any reasonable reference point, e.g., at least in the range $\boldsymbol{q}_{lo}$, $\boldsymbol{q}_{up}$, including both $\bar{\boldsymbol{q}} \notin Q_0$ and $\bar{\boldsymbol{q}} \in Q_0$. However, if $\bar{\boldsymbol{q}} \in Q_0$, then the minimal points of such a distance function are very rarely[13] Pareto-optimal.

For a general, nonconvex case, Pareto-optimality of the minimal points of a distance function can be secured only if we assume that the goals are in the displaced ideal set, as indicated in Figure 2(b), see [89]. However, such unrealistic reference points, very distant from the set of attainable objectives, cannot easily be used as the main parameters for changing or controlling the selection of Pareto-optimal outcomes. Thus, in the displaced ideal and other similar techniques, weighting coefficients were additionally used in a distance function in order to control the selection of its minima. However, the dependence of Pareto-optimal or efficient points on weighting coefficients is not transparent enough. Therefore, the use of reference points as the main parameters controlling the selection of Pareto-optimal points would be much more attractive; reference points usually provide better interpretations for the modeler than weighting coefficients. For this reason, instead of a distance function we must use a different, albeit related, function. This function is called an achievement scalarizing function or an order-consistent achievement scalarizing function which, as opposed to a distance function, preserves the monotonicity even if the point $\boldsymbol{q} = \bar{\boldsymbol{q}}$ is crossed.

[13] Only if $\bar{q}$ happens to be Pareto-optimal.

Simple forms of achievement functions We shall discuss here only the simplest forms of achievement functions. Suppose all objectives are maximized and are already rescaled to be dimension-free and aspiration levels $\bar{q}_i$ are used for these objectives. The simplest form of an order-consistent achievement function is then as follows:

$$\sigma(\boldsymbol{q}, \bar{\boldsymbol{q}}) = \min_{1 \leq i \leq k} (q_i - \bar{q}_i) + \varepsilon \sum_{i=1}^{k} (q_i - \bar{q}_i). \tag{6}$$

where $\varepsilon > 0$ is a small positive parameter. This function, a prototype of all order-consistent achievement scalarizing functions, is monotone with respect to all q_i, thus all its maxima are Pareto-optimal. Conversely, it can be shown, see e.g. [72], that for each Pareto-optimal decision $\hat{\boldsymbol{x}}$ defined in a specific[14] way there exist reference points[15] $\bar{\boldsymbol{q}}$ such that the maximization of (6) results in finding the selected Pareto-optimal decision $\hat{\boldsymbol{x}}$.

Other order-consistent achievement functions similar to (6) were also used in reference point methodology or other similar approaches to multi-objective optimization, see e.g., [72,58,65,75]. We give here an example of a general form of an order-consistent achievement function. Note that the terms $(q_i - \bar{q}_i)$ in (6) can be considered as the simplest example of more general functions $\sigma_i(q_i, \bar{q}_i)$ of the objective variable q_i and the reference or aspiration level $\bar{q}_i$. Such a function can also depend on other parameters, e.g. on a *reservation level* $\bar{\bar{q}}_i$. Examples of such functions are shown in Figure 3. Function σ_i must have special monotonicity properties and is often defined as a piece-wise linear function; this function is called the partial (often also the component) achievement function. By substituting the terms $(q_i - \bar{q}_i)$ in (6) by component achievement functions $\sigma_i(q_i, \bar{q}_i)$, we obtain the following more general form of the order-consistent achievement function:[16]

$$\sigma(\boldsymbol{q}, \bar{\boldsymbol{q}}) = \min_{1 \leq i \leq k} \sigma_i(q_i, \bar{q}_i) + \varepsilon \sum_{i=1}^{k} \sigma_i(q_i, \bar{q}_i). \tag{7}$$

The achievement function $\sigma(\boldsymbol{q}, \bar{\boldsymbol{q}})$ – in either form presented above, as well as other similar forms – is nondifferentiable. In the case of linear models, the nondifferentiability of the achievement function is unimportant, because in practice the component achievement functions $\sigma_i(q_i, \bar{q}_i)$ are concave, strictly monotone, piece-wise linear functions (increasing for maximized, decreasing for minimized criteria, respectively), and the maximization of the

[14] I.e. with some additional requirements which, however, are reasonable in applications.

[15] Typically, there is an infinite number of such points for each $\hat{\boldsymbol{x}}$.

[16] Sometimes – if the value 1 of a component achievement function has any importance, e.g. indicating that the aspiration level is reached while the value 0 corresponds to reservation level – the entire value of such a function $\sigma(\cdot)$ might be divided by $(1 + k\varepsilon)$, in order to give it the value 1 if all component functions have value 1.

order-consistent achievement function (7) can be equivalently expressed as a linear programming problem by introducing additional constraints and auxiliary variables.[17] However, in the case of nonlinear models, optimization algorithms for smooth functions are more robust (i.e., work more reliably without the necessity of adjusting their specific parameters to obtain results) than algorithms for nonsmooth functions. Therefore, there are two approaches to the maximization of such achievement functions. The first approach is to introduce additional constraints and auxiliary variables similarly as for linear models. The second one uses a smooth approximation of the achievement function. The choice depends on the characteristics of the underlying nonlinear model. Typically, for models with a large linear part, the first approach is more efficient, while for models with a dominating non-linear part the second one is likely to be better.

The reference point approach is related not only to the method of displaced ideal or goal programming, but also to many other techniques of multi-objective optimization. A popular technique [48] is to maximize the minimal of maximized objectives or of their increases above postulated reservation levels. Its use was indicated in Figure 2(c). It corresponds to the maximization of the achievement function (6) with $\varepsilon = 0$ which, however, has some disadvantages. This minmax method guarantees only weak Pareto-optimality. In Figure 2(c), weakly Pareto-optimal is the point $\bar{q}'$, which is clearly dominated by the Pareto-optimal point $\bar{q}''$. Note that the use of the achievement function (6) with $\varepsilon > 0$, as indicated in Figure 2(d), does not produce weakly Pareto-optimal points that could be improved in one component of the objective vector.

Finally, we briefly consider the concept of a neutral or compromise solution. Zeleny [89] introduced the concept of a compromise (Pareto-optimal) solution, where its objective outcome is situated "somewhere in the middle" of the Pareto set in objective space. The precise meaning of somewhere in the middle is specified either by defining the displaced ideal point or by using various weighting coefficients. When using the basic achievement function (6) and assuming that all objectives are already rescaled to be dimension-free, we might, however, define a neutral compromise solution more precisely: it corresponds to the result of maximizing function (6) while putting the reference point precisely at the ideal point and using weighting coefficients equal to each other (i.e., not using them at all, as in the function (6)) for the dimension-free objectives with equal ranges. Such a neutral compromise solution is a good starting point in an interactive process of analyzing a multi-objective model.

When using the reference point methodology, the process of analyzing a multi-objective model is simply constructed: first we let the decision maker or the modeler play freely with reference points and ask the DSS to respond to each reference point by maximizing the achievement function, e.g., of the form (6). This produces the Pareto-optimal/efficient decisions and their objective

[17] See Section 3.8 for details.

outcomes associated with this reference point. By doing this, the decision maker or modeler gains insight into the behavior of the multi-objective, substantive model of the decision situation. This initial phase might be supplemented, if necessary, with various other techniques of using reference points when supporting choice. Experience in multi-objective model analysis shows that the initial phase of learning by freely changing reference points might be most important, particularly for supporting the development of intuition of the decision maker or the modeler.

3.7 Sensitivity analysis

In mathematical programming sensitivity analysis is typically understood as analysis of changes of an optimal solution caused by change of the data in the model.[18]

For linear programming models, a traditional approach of such analysis is based on properties of an optimal solution. It typically consists of calculations of ranges of changes of parameters for which an optimal solution does not change, and on using a dual solution for calculations of changes of value of a goal function for changes of some parameters that are small enough for allowing such a simple evaluation procedure. These methodological topics, which all form the subject of post-optimal analysis, and the corresponding software tools have been extensively developed. However, their applicability is practically limited to rather small, linear models.

Moreover, there are several problems with applications of these approaches to real problems. We restrict the discussion to the three basic issues:

- The concept and tools for sensitivity analysis have been developed and implemented for analysis of rather small models. As the complexity of models applied in decision support grows, it is either cumbersome or practically impossible to use these tools.
- For mixed-integer and non-linear types of models, there are even less developed tools for post-optimal sensitivity analysis.
- In many models the quality of dual solution is rather questionable, often the dual solution is practically non-unique. This is due to the fact that most of large models are numerically badly conditioned. Moreover, one uses for them efficient presolve algorithms, which greatly decrease the resources (time and memory) needed for solving large problems. However, these algorithms guarantee the quality of the primal solution but often result in unreliable dual solution, which is the basis for classical sensitivity analysis. Therefore, such an analysis requires a good understanding of

[18] There exist also more general approaches to sensitivity analysis, concerned with the consequences of using conclusions such as an optimal solution or even an optimal feedback law derived from one model and applied to another model (e.g. with changed parameters), see [73], but the necessary tools for such generalized sensitivity analysis are not further developed yet.

various techniques and corresponding tools, which limits its applications to highly skilled specialists in mathematical programming.

These reservations concern sensitivity analysis in classical, scalar optimization (single-objective mathematical programming). In vector optimization and multiobjective decision support, basic concepts of sensitivity analysis must be changed. Mathematically, it is possible to analyze the sensitivity of the entire Pareto set, but it would require solutions of many more parametric optimization problems than is needed by more restricted single-criterion optimization approaches. Another solution would be to apply in a multiobjective analysis a generalized approach to sensitivity analysis as in [73]. However, even in such cases experiments with changed parameters for more complicated problems can grow very fast in volume.

On the other hand, vector optimization and multiobjective decision support offer better ways for providing some type of generalized sensitivity analysis. In classical single-criterion optimization only one criterion can be selected as a goal function, therefore typically several objectives are treated as constraints, for which one had to specify an acceptable value. However, in practice such values cannot be specified precisely, therefore their modifications are inevitable. This in turn requires sensitivity analysis of the impact of changes caused by specified constraining values for criteria that are treated as constraints. One needs to note that sensitivity analysis is typically based on dual solutions, which are often (especially for large and/or badly conditioned problems) not robust. Moreover, results of such an analysis are valid only in a (typically small) neighborhood of an optimal solution. These two facts commonly known for practitioners who deal with complex models explain why sensitivity analysis of optimal solutions provide very limited support for analysis of multicriteria problems that are analyzed through classical single-criterion optimization methods. These limitations and the demand for a truly multicriteria model analysis have been in fact one of the driving forces for the development of methods and tools for multiobjective model analysis.

3.8 Softly constrained inverse simulation

An important help for the modeler can be the *inverse simulation*, in which she/he assumes some desired model outcomes $\bar{\boldsymbol{y}}$ and checks – as in the classical goal programming – whether there exist admissible decisions which result in these outcomes. *Generalized inverse simulation* consists in specifying also some reference decision $\bar{\boldsymbol{x}}$, and in testing, whether this reference decision is feasible and results in the desired outcomes $\bar{\boldsymbol{y}}$. This can be written in the goal programming format of the minimization of a norm, for which it is useful to apply the augmented Chebyshev norm. However, because norms are minimized, and we want to keep to the convention that the achievements are usually maximized, we use the following form of the achievement function:

$$\sigma(\boldsymbol{y}, \bar{\boldsymbol{y}}, \boldsymbol{x}, \bar{\boldsymbol{x}}) = -(1-\rho)(\max_{1 \leq i \leq n} |x_i - \bar{x}_i| + \varepsilon \sum_{i=1}^{n} |x_i - \bar{x}_i|)$$
$$-\rho(\max_{1 \leq j \leq m} |y_j - \bar{y}_j| + \varepsilon \sum_{j=1}^{m} |y_j - \bar{y}_j|) \quad (8)$$

The coefficient $\rho \in [0; 1]$ indicates the weight given to achieving the desired output versus keeping decisions close to their reference values. It is assumed for simplicity sake that all variables are already re-scaled to be dimension-free.

A multi-objective optimization system based on reference point methodology can clearly be used also for inverse simulations. In such cases, we stabilize all outcomes and decisions of interest, and use for them partial achievement functions of the form $\sigma_i(y_i, \bar{y}_i)$ (or even $\sigma_i(y_i, \bar{y}_i, \bar{\bar{y}}_i)$), similar to those discussed in the Section 3.6 in terms of objectives q_i. Then an overall achievement function takes the form:

$$\sigma(\boldsymbol{y}, \bar{\boldsymbol{y}}, \boldsymbol{x}, \bar{\boldsymbol{x}}) = (1-\rho)(\min_{1 \leq i \leq n} \sigma_i(x_i, \bar{x}_i) + \varepsilon \sum_{i=1}^{n} \sigma_i(x_i, \bar{x}_i))$$
$$+\rho(\min_{1 \leq j \leq m} \sigma_j(y_j, \bar{y}_j) + \varepsilon \sum_{j=1}^{m} \sigma_j(y_j, \bar{y}_j)) \quad (9)$$

It is more convenient for the modeler, if such functions are defined inside the decision support system which also has a special function *inverse simulation,* prompting her/him to define which (if not all) decisions and model outputs should be stabilized and at which reference levels.

Even more important for the modeler might be another generalization of the above function, called *simulation with elastic constraints* or *softly constrained inverse simulation.* Common sense decisions might appear inadmissible for the model, because it interprets all constraints as *hard* mathematical inequalities or equations. On the other hand, we have already stressed that it is a good modeling practice to distinguish between *hard constraints* that can never be violated and *soft constraints* which in fact represent some desired relations and are better represented as additional objectives with given aspiration levels. Thus, in order to check actual admissibility of some common-sense decision $\bar{\boldsymbol{x}}$, the modeler should first answer the question which constraints in her/his model are actually hard and which might be softened and included in the objective vector $\boldsymbol{q}$. Thereafter, inverse simulation with soft constraints might be performed by maximizing an overall achievement function similar as above, but defined with respect to objectives q_i:

$$\sigma(\boldsymbol{q}, \bar{\boldsymbol{q}}, \boldsymbol{x}, \bar{\boldsymbol{x}}) = (1-\rho)(\min_{1 \le i \le n} \sigma_i(x_i, \bar{x}_i) + \varepsilon \sum_{i=1}^{n} \sigma_i(x_i, \bar{x}_i))$$

$$+\rho(\min_{1 \le j \le m} \sigma_j(q_j, \bar{q}_j) + \varepsilon \sum_{j=1}^{m} \sigma_j(q_j, \bar{q}_j)) \tag{10}$$

where we do not assume that all objectives are stabilized, but *include also maximized or minimized objectives e.g.* corresponding to inequalities in soft constraints. Again, it is better if the multi-objective optimization system has a special function called *elastic simulation*, which invokes the specific form of the achievement function and prompts the modeler to specify necessary data – in particular, to indicate constraints that might be softened.

We should stress here that, if either (9) or (10) is maximized with concave piece-wise linear partial achievement functions σ_i and for a linear model, then the underlying optimization problem can be converted to linear programming. In fact, if a partial achievement function – say, $\sigma_i(x_i, \bar{x}_i)$ – is piece-wise linear and concave, then it can be expressed as the minimum of a number of linear functions:

$$\sigma_i(x_i, \bar{x}_i) = \min_{l \in L_i} \sigma_{il}(x_i, \bar{x}_i) \tag{11}$$

where $\sigma_{il}(x_i, \bar{x}_i)$ are linear functions. Assume that a similar expression is valid for $\sigma_j(q_j, \bar{q}_j)$. The maximization of the function (10) can be then equivalently expressed as the maximization of the following function of auxiliary variables z, z_i, w, w_j:

$$(1-\rho)(z + \varepsilon \sum_{i=1}^{n} z_i) + \rho(w + \varepsilon \sum_{j=1}^{m} w_j) \tag{12}$$

with additional constraints:

$$\begin{aligned} \sigma_{il}(x_i, \bar{x}_i) &\ge z_i, \ \forall l \in L_i \\ z_i &\ge z, \ \forall i = 1, \dots n \\ \sigma_{jl}(q_j, \bar{q}_j) &\ge w_j, \ \forall l \in L_j \\ w_j &\ge w, \ \forall j = 1, \dots m \end{aligned} \tag{13}$$

Typically, component achievement functions $\sigma_i(\cdot)$ are concave, and therefore the maximization of an achievement function can be performed by the maximization of an auxiliary linear function with several additional (to the underlying model) linear constraints. Therefore the resulting optimization problem has similar characteristics as an optimization problem composed of a linear goal function and the constraints defined by the model. For linear models defined by only continuous variables this is a recommended approach.

Models with integer variables are typically much more difficult to be optimized than corresponding relaxed (where integer variables are replaced by continuous variables) models. For models with integer variables one can be tempted to consider non-concave piece-wise linear component achievement functions $\sigma_i(\cdot)$ because such functions can be represented by conversions that use auxiliary binary and continuous variables, which results in adding a small mixed-integer linear part to the underlying model that for medium size models does not increase remarkably the size of the original problem. However, the resulting optimization problem may be much more difficult to be solved than a problem resulting from concave component achievement functions. This is more likely for problems for which heuristic algorithms have to be applied for optimization. Therefore it is recommended to use concave piece-wise linear component achievement functions $\sigma_i(\cdot)$, which in practice provides a sufficiently good representation of user preferences. This approach does not remarkably increase the complexity of the resulting optimization problem. This recommendation is even much more stronger for non-linear models, because an introduction of integer variables to such models dramatically increases the complexity of the resulting optimization problem. However, as already discussed in Section 3.6, for some non-linear models smooth approximations of concave piece-wise linear functions $\sigma_i(\cdot)$ can be a more efficient approach that the use of original functions.

3.9 Relations between fuzzy sets and reference point methods

Theoretical background and a number of applications of fuzzy sets to decision making are discussed e.g. by Zimmermann [92] and Sakawa [57]. Both the fuzzy membership functions (discussed in detail e.g. in [90,91]), and the component achievement functions discussed in this chapter reflect the degree of satisfaction with a given set of criteria values. Therefore, it is interesting to consider some similarities and differences in applications of the Reference Point Optimization and Fuzzy Sets methods to model-based decision-making support.

Both approaches (fuzzy sets and reference point) assume, that for each criterion the user selects two values (called here the *aspiration* and the *reservation* level, respectively). The user considers solutions for which the criterion value is better than the aspiration level as good, and solutions for which this value is worse than the reservation level as bad. A goodness of solutions for which a criterion value is between the aspiration and reservation values depends on a measure of distance between the criterion value, and the aspiration and reservation levels.

In the fuzzy sets approach a value of a membership function (denoted[19] here by $\mu(q)$) is interpreted as a degree to which the corresponding solution

[19] For a simplification of the presentation, criteria indices in membership functions $\mu_i(\cdot)$, and in component achievement functions $\sigma_i(\cdot)$ are omitted in this subsection.

belongs to a set of *good* solutions. For maximized criteria $\mu(q)$ is a non-decreasing function conforming to the following requirements:

$$\begin{array}{lclll} \mu(q) & = & 0 & \text{for} & q \le R \\ 0 & < \mu(q) < & 1 & \text{for} & R < q < A \\ \mu(q) & = & 1 & \text{for} & q \ge A \end{array} \qquad (14)$$

where R and A denote the reservation and aspiration values, respectively.

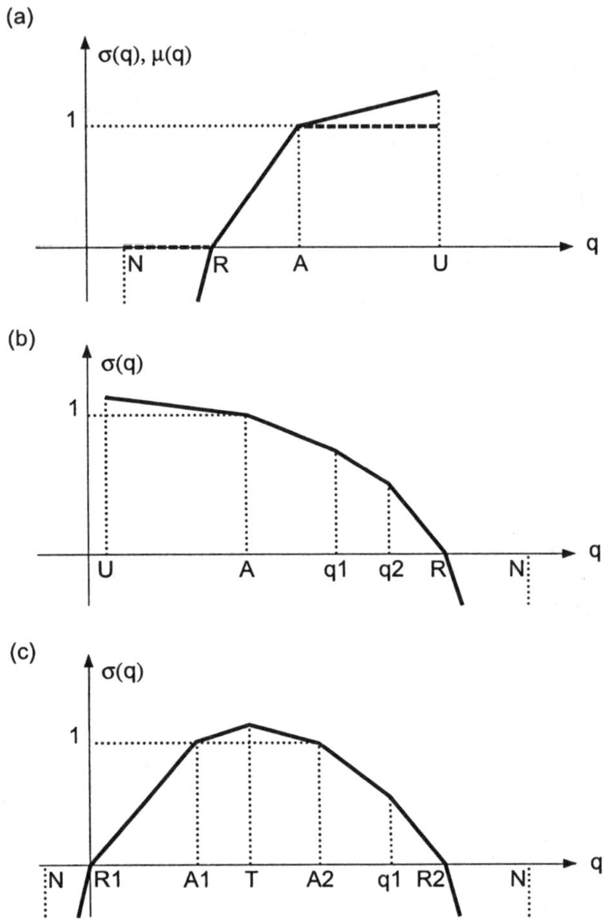

Fig. 3. Examples of component achievement functions $\sigma(q)$ for three types of criteria: (a) maximized (together with the corresponding membership function $\mu(q)$ shown by the dashed line), (b) minimized, (c) goal (stabilized), respectively. The letters N, R, A, U, and T denote values of nadir, reservation, aspiration, utopia, and target, respectively

An example of such a function is shown in Figure 3 (a) by the dashed line (which for the segment corresponding to values $R \leq q \leq A$ is covered by the solid line). The best and worst criterion values are called *Utopia* and *Nadir*, respectively, and are typically computed during the initial phase of the problem analysis. These values are denoted in Figure 3 (a) by letters U and N, respectively.

The reference point approach described in this chapter uses a so called extended-valued membership function as a component achievement functions $\sigma(\cdot)$. Such an extension of the membership function concept has been proposed in [21], who also suggested a method of constructing various forms of order-consistent component achievement scalarizing functions based on membership functions describing the satisfaction of the user with the attainment of separate objectives. For a maximized criterion $\sigma(q)$ is a strictly increasing function conforming to the following requirements:

$$\begin{array}{lcllr} \sigma(q) & < & 0 & \text{for} & q < R \\ \sigma(q) & = & 0 & \text{for} & q = R \\ 0 & < \sigma(q) < & 1 & \text{for} & R < q < A \\ \sigma(q) & = & 1 & \text{for} & q = A \\ \sigma(q) & > & 1 & \text{for} & q > A \end{array} \tag{15}$$

An example of the most simple (defined by only four points: utopia, aspiration, reservation, and nadir) PWL (piece-wise linear) function $\sigma(\cdot)$ for a maximized criterion is shown in Figure 3 (a) by the solid line.[20]

An example of a PWL function $\sigma(\cdot)$ for a minimized criterion is shown in Figure 3 (b). For this example the user has specified (in addition to the aspiration and reservation levels denoted by A, and R, respectively) values of his/her degree of satisfaction for two more values of criteria, denoted by q1, and q2, respectively. Note that the concavity requirement corresponds well to the nature of the problem since one accepts small changes of $\sigma(\cdot)$ when a criterion value is better or close to an aspiration level. The speed of such a change should increase along with moving away from an aspiration level towards a reservation level, and should be greatest for the criteria values worse than the corresponding reservation level. Such features of $\sigma(\cdot)$ are justified not only by a theoretical reasoning but also by experience from real-world applications, which shows that users are primarily interested in improving those criteria whose values are close to the corresponding reservation levels (especially, if the values are much worse than these levels).

In some (e.g. engineering) applications a criterion should neither be maximized nor minimized, but it is desired that its value is possibly close to a *target*. For such cases a *goal type* (also called *stabilized*) criterion is suitable. An example of a component achievement function $\sigma(\cdot)$ for such criterion is

[20] For practical reasons (the value of $\sigma(N)$ is typically much smaller than -10, see e.g. [38] for details) only a small part of the segment for criteria values smaller than the reservation level is shown.

shown in Figure 3 (c). The goal type criterion requires a specification of a target value (denoted by T), and of two pairs of aspiration and reservation levels (for the criterion values smaller and greater than the target value, respectively), which are denoted in Figure 3 (c) by pairs (A1, R1) and (A2, R2), respectively. For the goal-type criterion a component achievement function is composed of two parts, defined respectively for the criterion values smaller and larger than the given target. Obviously, the first (second) part has properties of $\sigma(\cdot)$ for maximized (minimized) criteria, respectively. Optionally, the user can specify a level of his/her satisfaction with the criterion values between aspiration and reservation levels, as illustrated by the value q1 in Figure 3 (c).

The minimized, maximized, and goal types of criteria are the most commonly known and used. However, the Reference Point Optimization method has been successfully applied also for other types of criteria. Several of them are discussed in [38]. Moreover, we would like to point out that modular software tools are available for easy implementations of user-friendly interfaces to the Reference Point Optimization method (including an interactive definition of criteria, components achievement functions $\sigma(\cdot)$, comparisons of solutions, etc.), see e.g. [40] for details.

The above summary shows that there are many similarities, and many differences between the Reference Point and the *Fuzzy Multi-objective Programming* approaches. The main similarity is in the interpretation of values of the membership function $\mu(\cdot)$ and the component achievement function $\sigma(\cdot)$ for the criteria values between the aspiration and reservation levels: both functions reflect a goodness of the corresponding solution in respect to the corresponding criterion. There are however two types of differences between these two approaches.

First, functions $\sigma(\cdot)$ for minimized and maximized criteria are strictly monotone in their whole domains, while membership functions $\mu(\cdot)$ are constant for the criterion values better than the aspiration level, or worse than the reservation level. Thus the maximization of functions $\mu(\cdot)$ does not force improvements of criteria, if their reservation and/or aspiration levels were selected too optimistically and/or too pessimistically, respectively.

The second difference is due to the specification and use of the $\mu(\cdot)$ and $\sigma(\cdot)$ functions.

The fuzzy set approaches assume that the membership functions are elicited before the interactive analysis of the problem. The interactive fuzzy multi-objective programming as proposed in [59,57] uses given membership functions (one function for each criterion) for the interactive procedure in which the user specifies the aspiration levels of achievement of the membership values for all of the membership functions, called the reference membership level. Hence, this method requires prior specification of aspiration and reservation levels which are used for the definitions of the membership function. Moreover, it is implicitly assumed that the criteria values for all interesting solutions are between the corresponding aspiration and reservation levels (because the applied membership function does not differentiate between solutions with values better than the aspiration level and between those with values worse than the reservation level). The user interactively specifies the reference membership levels for each membership function, which can be interpreted as a degree of achievements of the aspiration for each criterion (scaled by the difference between aspiration and reservation). Therefore, in this approach the user cannot change aspiration levels in terms of criteria values, because they have to be specified a priori for the definition of the membership function.

The reference point method does not use the membership function directly. It assumes that the user may change aspiration and reservation levels (and optionally specify her/his preferences for selected values of criteria between the aspiration and reservation levels) during the interaction upon the analysis of previously obtained solutions. The user interactively specifies the preferences in the space of the criteria values which seems to be more natural than a specification of preferences in terms of degrees of achievements of membership function values. A selection in the criteria space can, however, be interpreted in terms of fuzzy sets by a definition of a membership function for a linguistic variable (e.g. *good solution*) for each criterion, and an ex post interpretation to which degree a solution belongs to a set of *good* solutions. Moreover, there is no need for restrictions for the specification of aspiration and reservation levels in the criteria space. This is important for the analysis of large-scale complex problems for which it might be difficult to specify attainable/non-attainable reservation/aspiration levels, respectively.

Namely, the membership functions $\mu(\cdot)$ have to be elicited before an initial iteration, while the component achievement functions $\sigma(\cdot)$ are interactively changed by the user upon analysis of obtained solutions, which allows for progressive changes of the achievement functions during a learning process.

4 Decision Support Systems

Given the complexity of the DMP, and of multi-paradigm modeling, discussed in Sections 2 and 3, respectively, an appropriate computerized support for decision-making related to any complex problem can only be provided by a customized computing environment, which is conventionally called a Decision Support System (DSS).

The concept of a DSS was first used in the late 1970s, see e.g. [67], in opposition to data processing systems developed in 1950s. A DSS was supposed to have more functions than Data Base Management Systems (DBMS), or Management Information Systems (MIS), see e.g. [14] and [17], respectively. One of the early definitions of DSS given by Sprague [63] stresses that a DSS is an interactive computer-based system designed to help DMs use data and models to find solutions to unstructured[21] problems. While the classical definition of DSS can still be considered valid, the meaning of "use data and models" has changed substantially along with the change of context of modern decision-making discussed in Section 2. Thus also the DSS functionality and architecture has changed considerably in comparison to DSSs developed earlier than even a decade ago. These developments have been possible due to fast changes in hardware and software technologies, which have made it possible to address the needs of DSS users in a far better way than previously.

Models developed for decision-making support represent this part of knowledge relevant to a decision making process, which can be used in a more efficient way in a computerized form than provided by other means. The meaning of this statement implies that the models, and the way they are used cannot be defined a priori. The knowledge useful for model-based representation is typically heterogeneous, therefore it is often initially represented by several models developed and tested separately before they are combined into one model, or in several larger models that are used as a system of models. In fact most of successful applications show that the requirements for model-based support change along a (typically long) process of development and application of a model. In some cases the modeling process (which starts with structuring the modeled problem) is even more important than the final version of a model, and the end-product of this process, which is a DSS into which the developed model is included. Moreover, support of different phases of DMP (discussed in Section 2) may require different models.

A modeling process is typically composed of the following interlinked elements:

- model (symbolic) specification,
- collection and verification of data to be used for model instantiation,
- model verification and testing,
- generation of various instances of the model,
- comparative analysis of each of these instances.

The modeling process is in fact also a learning process for both the model developers and future users. During this adaptive process the assumptions and requirements for the model are modified until a best possible fit between possibilities offered by the modeling technology, available knowledge, and data on one side, and the needs of the users on the other side is achieved. A more detailed discussion of these elements of the modeling process (related to the RAINS model outlined in Section 3.2 but applicable also to many other complex models) can be found in [37]. Different methods and tools are used for different elements of modeling process, which typically requires a substantial amount of resources. Therefore it is typically rational to reuse

[21] Unstructured in this context means that a problem cannot be adequately represented by a mathematical programming problem, whose unique solution provides a satisfactory advice for making a decision.

software that was developed (and tested) for other applications. However, for making reusing software components practicable, the modeling environment has to be composed of modular tools that can be configured according to various needs.

In order to illustrate the advantages of a modular structure of a DSS, we outline here the architecture of two DSSs applied in two different areas, namely regional water quality management, and land use planning. Each of these applications was developed for simpler (than that for which the the RAINS model was developed) problems. However, they illustrate well a modular structure of a DSS necessary when using various modeling paradigms. Moreover, they also show how easily one can use in practice various modular modeling tools to develop a DSS which supports the reference point methodology to a model analysis.

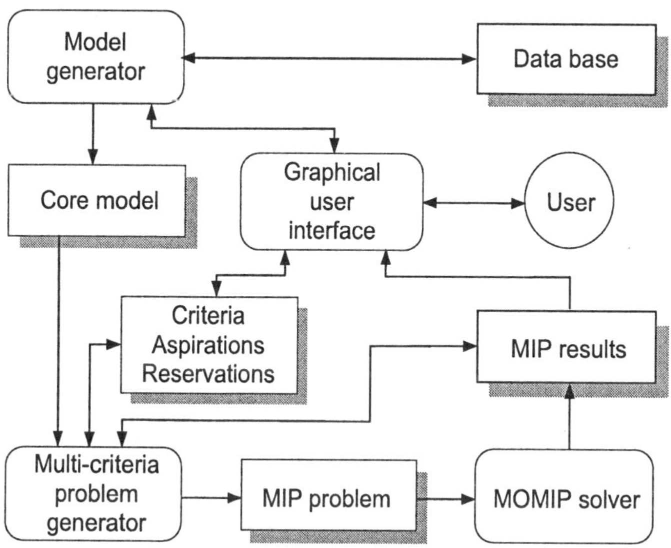

Fig. 4. The structure of a Decision Support System for the water quality management in the Nitra River River basin

The structure of the DSS used for the water quality management in the Nitra Basin, see [43] for details, is illustrated in Figure 4. The implementation of this DSS is based on modular and portable software tools that support the following functions of a DSS:

- A problem-specific model generator for generating the core model that relates wastewater emissions, treatment decisions, and the resulting ambient water quality. It is important to stress again that the core model includes only physical and logical relations between the decision and output variables, and not the preferential structure of the DM.

- A graphical user interface (GUI) for handling all the interaction with the user, who uses the GUI to first select an instance of the model he/she wants to analyze, and then specifies during an interactive process his/her preferences. The latter is done by selecting outcome variables that are used as criteria. Then for each selected criterion the user defines (through a specialized graphical interface) a component achievement function $\sigma_i(\cdot)$, which has properties discussed in Section 3.6.
- The GUI is linked with the MCMA package, which handles the definitions of criteria, aspiration, and reservation levels, as well as the generation of a multi-criteria problem. The MCMA includes ISAAP (Interactive Specification and Analysis of Aspiration-based Preferences by Granat and Makowski [20], which is directly linked with LP-MULTI by Makowski [36], which in turn is a modular tool for handling multi-criteria problems using the reference point methodology presented in more detail in Section 3.6. The resulting parametric mixed-integer programming (MIP) problem is based on the core (substantive) model and on an interactively specified preferential structure of a DM.
- A modular solver for mixed integer programming problems, MOMIP, developed by Ogryczak and Zorychta [45].
- A data interchange tool LP-DIT [35]. This tool provides an easy and efficient way for defining and modifying MIP problems, as well as interchanging of data between a problem generator, a solver, and software modules that serve for problem modification and solution analysis.

The structure of the DSS for land-use planning described in [4] is the same as that presented above. Only one substantial module, namely, another problem-specific model generator for generating the land-use core model, had to be developed. Another difference is of a technical nature. The resulting optimization model is of the LP type, therefore another solver is automatically selected by the DSS instead of the MOMIP solver.

In addition to problem-specific model generators (and associated utilities for data management), each of these two DSSs has problem-specific utilities for analysis of the results, but the kernel of the software for both DSSs is the same. This illustrates well the power of modular modeling tools, and the associated efficiency resulting from the structured design of software.

A more detailed discussion of the DSS architecture is beyond the scope of this chapter. However, the relations between the decision-making process and the DSS architecture, its basic components, and implementation issues, all illustrated by a modular DSS actually implemented for environmental case studies, are presented in more detail by four case studies described in [41]. The fast developments in various areas that contribute to the decision support methods and technology call for new approaches to knowledge management and representation by analytical models, and its use for decision-making support by utilization of also fast developments of global information networks. The implications of these developments on future DSS are discussed in [42].

5 Conclusions

This chapter focused on the use of models of knowledge in decision support. All these concepts – of knowledge, models, decisions and support – can be used in very general but also in a very specific sense. We have discussed some more specific interpretations of these concepts that resulted from our experience in diverse applications of decision support.

Because of the variety of decision problems and of habitual domains of DMs, there will probably never be just one method of model-based decision support. In fact, no single modeling paradigm alone is sufficiently good enough to identify and analyze various enough policy options for any complex decision problems that are necessary for making rational decisions. Rather, an integration of various modeling methods and tools is needed to provide the best available support possible to analyze complex problems.

It is worth to note that a number of diverse modeling paradigms has been developed to a large extent in parallel by researchers who tried their best to find better (than those offered by traditional mathematical programming) methods for solving various real-world problems. For example, the fuzzy set approaches were motivated by problems that could be adequately represented neither by traditional (crisp) models nor by stochastic models. Similarly, multicriteria model analysis methods were developed in response to limitations of traditional single-criterion optimization approaches. For a quite long time these two streams have been nurtured by research communities that were to a large extent separated. In this chapter we have shown that in fact some concepts in both approaches are rather similar (although named differently), and that both approaches belong to the class of methods that are now described as *soft computing*. Both of them have contributed greatly to model-based decision support, because *support* implies that only *soft computing* (as opposed to approaches based on providing an optimal solution of a precisely defined mathematical programming problem) can be helpful in a better understanding of a complex problem, which is a prerequisite for finding a satisfactory solution to it.

Lessons learned from the applications of various modeling paradigms to diverse types of real-world problems, and the recent abundance of computing hardware and software tools makes it possible to integrate several methods of a model specification and analysis, and to apply them to large and complex problems. Such an integration calls for a collaboration of specialists, who have concentrated their efforts – and therefore have substantial experience – in a particular method. Therefore, one should expect that diverse integrations of various modeling paradigms will be used more broadly to improve decision-making support in a wide range of practical problems.

However, the key role in actual decision making will stay with human decision makers. Thus, we have also stressed the role of human intuition – together with a rational theory of intuition that makes it possible to understand how to enhance intuition by computerized decision support. We have also stressed the interactive, learning aspects of decision support processes. All these aspects can be summarized by one concluding postulate of *user sovereignty*. This postulate includes the postulate of *user friendliness* of information systems, but is further reaching, and describes the dominant role of a human decision maker in human-computer interactions.

References

1. Ackoff, R.: 1967, Management misinformation systems, *Management Science* **14**(4), 43–89.
2. Ackoff, R.: 1979, The future of operational research is past, *Journal of OR Society* **30**(2), 93–104.
3. Amann, A. and Makowski, M.: 2000, Effect-focused air quality management, in [81], pp. 367–398. ISBN 0-7923-6327-2.
4. Antoine, J., Fischer, G. and Makowski, M.: 1997, Multiple criteria land use analysis, *Applied Mathematics and Computation* **83**(2–3), 195–215 (available also as IIASA's RR-98-05).
5. Axelrod, R.: 1984, *The Evolution of Cooperation*, Basic Books, New York.
6. Bergson, H.: 1903, Introduction à la métaphysique, *Revuc de la métaphysique et de morale* **11**, 1–36 (translated by T.E. Hulme as *Introduction to Metaphysics*, New York, 1913 and 1949).
7. Bertalanffy, L. von: 1968, *General Systems Theory: Foundations, Development, Applications*, Braziller, New York.
8. Bosc, P. and Kacprzyk, J.: 1995, *Fuziness in Database Management Systems*, Springer Verlag, Berlin, New York.
9. Carlsson, C. and Fullér, R.: 2002, *Fuzzy Reasoning in Decision Making and Optimization*, Physica Verlag, New York.
10. Cartwright, N.: 1999, *The Dappled World. A Study of the Boundaries of Science*, Cambridge University Press, Cambridge.
11. Chapman, C.: 1988, Science, engineering and economics: OR at the interface, *Journal of Operational Research Society.*
12. Charnes, A. and Cooper, W.: 1967, *Management Models and Industrial Applications of Linear Programming*, J. Wiley & Sons, New York, London.
13. Charnes, A. and Cooper, W.: 1977, Goal programming and multiple objective optimization, *J. Oper. Res. Soc.* **1**, 39–54.
14. Codd, E.: 1970, A relational model for large shared data banks, *Comm. ACM* **13**(6), 377–387.
15. Dreyfus, H. and Dreyfus, S.: 1986, *Mind over Machine: The Role of Human Intuition and Expertise in the Era of Computers*, Free Press, New York.
16. Edwards, P.: 2001, Representing the global atmosphere: Computer models, data, and knowledge about climate change, *in* C. Miller and P. Edwarrds (eds), *Changing the Atmosphere. Expert Knowledge and Environmental Governance*, The MIT Press, Cambridge, London, pp. 31–65.
17. Emery, J.: 1987, *Management Information Systems, The Critical Strategic Resource*, Oxford University Press, New York.
18. Fink, E.: 2002, *Changes of Problem Representation*, Springer Verlag, Berlin, New York.

19. Glushkov, V.: 1972, Basic principles of automation in organizational management systems, *Upravlayushcheye Sistemy i Mashiny.*
20. Granat, J. and Makowski, M.: 2000, Interactive Specification and Analysis of Aspiration-Based Preferences, *EJOR* **122**(2), 469–485 (available also as IIASA's RR-00-09).
21. Granat, J. and Wierzbicki, A. P.: 1994, Interactive specification of DSS user preferences in terms of fuzzy sets, *Working Paper WP-94-29*, International Institute for Applied Systems Analysis, Laxenburg, Austria.
22. Grauer, M., Thompson, M. and Wierzbicki, A. (eds): 1985, *Plural Rationality and Interactive Decision Processes*, Vol. 248 of *Lecture Notes in Economics and Mathematical Systems*, Springer Verlag, Berlin, New York.
23. Hacking, I.: 1964, *Scientific Revolutions*, Oxford University Press, Oxford.
24. Hloyningen-Huene, P.: 1993, *Restructuring Scientific Revolutions*, The University of Chicago Press, London.
25. Isermann, H. and Steuer, R. E.: 1987, Computational experience concerning payoff tables and minimum criterion values over the efficient set, *European J. Oper. Res.* **33**, 91–97.
26. Kahneman, D. and Tversky, A.: 1982, The psychology of preferences, *Scientific American* **246**, 160–173.
27. Keeney, R. and Raiffa, H.: 1976, *Decisions with Multiple Objectives: Preferences and Value Tradeoffs*, J. Wiley & Sons, New York.
28. Knight, J.: 2002, Statistical error leaves pollution data up in the air, *Nature* **417**(13), 677.
29. Kuhn, T.: 1964, A function of thought experiments, *in* I. Hacking (ed.), *Scientific Revolutions*, Oxford University Press, Oxford (originally published in *L'aventure de la science, Melanges Alexandre Koyre*, Vol. **2**, pp. 307-334, Hermann, Paris 1964).
30. Kuhn, T.: 1970, *The Structure of Scientific Revolutions*, The University of Chicago Press, Chicago.
31. Lewandowski, A. and Wierzbicki, A. (eds): 1989, *Aspiration Based Decision Support Systems: Theory, Software and Applications*, Vol. 331 of *Lecture Notes in Economics and Mathematical Systems*, Springer Verlag, Berlin, New York.
32. Liu, B.: 2002, *Theory and Practice of Uncertain Programming*, Springer Verlag, Berlin, New York.
33. Lorentz, K.: 1965, *Evolution and Modification of Behavior: A Critical Examination of the Concepts of the "Learned" and the "Innate" Elements of Behavior*, The University of Chicago Press, Chicago.
34. Maclean, D.: 1985, Rationality and equivalent redescriptions, *in* M. Grauer, M. Thompson and A. Wierzbicki (eds), *Plural Rationality and Interactive Decision Processes*, Vol. 248 of *Lecture Notes in Economics and Mathematical Systems*, Springer Verlag, Berlin, New York, pp. 83–94.
35. Makowski, M.: 1994a, LP-DIT, Data Interchange Tool for Linear Programming Problems, (version 1.20), *Working Paper WP-94-36*, International Institute for Applied Systems Analysis, Laxenburg, Austria (available on-line from `http://www.iiasa.ac.at/ marek/pubs`).
36. Makowski, M.: 1994b, Methodology and a modular tool for multiple criteria analysis of LP models, *Working Paper WP-94-102*, International Institute for Applied Systems Analysis, Laxenburg, Austria (available on-line from `http://www.iiasa.ac.at/ marek/pubs/`).

37. Makowski, M.: 2000, Modeling paradigms applied to the analysis of European air quality, *EJOR* **122**(2), 219–241 (available also as IIASA's RR-00-06).
38. Makowski, M.: 2001, Modeling techniques for complex environmental problems, *in* M. Makowski and H. Nakayama (eds), *Natural Environment Management and Applied Systems Analysis*, International Institute for Applied Systems Analysis, Laxenburg, Austria, pp. 41–77.
39. Makowski, M.: 2003, Structured modeling technology, *EJOR* (to appear).
40. Makowski, M. and Granat, J.: 2000, Interfaces, in [81], pp. 283–307.
41. Makowski, M. and Wierzbicki, A.: 2000, Architecture of decision support systems, in [81], pp. 48–70.
42. Makowski, M. and Wierzbicki, A.: 2002, Modeling knowledge in global information networks, *in* KBN (ed.), *Importance of ICT for Research an Science*, KBN, Warsaw, p. 14 (draft version available from `http://www.iiasa.ac.at/ marek/pubs/prepub.html`).
43. Makowski, M., Somlyódy, L. and Watkins, D.: 1996, Multiple criteria analysis for water quality management in the Nitra basin, *Water Resources Bulletin* **32**(5), 937–951.
44. Ogryczak, W.: 1996, A note on modeling multiple choice requirements for simple mixed integer programming solvers, *Computers & Operations Research* **23**, 199–205.
45. Ogryczak, W. and Zorychta, K.: 1996, Modular optimizer for mixed integer programming, MOMIP version 2.3, *Working Paper WP-96-106*, International Institute for Applied Systems Analysis, Laxenburg, Austria.
46. Paczyński, J., Makowski, M. and Wierzbicki, A.: 2000, Modeling tools, in [81], pp. 125–165.
47. Pawlak, Z.: 1991, *Rough Sets. Some Aspects of Reasoning about Knowledge*, Kluwer Academic Publishers, Dordrecht.
48. Polak, E.: 1976, On the approximation of solutions to multiple criteria decision making problems, *in* M. Zeleny (ed.), *Multiple Criteria Decision Making*, Springer-Verlag, New York (originally published in *L'aventure de la science, Melanges Alexandre Koyre*, Vol. **2**, pp. 307-334, Hermann, Paris 1964).
49. Popper, K.: 1959, *The Logic of Scientific Discovery*, Hutchinson, London.
50. Popper, K.: 1975, The rationality of scientific revolutions, *in* R. Harre (ed.), *Problems of Scientific Revolution*, University Press, Oxford, pp. 72–101.
51. Popper, K.: 1983, *Realism and the Aim of Science*, Hutchinson, London.
52. Pospelov, G. and Irikov, V.: 1976, *Program- and Goal-Oriented Planning and Management*, Sovietskoye Radio, Moscow.
53. Radermacher, F.: 1994, Decision support systems: Scope and potential, *Decision Support Systems* **12**(4/5), 257–265.
54. Raiffa, H.: 1997, *Decision Analysis: Introductory Lectures of Choices Under Uncertainty*, MacGraw-Hill Companies, New York, Tokyo, Toronto.
55. Rapoport, A.: 1989, *Decision Theory and Decision Behaviour, Normative and Descriptive Approaches*, Vol. 15 of *Theory and Decision Library, Mathematical and Statical Methods*, Kluwer Academic Publishers, Dordrecht, Boston, London.
56. Ruan, D., Kacprzyk, J. and Fedrizzi, M. (eds): 2001, *Soft Computing for Risk Evaluation and Management*, Physica–Verlag, Heidelberg and New York.
57. Sakawa, M.: 1993, *Fuzzy Sets and Interactive Multiobjective Optimization*, Plenum Press, New York, London.

58. Sawaragi, Y., Nakayama, H. and Tanino, T.: 1985, *Theory of Multiobjective Optimization*, Academic Press, New York.
59. Seo, F. and Sakawa, M.: 1988, *Multiple Criteria Decision Analysis in Regional Planning: Concepts, Methods and Applications*, D. Reidel Publishing Company, Dordrecht.
60. Simon, H.: 1955, A behavioral model of rational choice, *Quarterly Journal of Economics* **69**, 99–118.
61. Simon, H.: 1957, *Models of Man*, J. Wiley & Sons, Chichester, New York.
62. Simon, H.: 1958, *Administrative Behavior, a Study of Decision Making Process in Administrative Organization*, Macmillan, New York.
63. Sprague, R.: 1983, A framework for the development of decision support systems, *Decision Support Systems: A Data Based, Model-Oriented, User-Developed Discipline*, Petrocelli, Princeton, N.J.
64. Springer, S. and Deutsch, G.: 1981, *Left Brain – Right Brain*, Freeman, San Francisco.
65. Steuer, R.: 1986, *Multiple Criteria Optimization: Theory, Computation, and Application*, J. Wiley & Sons, New York.
66. Stewart, T.: 1992, A critical survey on the status of multiple criteria decision making theory and practice, *OMEGA, International Journal of Management Science* **20**(5/6), 569–586.
67. Tukey, J.: 1977, *Exploratory Data Analysis*, John Wiley & Sons, New York.
68. Tversky, A. and Kahneman, D.: 1985, The framing of decisions and philosophy of choice, *in* G. Wright (ed.), *Behavioral Decision Making*, Plenum, New York, pp. 25–42.
69. Vollmer, G.: 1984, Mesocosm and objective knowledge, in [83].
70. Wierzbicki, A.: 1977, Basic properties of scalarizing functionals for multiobjective optimization, *Mathematische Operationsforschung und Statistik, s. Optimization* **8**, 55–60.
71. Wierzbicki, A.: 1980, The use of reference objectives in multiobjective optimization, *in* G. Fandel and T. Gal (eds), *Multiple Criteria Decision Making, Theory and Applications*, Vol. 177 of *Lecture Notes in Economics and Mathematical Systems*, Springer Verlag, Berlin, New York, pp. 468–486.
72. Wierzbicki, A.: 1982, A mathematical basis for satisficing decision making, *Mathematical Modelling* **3**(5), 391–405.
73. Wierzbicki, A.: 1984, *Models and Sensitivity of Control Systems*, Elsevier-WNT, Amsterdam, Warsaw.
74. Wierzbicki, A.: 1986, On the completeness and constructiveness of parametric characterizations to vector optimization problems, *OR Spektrum* **8**, 73–87.
75. Wierzbicki, A.: 1992a, Multi-objective modeling and simulation for decision support, *Working Paper WP-92-80*, International Institute for Applied Systems Analysis, Laxenburg, Austria.
76. Wierzbicki, A.: 1992b, Multiple criteria games: Theory and applications, *Working Paper WP-92-79*, International Institute for Applied Systems Analysis, Laxenburg, Austria.
77. Wierzbicki, A.: 1992c, The role of intuition and creativity in decision making, *Working Paper WP-92-78*, International Institute for Applied Systems Analysis, Laxenburg, Austria.
78. Wierzbicki, A.: 1993, On the role of intuition in decision making and some ways of multicriteria aid of intuition, *Multiple Criteria Decision Making* **6**, 65–78.

79. Wierzbicki, A. and Makowski, M.: 2000, Modeling for knowledge exchange: Global aspects of software for science and mathematics, *in* P. Wouters and P. Schröder (eds), *Access to Publicly Financed Research*, NIWI, Amsterdam, the Netherlands, pp. 123–140.
80. Wierzbicki, A. and Wessels, J.: 2000, The modern decision maker, in [81], pp. 29–46.
81. Wierzbicki, A., Makowski, M. and Wessels, J. (eds): 2000, *Model-Based Decision Support Methodology with Environmental Applications*, Series: Mathematical Modeling and Applications, Kluwer Academic Publishers, Dordrecht.
82. Wuketits, F.: 1984a, Evolutionary epistemology - a challenge to science and philosophy, in *Concepts and Approaches in Evolutionary Epistemology* [83].
83. Wuketits, F. (ed.): 1984b, *Concepts and Approaches in Evolutionary Epistemology*, D. Reidel Publishing Co., Dordrecht.
84. Yu, P.L.: 1985, *Multiple-Criteria Decision Making: Concepts, Techniques, and Extensions*, Plenum Press, New York, London.
85. Yu, P.L.: 1990, *Forming Winning Strategies, An Integrated Theory of Habitual Domains*, Springer Verlag, Berlin, New York.
86. Yu, P.L.: 1995, *Habitual Domains: Freeing Yourself from the Limits on Your Life*, Highwater Editions, Shawnee Mission, Kansas.
87. Zadeh, L.A.: 1965, Fuzzy sets, *Information and Control* **8**, 338–353.
88. Zadeh, L.A. and Kacprzyk, J.: 1999, *Computing with Words in Information/Intelligent Systems: Foundations*, Springer Verlag, Berlin, New York.
89. Zeleny, M.: 1974, A concept of compromise solutions and the method of the displaced ideal, *Comput. Oper. Res.* **1**, 479–496.
90. Zimmermann, H.: 1978, Fuzzy programming and linear programming with several objective functions, *Fuzzy Sets and Systems* **1**, 45–55.
91. Zimmermann, H.: 1985, *Fuzzy Set Theory – and Its Applications*, Kluwer Academic Publishers, Boston, Dordrecht, Lancaster.
92. Zimmermann, H.: 1987, *Fuzzy Sets, Decision Making, and Expert Systems*, Kluwer Academic Publishers, Boston, Dordrecht, Lancaster.

Benefits of Decision Support Using Soft Computing

William Cheetham, Kai Goebel and Piero Bonissone

General Electric Company
1 Research Circle
Niskayuna, NY 12309, USA

E-mail: {cheetham,goebelk,bonissone}@crd.ge.com)

Abstract. Soft Computing allowed developers at General Electric to maximize the benefits and minimize the costs of the Decision Support Systems they have created. Soft Computing techniques can be used for designing, automating, maintaining, and maximizing the value of the decision process. These benefits were shown in two applications. The first application used fuzzy rule-based and case-based reasoning to create a decision system that is able to automatically determine risk categories for insurance applications. This second application used fuzzy information fusion for selecting replacement parts required to repair medical imaging machines, such as Magnetic Resonance Imaging system.

1. Introduction

General Electric's Research Center has been using soft computing (SC) and related techniques to create decision support systems (DSS) since the early 1980's. After fielding and maintaining multiple applications we have found many benefits from using soft computing to create knowledge based decision support applications. This chapter will describe the benefits of using SC for creating a DSS and then show two recently fielded applications. The applications are a classification tool for determining the risk of financial transactions and a parts advisor for servicing medical equipment.

1.1 Background

DSSs are interactive computer-based systems intended to help decision makers use data and models to identify and solve problems and make decisions [35]. The term DSS was created in the 1970's to represent information systems designed to help managers in making choices. These systems used spreadsheets, integer programming, and expert systems. In the 1980's, vendors created decision support tools called executive information systems (EIS). An EIS was used to view organizational information and predict the future state of the organization. In the 1990's, on-line analytical processing (OLAP) was introduced as a method for manipulating enterprise data models across many dimensions to provide a DSS. However, the 1990's also brought a realization that non-mangers are performing knowledge-based work that can benefit from the assistance of a DSS [36,16]. The term DSS was also applied to complex domains with uncertain information. The needs of the DSS of the future that are used by non-managers in uncertain domains are a good match with the capabilities provided by the field of Soft Computing (SC). SC provides a synthesis of methodologies from fuzzy logic, probability theory, neural networks and evolutionary computing. SC differs from conventional (hard) computing in that it takes advantage of imprecision, uncertainty and partial truth to achieve tractability, robustness and low solution cost [41,42].

2. Benefits of Soft Computing to Knowledge Engineering

Knowledge engineering (KE) deals with the development of knowledge-based systems including DSS. The phases of the KE process are planning, design, knowledge acquisition, system building, and system maintenance [37]. KE tries to turn the creation of knowledge-based systems from an art into a science, similar to the way software engineering turned software development from an art to a science [30,8]. KE traditionally augmented software engineering principles with guidelines for creating expert systems. However, the symbolic manipulation and predicate logic of expert systems is not sufficient for creating complex DSSs [43]. We have found that KE can be expanded to include SC and related techniques. This section outlines the benefits of using SC as part of the KE process for creating DSSs.

2.1 Planning and design

Planning and design are the first phases of the traditional KE lifecycle. An understanding of SC techniques is useful in planning and designing a DSS that adds the maximum possible value to the decision process. The planning and design phases include setting the goals of the project and determining the approach to reach those goals. An understanding of SC can help with both of these tasks

2.1.1 Setting project goals

SC techniques can allow a DSS to have a level of decision automation that is more valuable than is possible without SC. DSSs can provide information for making a decision, suggest possible solutions, or completely automate the decision process.

Traditional DSSs and knowledge management systems [13,32] only provide information for making a decision. SC techniques are often needed to correctly suggest possible solutions or automate the decision process. When SC or other related techniques are used to produce a DSS that provides a higher level of automation the DSS can provide greater productivity benefits to the decision process. A DSS with the higher level of automation will be a more valuable DSS. Thus, the goal of a DSS should be to automate as much of the decision process as possible, which would maximize the need for use of SC techniques. The two applications described in the following sections will provide examples of the higher levels of automation. The parts advisor suggests possible solutions and the financial risk classifier automates the decision process.

2.1.2 Technology selection

Another important task in planning and design of a DSS is selecting the appropriate technique of supporting the decision process. There are many technologies that can be used to create a DSS and the correct selection of a technology has a large impact on the project. SC includes a variety of techniques that can be used directly or combined with other techniques to produce a DSS. Each of the SC techniques has advantages and disadvantages so no one technique would be appropriate for all DSSs. We will briefly discuss these advantages and disadvantages.

Fuzzy logic is a methodology for the representation and manipulation of imprecise and vague information [33]. One common way of using fuzzy logic in a DSS is creating a fuzzy rule based system. A fuzzy rule based system has the advantages that it can represent domain knowledge in the form or rules, similar to an expert system, but it can also reason with uncertain information and perform numerical calculations. A disadvantage of the fuzzy rule based system is that the rules can be difficult to acquire and tune, as is true with expert systems.

An artificial neural network (ANN) is a method of reasoning that is inspired by principles observed in neurons of the central nervous system of humans and animals [19]. An ANN is a set of one or more layers of nodes interconnected by directed, weighted links. Given examples of input patterns and desired outputs the ANN can learn weights on the links that provide the correct outputs for the training inputs and other similar inputs. The ANN can be used to create a DSS when sufficient training examples exist. However, it is difficult to incorporate heuristic or rule based domain knowledge in an ANN.

Evolutionary computation (EC) is a method of optimization that is based on simulating evolution on a population of possible solutions to a given problem [3]. Simulated chromosome crossover and mutations are used with survival of the fittest to evolve the population of solutions so that after many generations the population consists of the best solutions found. A DSS that is created using evolutionary computing can evaluate a large population of potential solutions and suggest the best one that was found. Some disadvantages of EC are that there must be a fitness function that can evaluate the fitness of any possible solution and that all possible solutions are not evaluated so that the optimal solution is not always found.

Bayesian belief nets (BBN) have nodes that represent propositions and arcs between nodes that represent probabilistic relationships to other propositions [34]. The BBN can be used in a DSS to reason about the propositions that are represented by the nodes. The disadvantages of a BBN are that the choice of propositions to represent limits the reasoning to the set selected and the relationships between nodes can be difficult to specify precisely.

Hybrid techniques can combine multiple SC techniques or some SC techniques with other related techniques such as database theory, information retrieval, machine learning, or case based reasoning.

2.2 Knowledge acquisition and system building

Knowledge acquisition includes collecting the data and information that will be used to make decisions plus capturing the process that will be used to create a decision from the data and information. A traditional DSS would rely on a human's undocumented intuition or experience to provide the tacit knowledge [25] needed to create a decision. This tacit knowledge can be difficult to communicate or share with others. SC provides a variety of techniques for capturing the tacit knowledge. When a humans' tacit knowledge is used in a decision process that process is an art, but if the knowledge is captured the process can become a science. Once a decision process is captured it can be made more consistent, be shared, evaluated, and improved. Consistency for a single decision maker or a group of decision makers can be achieved by having humans specify how to make a decision and using a technique from SC to create a tool that performs the decision process. Since the decision process is captured in a deterministic tool it will be consistently repeatable. The humans' job is transformed from making decisions to specifying how they are made. Once a decision process is captured in a tool it can be easily shared with others. Instructions on how to use the tool are all that is needed for others to execute the decision process. The tool can be evaluated to establish how well it performs. Some evaluations include determining the error rate (percentage of times the tool makes an incorrect decision), false positive rate (percentage of times the tool determine a positive decision should be made when actually a negative decision should be made), and false negative rate (percentage of time a negative decision is made and a positive decision should be made). The tool can be improved by making modifications and determining the modified tools error rate, using the same data set. Modifications that reduce the error rate are retained and ones that do not reduce the rate are discarded. If costs are associated with false positives and false negatives then the tool can be tuned to minimize the cost of making a mistake. Furthermore, statistics on the tool can be used to determine if the error rate for a given type of decision are higher or lower than the typical error rate. If the error rate is low then there can be high confidence in a solution the tool creates. The high confidence solutions can be automated even if all the decisions cannot be automated. This relieves the burden from decision makers to perform easy but repetitive decisions and allows them to concentrate on more challenging decisions or improving the way that decisions are made.

2.3 System maintenance

The world is constantly changing. Decisions that were correct in the past are not always correct today, and may be different in the future. The DSS is only as good as the information on which the decision is based and reasoning process that uses the information. Determining the quality of information and updating the reasoning process are tasks that a traditional DSS left for the end user. SC techniques can be used to deal with changing information and updating the reasoning process.

Quality information has always been important to a DSS. The user of a traditional DSS often played the role of the judge of information quality. If the information was not current, complete, correct, and consistent they would have made a decision with uncertain information. SC techniques are specifically designed to be used with uncertain information.

A DSS must constantly have its reasoning process updated to represent the current reality. The user of a traditional DSS was often assumed to update their reasoning process as needed. This updating was often a hidden task of the user and was not done in a formal way if it was done at all. However, when the decision process is automated the users primary task changes from making decisions to updating the reasoning process. SC techniques can be used for both automating the decision process and maintaining the process. For example, the financial risk classifier will describe how fuzzy rules are used to automate the process and an evolutionary algorithm is used to update the rules when needed.

2.4 Value proposition

Maximizing the value of the DSS should be the goal of everyone who creates a DSS. In order to do this, it is imperative to understand the benefits that are derived from the DSS and the cost of creating it. The cost and benefits of DSSs are a little different than traditional software applications. DSSs add value to the intellectual capital of an organization [17]. The following is a list of some of the benefits of a DSS:

- Increased productivity
- Quality of decisions (correctness)
- Speed with which a decision can be made (as seen by decision maker or customer)
- Consistency of decisions

- Reduced training time needed by decision maker (staff flexibility)

The total benefit of a DSS is a sum of the benefits of each of these items. Some of these benefits are realized every time the DSS is used (B_1), an example is increased productivity. Some of the benefits are realized in a reduction of yearly costs (B_2), an example is reduced training time.

There is also a list of costs associated with a DSS that include:

- Cost to create the system (C)
- Cost to maintain the information and reasoning (M)
- Cost to use the system (U)
- Cost to apply the result of the system (A)

A formula that determines the value of a DSS, where N is the number of times the DSS is used and T is the time that the DSS remains useful in years, is given in equation 1.

$$\text{Value} = (B_1 * N) + (B_2 * T) - C - (M * T) - ((A + U) * N) \tag{1}$$

When a DSS is being planned, designed, and created this formula should be used to make sure the DSS gives the maximum possible value. An evaluation of the formula shows the importance of maximizing B_1 and minimizing (A + U). These benefits and costs are multiplied by the number of times the DSS is used so they can have a large effect on the value of the DSS. SC can increase the benefits and decrease the costs so that the value is optimized. The remainder of this chapter will show how SC was used to optimize the value of two fielded DSSs.

3. Automated Decision System for Financial Transactions

3.1 Problem description

In many transaction-oriented processes, human decision makers evaluate new applications for a given service (mortgages, loans, credits, insurances, etc.) and assess their associated risk and price. The automation of these key business processes is likely to increase throughput and reliability while reducing risk. The success of these ventures is highly dependent on the availability of generalized decision-making systems that are not just able to reliably replicate the human decision-making process, but can do so in an *explainable*, *transparent* fashion. Insurance underwriting is one such high-volume application domain where intelligent automation can be highly beneficial, and reliability and transparency of decision-making are critical.

A highly trained individual traditionally performs insurance underwriting. A given insurance application is compared against several standards put forward by the insurance company and classified into one of the risk categories (rate classes) available for the type of insurance requested. The risk categories then affect the premium paid by the applicant--the higher the risk category, the higher the premium. Note that the accept/reject decision is also part of this risk classification, since risks above a certain tolerance level set by the company will simply be rejected.

There can be a large amount of variability in the underwriting process when performed by human underwriters. Typically the underwriting standards cannot cover all possible cases, and sometimes they might even be ambiguous. The subjective judgment of the underwriter will almost always play a role in the process. Variation in factors such as underwriter training and experience, and a multitude of other effects can cause different underwriters to issue inconsistent decisions. Sometimes these decisions can fall in a "gray area" not explicitly covered by the standards. In these cases, the underwriter uses his/her own experience to determine whether the standards may or should be adjusted. Different underwriters could make different determinations about when these adjustments are allowed, as they might apply stricter or more liberal interpretations of the standards.

3.2 Approach

We will describe a system to automate the application placement process for cases of low or medium complexity (using a fuzzy rule-based and a fuzzy case-based reasoning engine.) Alternatively, the system can provide the underwriter with partial analysis and conclusions for more complex cases. With this kind of automation, variability of the risk category decision would be greatly reduced or eliminated. This will eliminate a source of risk exposure for the company, allowing it to operate more competitively and profitably. A smart automation process, capable of determining its applicability to each new case will increase the underwriting capacity for handling larger volume of applications.

3.3 Prior work

Reported research in the area of automated insurance underwriting is quite sparse. However, there are a few documented approaches. Collins et al. [12] describe the application of a neural network to replicate the decision-making of mortgage insurance underwriters by training the system on a database of certified cases. Insurance underwriting based on neural-networks or similar modeling approaches leads to *opaque* decision-making, wherein the learned and encoded interrelationships between decision variables that are used to arrive at decisions is not well defined and explainable. Nikolopoulos et al. [24] describe the application of evolutionary learning and classification tree techniques to build a knowledge base that determines the termination criteria for an insurance policy. DeTore, et al. [15] describe a different approach, based on the assignments of weights to elements on the basis of predetermined relationships. A risk classification is determined for the potentially insurable risk from the weights assigned.

3.4 Solution description

Our approach differs from the ones above described because it uses transparent, interpretable models that can represent partial degree of constraint satisfaction in a flexible, natural manner.

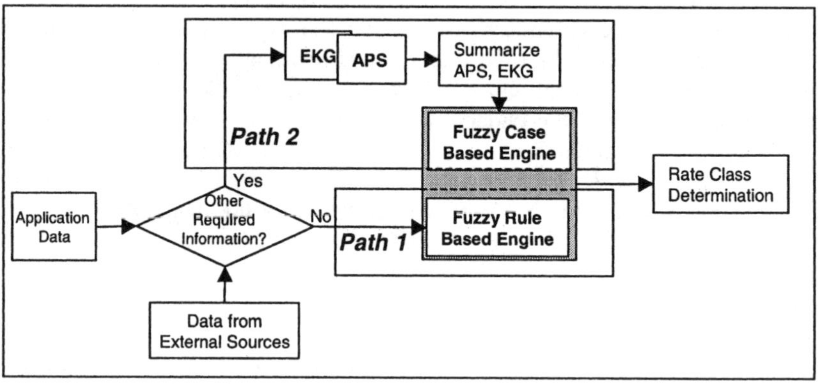

Figure 1: Decision engine selection as a function of case complexity

3.4.1 Rule-based decision system

The fuzzy rule-based decision system is based on rule-sets that encode underwriting standards. Each rule-set represents a set of constraints with fuzzy regions at the boundaries of different rate classes. The decision system is derived from knowledge engineering sessions with expert underwriters that helped identify factors such as blood pressure levels and cholesterol levels that are critical in defining the applicant's risk and corresponding premium. The goal is to assign an applicant to the most competitive rate class, providing that the applicant's vital data meet all of the constraints of that particular rate class to a minimum degree of satisfaction. Each constraint is represented by a fuzzy set A(x), which can be interpreted as the degree of preference induced by a value x for satisfying condition A. An inference engine will determine the value of satisfaction of each condition and the overall degree of satisfaction of the rule. The definition of each fuzzy constraint A(x) and the determination of the minimal degree of satisfaction are design parameters that are initialized via these knowledge engineering sessions. Figure 2 illustrates the evaluation of the data from a hypothetical applicant using the rule-based constraints of a given rate class.

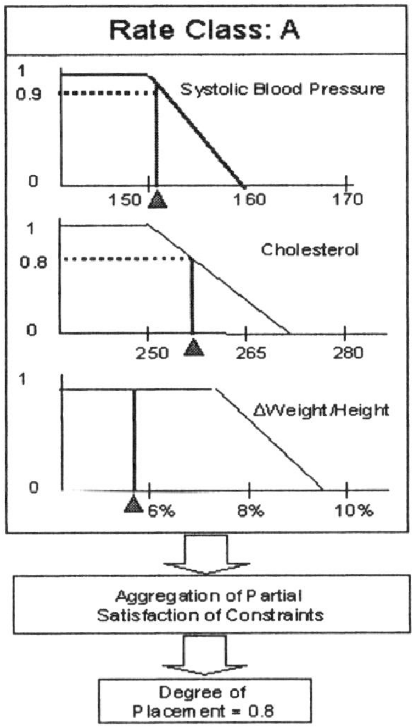

Figure 2: Evaluation of hypothetical applicant data (triangular markers) using fuzzy rule-based constraints

3.4.2 Case-based decision system

Unlike rule-based decision-making, which provides a generative approach to the representation of knowledge, case-based decision-making represents an analogical approach to the solution of the same problem. For instance, in the judicial system, statutory law is interpreted by a judge and applied to specific instances, which then becomes part of jurisprudence. When the law is not clear or unambiguous, legal precedents are used to help the judge make consistent decisions on new but similar cases. In this analogy, statutory law corresponds to a set of underwriting rules, while the set of precedents corresponds to past applications for which a rate class was determined. In this endeavor, complex cases require the aid of case-based reasoning (CBR) to determine the correct rate class for an applicant.

The selection of the appropriate rate class (solution) for a new applicant (probe) is done by using a set of previous problem / solution pairs that are stored in a Case

Base (CB). The CBR process, which uses this CB, consists of the following steps [1]:

- Retrieval of similar cases,
- Reuse of the retrieved case for the new application,
- Revision based on a underwriters evaluation, and
- Retention of the newly created application and category in the case base.

The retrieval step involves finding all cases in the CB that have a problem similar to the new decision (probe). The similarity is determined by comparing each feature characteristic of probe with the corresponding feature in a case. If all features in an existing case are within a pre-specified distance from the ones in the probe, then the case is retrieved. The similarity of the retrieved cases to the new application is calculated using a fuzzy membership function for each feature characterizing the probe [5,9].

The retrieved cases are reused to determine a solution for the new decision. Each retrieved case is adapted to more accurately reflect the category or class that should be used for the new decision. The adapted solutions define a *distribution* over the space of decisions. Then, we use the mode of the distribution (similar to the use of the Mean of Maxima in the defuzzification of a Fuzzy Controller) to select the appropriate risk for the new probe.

Knowledge engineering is used to determine the most efficient indices that represent cases, to define the fuzzy similarity metric that induces the best ranking upon the retrieved cases, and to adapt the solution of the closest cases to the application on hand. The parameters used by the case-based decision system to search for similar cases and to rank them are also design choices that need to be determined and maintained over time for optimal performance. Figure 3 illustrates an example where the applicant data is represented by a point in a multi-dimensional index space, and the points in the surrounding sphere represent potentially similar cases, each with a corresponding color-coded rate class. The corresponding histogram illustrates the distribution of the retrieved neighboring cases, and the information conveyed by this histogram is used by the case-based decision system to make the final decision for a given application. While the mode of the histogram determines the output, the shape and cardinality of the histogram can be used to determine the confidence value in such output [6].

3.5 Results

The rule-based and case-based engines have been tested with about 3,000 reference cases, i.e. cases for which the outcome has been independently validated. Both systems outperformed the human decision maker in terms of accuracy and consistency of the decision.

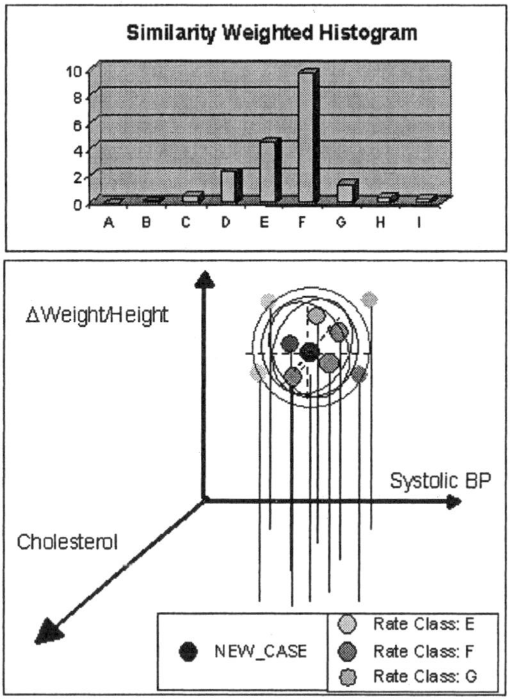

Figure 3: Neighbors closest to a new case in case-based decision-making

However, there was no guarantee that these parameters were optimal. Therefore, we decided to define a process to optimize the decision engines according to some predefined metrics. We developed a method whereby the decision thresholds and internal parameters of these decision-making systems were tuned using a multi-stage mutation-based evolutionary algorithm. The fitness function used selectively penalized different degrees of misclassification, and served as a forcing function to drive correct classifications.

We evaluated the performance of the decision systems based on the three metrics below:

- *Coverage*: number of cases whose rate class assignments are decided by a decision system as a fraction of the total number of input cases.
- *Relative Accuracy*: number of cases with a correct rate class assignment as a fraction of the number of cases covered by the system.
- *Global Accuracy*: number of cases with a correct rate class assignment as a fraction of the total number of input cases.

The following two tables describe the performance enhancement for each decision engine, obtained via the evolutionary optimization methods. This effort is further described in reference [7].

Table 1: Performance of the un-tuned and tuned rule-based decision system

METRIC	SUB-OPTIMAL PARAMETERS	OPTIMIZED PARAMETERS
Coverage	90.38%	91.71%
Relative Accuracy	92.99%	95.52%
Global Accuracy	90.07%	93.63%

Table 2: Performance of the un-tuned and tuned case-based decision system

METRIC	SUB-OPTIMAL PARAMETERS	OPTIMIZED PARAMETERS
Coverage	47.97%	98.86%
Relative Accuracy	92.10%	90.80%
Global Accuracy	44.18%	89.77%

3.6 Discussion

We developed two complementary methodologies based on fuzzy rule-based and case-based reasoning to create decision systems that are able to automatically determine risk categories for insurance applications. In the design phase of these engines we specified initial parameters, such as the membership functions describing the constraints and the thresholds determining the minimum degree of satisfaction required to qualify for a given rate class. The decision thresholds and internal parameters of these decision-making systems are tuned using a multi-stage mutation-based evolutionary algorithm. The fitness function used selectively penalizes different degrees of misclassification, and serves as a forcing function that drives correct classifications. The tunable parameters have a critical impact on the coverage and accuracy of decision-making, and a reliable method to optimally tune these parameters is critical to the quality of decision-making and the maintainability of these systems.

4. Medical Repair Parts Selection DSS

This section describes a DSS for selecting replacement parts required to repair a medical imaging machine, such as MRIs, CT scanners, and the like. Central to the repair action is the notion of a dispatch. A dispatch is a type of response to a machine malfunction where a field engineer is sent from a central location to the machines location where he is supposed to correct the machine malfunction. Typically, after a fault has been logged, an on-line center engineer attempts to perform corrective action remotely. If that fails or if the malfunction is rated severe, a field engineer is contacted and a dispatch log is created. This dispatch log will contain information about the circumstances of the malfunction (such as symptoms) and the repair attempts made (including the parts used). It is the field engineers job to select the parts that will be needed to repair the machine. The DSS described herein assists the field engineer in making that selection.

4.1 Problem description

It is often necessary to replace parts during equipment service. This is true for both scheduled maintenance activities and troubleshooting. For the latter, it is imperative to replace the right part. If the wrong part is replaced, the troubleshooting cycle goes through another iteration, typically requiring ordering another part. This is undesirable because of the time involved for each cycle. For complex machinery, it is infeasible for maintenance personnel to carry even a small percentage of replacement parts. Not only may a particular piece of equipment consist of thousands of individual parts, some of which may be rather large, but, it is also quite common for a field engineer to service different types of equipment. When a potentially broken part is identified, it is ordered from a warehouse. Usually, shipment of the part takes at least a day. The field engineer will return upon confirmed arrival of the part and attempt the correction of the problem. If the particular part did not solve the problem, another part is ordered and the cycle starts over. The implication on the status of the machine is that there is at best reduced functionality and at worst a complete shutdown until the machine can be fixed. From the customer's perspective, this is highly undesirable. Depending on the service contract, this may also have repercussions on the service provider, if, for example, a certain amount of up-time is guaranteed.

4.2 Approach

This application addresses how to create a DSS for ordering the right part the first time, thus reducing the elapsed time to repair (ETTR) and the demand/use ratio for a given part. Making an informed decision about which part to order is not a trivial task. Ideally, the decision maker – typically the field engineer tasked with fixing some equipment – has only partial information that would be required to make an optimal decision. This is in part true because the information resides in various disjoint databases, which may physically not be in the same place. However, even if the decision maker had all information readily available, the diverse nature as well as partially conflicting information may impede a proper decision. This section shows how such information can be combined and aggregated, uncertainty reduced and conflicts resolved.

4.3 Prior work

Information fusion is a field that has recently spawned interest in variety of fields ranging from track association in radar images, to image fusion to classifier output fusion. Information fusion is related to Soft Computing because it leverages the information contained in a variety of possibly uncertain sources [20]. Moreover, all elements of Soft Computing permeate Information Fusion [39, 14, 4], ranging from Dempster Shafer [2], Bayesian Belief Nets [10], Neural Nets [31], and Fuzzy Logic [21, 27], to Genetic Algorithms [29].

4.4 Solution description

The DSS shown here demonstrates how a field engineer is aided in making a decision about which parts to replace on defective medical equipment. Input to the DSS tool is dispatch number, system (equipment) ID, a symptom keyword description, and the field engineer's proposed part selection (if any). Figure 4 shows a typical user input screen where the symptom keyword is "archive" and the part selection is "MAX OPTICAL DRIVE (MOD557431)."

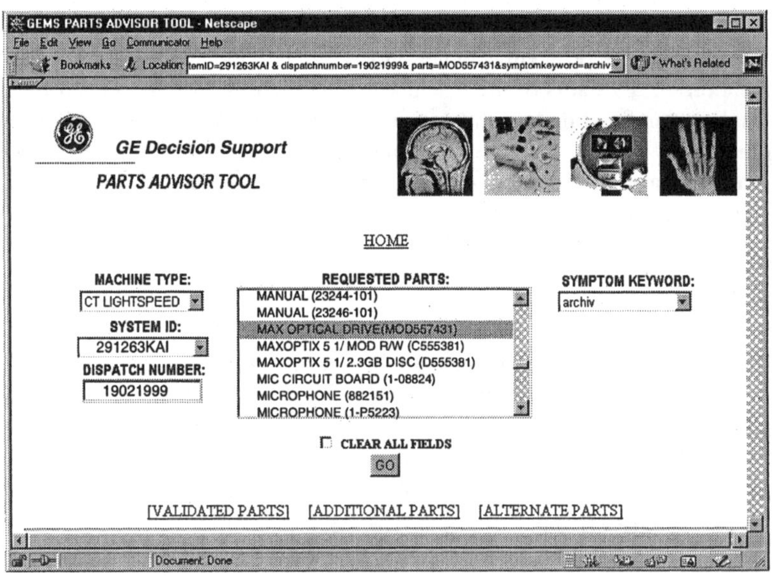

Figure 4: Parts suggestion input screen

The output is tri-modal: 1.) a validation rating for the part(s) selected, 2.) a recommendation for additional parts that will be used along with the parts selected and 3.) alternate parts that will be used instead of the parts selected. Each output is assigned a confidence rating. Figure 5 and Figure 6 show the output screens where the confidence in the validation for the parts selected is 100%, the confidence in additional parts ranges from 75% to 7%, and the confidence for the alternate part ranges from 50% to 91%.

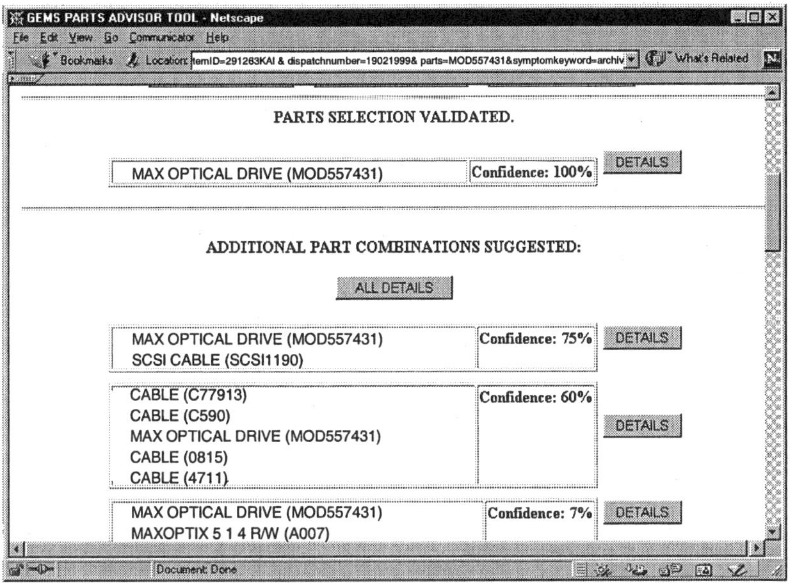

Figure 5: Validate parts and additional parts suggestion screen

The DSS uses the process shown in Figure 7 to determine the three confidence values. The DSS tool reads in information from several databases, in particular dispatch database, machine history database, demand/use database, and error log database. The dispatch database contains specific equipment history. The machine history database contains successful repairs of the entire installed base. The demand/use database contains the demand/use ratio for all parts. Finally, the error log database contains equipment error codes generated during system malfunction and associated successful repair solutions.

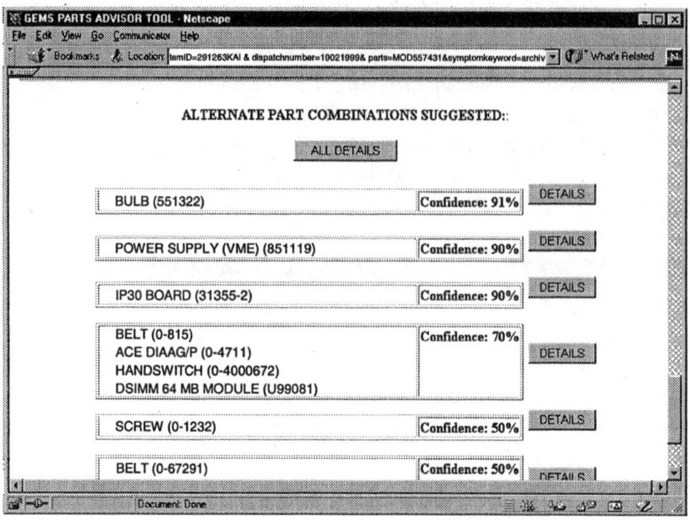

Figure 6: Alternate parts suggestion screen

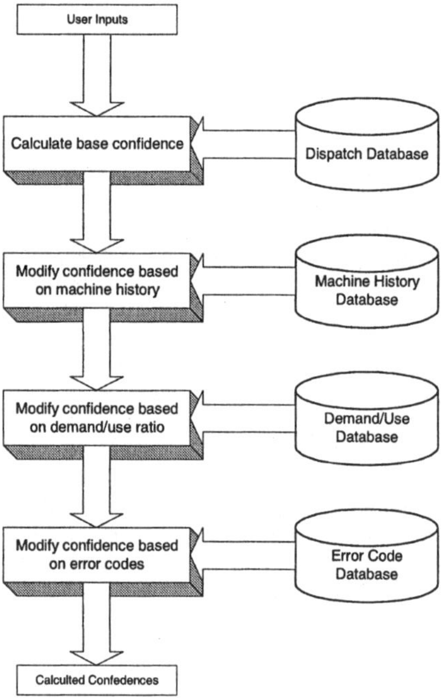

Figure 7: Confidence Calculation Process

4.4.1 Base confidence

The first stage of the DSS tool calculates a base confidence in the three output values using the dispatch database; validated, additional, and alternate. The DSS does these values by through all dispatches. The validated confidence, i.e., the part selection made by the field engineer, is calculated as follows. If a match between the field engineer's parts selection and a previous successful dispatch with the same identical part(s) is found, a confidence is assigned to the validated part. The magnitude of the confidence depends on whether the same symptom was present, or none was selected. The match is established based on the ratio of keywords that are found in the current symptom description and symptoms stored with prior dispatches. Keywords are strings found in symptom description that have high correlation to repair action. A list of meaningful keywords is established a priori in cooperation with field engineers and online engineers. An example symptom description might be "cannot archive". The keyword in that symptom description is "archive".

The confidence for the validated part(s) is calculated by

$$confidence_{validated} = \begin{cases} \min\left(\dfrac{\#\,\text{matching keywords in current dispatch}}{\#\,\text{keywords in retrieved dispatch}},\ \dfrac{\#\,\text{keywords in retrieved dispatch}}{\#\,\text{matching keywords in current dispatch}} \right) & \text{if keyword selection exists} \\ \text{confidence}_{\text{validated0}} & \text{if no keyword is selected in either dispatch} \end{cases} \quad (2)$$

As an illustrative example, consider the initial part selection of an optical drive, "MAX OPTICAL DRIVE (MOD557431)," with the choice of symptom keywords "archive". The DSS system retrieves several instances of the same parts and symptom selections and assigns the initial confidence for that part selection per equation (2) as

$$confidence_{validated} = 1$$

The additional parts confidence is calculated as follows. If only some parts in previous successful dispatches match the initial part selection, then the set of parts that were not selected by the user is considered an additional part combination. A confidence is assigned to each of the additional part offerings i based on whether the same symptom was observed in the current situation and the previous dispatch (or whether none was selected).

Here,

$$confidence^{i}_{additional} = \begin{cases} \min\left(\dfrac{\#\,\text{matching keywords in current dispatch}}{\#\,\text{keywords in retrieved dispatch}}, \dfrac{\#\,\text{keywords in retrieved dispatch}}{\#\,\text{matching keywords in current dispatch}} \right) & \text{if keyword selection exists} \\ confidence_{additional_0} & \text{if no keyword is selected in either dispatch} \end{cases} \quad (3)$$

where

$confidence_{additional_0}$ = initial confidence for additional part selection (0 < base confidence <1). The initial confidence is subject to tuning.

For example, if the additional part retrieved is a SCSI cable (while the symptom keyword is also "archive"), then, with predetermined $confidence_{additional_0}$ =0.8, the parts combination attains the value

$$confidence^{i}_{additional} = 0.8.$$

The alternate parts confidence is calculated as follows. For retrieval of parts that do not match any of the initial part selections, for a given symptom, the confidence for these alternate parts offering is assigned based on how many instances for that particular symptom-part combination are retrieved from the historical database. A symptom-part combination is a constellation where previous dispatches are retrieved that have the same symptom keyword but different parts were used in the repair. The confidence of these alternate parts is calculated as:

$$confidence^{i}_{alternate} = {confidence_{0_{alternate}}}^{exponent_{alternate}} \quad (4)$$

where

$confidence_{0_{alternate}}$ = base parameter for additional parts (0 < parameter < 1)

$$exponent_{alternate} = \frac{1}{\sum_i \text{matching symptom part selections}} \quad (5)$$

Note that the exponent is equal to the inverse of sum of the matching instances.

For example, if four other part combinations were found for the same symptom, then, with predetermined base initial $confidence_{0_{alternate}} = 0.8$, $confidence_{alternate}$ is calculated as:

$$confidence^{i}_{alternate} = 0.8^{\frac{1}{4}} = 0.9457\,.$$

At this point the base confidences are all calculated. These base confidences are further revised through modifiers to account for a number of factors that change the base confidences' support. The calculation of these modifiers is described in the next sections. Information required for calculation of these modifiers is extracted from databases pertaining to machine history, demand/use information, and error codes.

4.4.2 Machine history

Extraction of machine history involves the computation of discount values relating to the part selection confidences based on the history of that particular machine. To that end, the machine history is compiled from the overall dispatch database by extracting the related dispatches with matching machine ID. Next, the algorithm determines whether the same dispatch number is already part of the machine history. If it is, it assigns discounts to the validated part selection, the additional part selection, and the alternate part selection if the same parts were already used in the same dispatch.

Here,

$$discount^{i}{}_{X_{history}} = \begin{cases} discount_{X_{history0}} & \text{if part was already used in same dispatch} \\ 1 & \text{otherwise} \end{cases} \tag{6}$$

where

$discount^{i}{}_{X_{history}}$ is the machine history discount where X is the placeholder for "validated parts", "additional parts", or "alternate parts" and i is the ith part suggestion

$discount_{X_{history0}}$ is the base history discount for part combination X

Next, the algorithm establishes whether the same parts were used on the same machine in the past (outside the same dispatch number) and computes a discount as a function of time passed since the last use of that part. The function used is

$$discount^{i}{}_{X_{history'}} = discount^{i}{}_{X_{history}} \cdot \frac{1}{1 + e^{-time\_passed \cdot modifier_{time}}} \tag{7}$$

where

> *time_passed*= time passed since same parts were used on machine (note: time measured in days)
> $modifier_{time}$= tunable parameter for machine history discount (0 <= parameter <= 1)

For example, if the part was not used in the same dispatch but was installed on the same machine 32 days ago, then the discount for the validated part selection for the optical drive is computed (with a time modifier of $modifier_{time}$=0.05) as

$$discount^{i}{}_{validated_{history}} = 1 \cdot \frac{1}{1+e^{-32 \cdot 0.05}} = 0.832$$

4.4.3 Demand use ratio

This module uses demand and use data for parts to manipulate discount variables. The idea is that parts with a high demand/use ratio are parts that are often times incorrectly requested. This is in essence an a priori likelihood that the part will be used. This information factored into a discount value that penalizes parts selection that have a high demand/use ratio. The discount is modified by the ratio of minimum of overall consumption and 97th consumption percentile divided by the 97th percentile of parts consumption. This modifier deals with nonlinear exponential changes in the usage distribution that can potentially skew the demand/use ratio information. In addition, this modifier is further scaled by a tunable power operator. In particular, the discount for the ith part combination X is calculated using the formula

$$discount^{i} X_{demand/use} = \frac{use}{demand} \cdot \left(\frac{\min(overall\_consumption, value_{97})}{value_{97}} \right)^{exponent_{demand/use}} \qquad (8)$$

where

> *consumption*=overall consumption over a fixed period (here 12 months)
> $value_{97}$=value of 97th percentile of max value
> $exponent_{use\_demand}$=tunable parameter for use demand discount (0 <= parameter <= 1)

For the illustrative example, let the demand/use ratio for the optical drive be demand/use=2.1, overall consumption of optical drives is *overall_consumption*=2250, 97th percentile of consumption is $value_{97}$=9756 and demand/use exponent is $exponent_{demand/use}$=0.8, then the demand/use discount is calculated as

$$discount^{i}_{additional_{demand/use}} = 0.4762 \cdot \left(\frac{\min(2250,9756)}{9756} \right)^{0.1} = 0.4112$$

4.4.4 Error code

This sub-module integrates information from error codes. Error codes are automatically created by the machine and logged in a central database. Error codes can also be observed and recorded by the field engineer through additional diagnostic tasks. The error codes were previously matched with successful solutions obtained for cases where the same error codes where observed. The relevant parts are then used in the evaluation and confidences are assigned according to the match. The output from this sub-module is a booster value that is used to strengthen the opinion of a part selection (as opposed to the discount values that weaken a confidence in a part selection).

Here,

$$booster^{i}{}_{X_{error\_code}} = \begin{cases} booster_{X_{error\_code_0}} & \text{if error codes match} \\ 1 & \text{otherwise} \end{cases} \tag{9}$$

where

$booster_{X_{error\_code_0}}$ is the base booster value for the ith part combination X

4.4.5 Information fusion

The fusion module aggregates the information calculated in the previous sub-modules. In particular, the confidence and discount values for the ith part combination X are coalesced using the following formula:

$$confidence^{i}{}_{X} = \left(\left(confidence^{i}{}_{X_0}{}^{discount^{i}{}_{X_{history}}} \right)^{discount^{i}{}_{X_{demand\_use}}} \right)^{\frac{1}{booster^{i}{}_{X_{error\_code}}}} \tag{10}$$

Finally, a confidence value is calculated for the overall parts choices. This can be accomplished by taking the mean of the individual part confidences:

$$confidence_{X_{overall}} = mean\left(confidence^{part1}{}_{X}, confidence^{part2}{}_{X}, \ldots confidence^{partn}{}_{X}\right) \tag{11}$$

The output is piped back to a screen and allows the user to assess the selected parts selection and evaluate additional part combinations as well as alternate part combinations. Sample output screens were shown in Figure 2 and Figure 3.

Further drill down capability allows to view symptoms and other pertinent information associated with the suggested parts.

4.5 Results

Test data were obtained from several GEMS databases containing information about part usage, machine history, parts availability, etc. The information processed was selected for "CT Lightspeed" CT scanners for a period of 13 months. The tool was implemented in JAVA using Servlet technology and ODCB Oracle database connectivity. A comprehensive quantitative evaluation will be performed once the system has been in use sufficiently long. Feedback from users so far justifies initial claims to the usefulness of this tool.

4.6 Discussion

The value of this project will ultimately be calculated through the reduced downtime of the equipment, the smaller number of parts that are used. It is estimated that the savings will be several hundred thousand dollars per year. The system captures formerly tacit knowledge in the four databases. It then converts that data into useful information for decision support through explicit knowledge embedded into the reasoning engine.

The quality of the data is not very high. Missing data, incorrect entries, and false labels make reasoning with these constraints hard. However, this system exhibits a ductility to poor data quality that allows an assessment even under these adverse and uncertain conditions.

5. Conclusion

This chapter showed the benefits of using SC techniques to create a DSS. SC allowed us to maximize the benefits and minimize the costs of each of the DSSs described. The financial risk classification tool had the following benefits:

- Increased productivity of the insurance underwriters
- Improved quality of decisions (more correct risk classifications)
- Faster risk classifications (as seen by decision maker or customer)
- More consistency in risk classifications
- Reduced training time needed by classifiers using the tool

The costs associated with the financial risk classifier include:

- A significant initial creation cost (multiple person years of effort)

- A low cost to maintain and reasoning due to the evolutionary algorithm, which periodically updates the system.
- A very low cost to use the system since it is completely automated
- A very low cost to apply the result of the system

The parts advisor had the following benefits:

- Increased productivity of the field engineer, because they have the correct parts
- Improved quality of decisions (more equipment repaired on first attempt)
- Faster repair of equipment (as seen by the customer)
- More consistency in parts selected by different field engineers
- Reduced training time needed for field engineers

The costs associated with the risk classifier include:

- A large initial creation cost (about one person year of effort)
- A low cost to maintain the reasoning since the equations do not change. Only the data in the databases change.
- A very low cost to use the system, since data entry is web based
- A very low cost to apply the result of the system, since the output is immediately displayed to the user.

We have found great value in digitizing many of General Electric's decision processes. The early DSSs we created simply assisted a user in making a manual decision. However, we have found that SC techniques can be used to partially or completely automate many decision processes. Changing from a manual to automated process using SC is specifically useful when a process is performed frequently, needs improved accuracy, should be performed quicker, requires consistency, or when there are not enough people qualified to perform the process. Automating processes changes the way people work. People are no longer performing repetitive decision processes. They are spending their time deciding how the decision process should take place and then creating a DSS that automates it.

References

[1] Aamdot, A. (1994). Case-Based Reasoning: Foundational Issues, Methodological Variations, and System Approaches, AICOM, (7)1

[2] Azcarraga, A. P., Marcos, N., (2001), Successive Refinement towards Belief-Evidence Fusion, Proceedings of 4th Annual Conference on Information Fusion, ppTuA1-3 – TuA-10

[3] Back, T., Fogel, D., Michalewicz, Z., (1997) Handbook of Evolutionary Computation, IOP Publishing, Bristol

[4] Behloul, F.; Janier, M.; Croisille, P.; Poirier, C.; and others, (1998), MRI-PET Data Fusion using Soft Computing for the Automatic Assessment of Myocardial Viability, Computers in Cardiology 1998. Vol. 25. pp. 29-32

[5] Bonissone, P., Cheetham, W., (1997), Financial Applications of Fuzzy Case-Based Reasoning to Residential Property Valuation, Proc 6th IEEE Conf. on Fuzzy Systems, vol. 1, Barcelona, Spain, pp 37-44

[6] Bonissone, P., Cheetham W., (2001). Fuzzy Case-Based Reasoning for Decision Making, Proceedings of the IEEE International Conference on Fuzzy Systems, Melbourne, Australia

[7] Bonissone, P., Subbu, R., and Aggour K, (2002) Evolutionary Optimization of Fuzzy Decision Systems for Automated Insurance Underwriting Proceedings of the IEEE International Conference on Fuzzy Systems, Honolulu, Hawaii

[8] Chang, S. K., ed, (2001), Handbook of Software Engineering and Knowledge Engineering Vol. 1 Fundamentals, World Scientific Pub. Co

[9] Cheetham, W., (1997). Case-Based Reasoning for Color Matching, Second Int. Conf. Case-Based Reasoning, Providence, RI, Aug. 1997

[10] Chen, B. A., (2001) A Bayesian Sampling Approach to Decision Fusion, Proceedings of 4th Annual Conference on Information Fusion, pp. TuA1-11 – TuA1-16

[11] Cholvy, L., (2000), Applying Theory of Evidence in Multisensor Data Fusion A Logical Interpretation, Proceedings of 3rd International Conference on Information Fusion, pp. TuB4-17 – TuB4-24

[12] Collins, E., Ghosh, S., and Scofield, C. (1988). An Application of a Multiple Neural Network Learning System to Emulation of Mortgage Underwriting Judgments, Proceedings of the IEEE International Conference on Neural Network, Vol. 2, San Diego, CA s, pp 459-466

[13] Davenport, T., (1998), Working Knowledge: How organizations manage what they know, Harvard Business School Press, Boston Mass

[14] Desachy, J.; Roux, L.; Zahzah, E.-H., (1996), Numeric and Symbolic Data Fusion: a Soft Computing Approach to Remote Sensing Images Analysis, Pattern Recognition Letters, 25 Nov. 1996, vol.17, (no.13):1361-78

[15] DeTore, et al. (1990). Method and apparatus for evaluating a potentially insurable risk, US Patent 4,975,840, December 4, 1990

[16] Drucker, P., (2001), The Essential Drucker, HarperBusiness, New York

[17] Edvinsson, L., (1997), Intellectual Capital, HarperBusiness, New York

[18] English, L., (1999), Improving Data Warehouse and Business Information Quality, Wiley, New York

[19] Fiesler, E., Beale, R., (1997) Handbook of Neural Computation, IOP Publishing, Bristol

[20] Goebel, K., (2001), K. Architecture and Design of a Diagnostic Information Fusion Tool, AIEDAM: Special Issue on AI in Equipment Service, pp. 335-348, April 2001

[21] Glockner, I., Knoll, A., (2001), Fuzzy Quantifiers: A Natural Language Technique for Data Fusion, pp. Proceedings of 4th Annual Conference on Information Fusion, TuA3-3 – TuA3-8

[22] Gray, P., Watson, H., (1998) Decision Support in the Data Warehouse, Prentice Hall, New Jersey

[23] Morey, D., (2000), Knowledge Management: Classic and Contemporary Works, MIT Press, Cambridge, Mass

[24] Nikolopoulos, C and Duvendack, S. (1994). A Hybrid Machine Learning System and its Application to Insurance Underwriting, Proceedings of the IEEE International Congress on Evolutionary Computation, Vol. 2, Orlando, FL, pp 692-695
[25] Nonaka, I., (1995), The Knowledge-Creating Company, Oxford University Press, New York
[26] O'Dell, C., (2000), Knowledge Management: A guide for your journey to best-practice process, American Productivity and Quality Center, Houston Texas
[27] Pigeon, L., Inglada, J., and Solaiman, B., (2001) Genetic Algorithms fir multiagent fusion system learning, Proceedings of SPIE: Sensor Fusion: Architectures, Algorithms, and Applications V, pp.87-95
[28] Power, D.J., (2001), Decision Support Systems Web Tour. World Wide Web, http://dssresources.com/tour, version 4.1
[29] Prajitno, P, and Mort, N., (2001), Fuzzy Model-Based Multisensor Data Fusion System, Proceedings of SPIE: Sensor Fusion: Architectures, Algorithms, and Applications V, pp. 301-312
[30] Pressman, R., (1987) Software Engineering: A Practitioner's Approach, McGraw-Hill, United States
[31] Rogova, G. L. and Kasturi, J., (2001), Reinforcement Learning Neural Network for Distributed Decision Making, Proceedings of 4th Annual Conference on Information Fusion, pp. TuA2-15 – TuA2-20
[32] Rumizen, M., (2002), The Complete Idiot's Guide to Knowledge Management, CWL Publishing, Madison, WI
[33] Ruspini, E., Bonissone, P., Pedrycz, W., (1998) Handbook of Fuzzy Computation, Institute of Physics Publishing
[34] Russell, S., Norvig, P., (1995) Artificial Intelligence: A Modern Approach, Prentice Hall, New Jersey
[35] Sage, A., (1991), Decision Support Systems Engineering, Wiley, New York
[36] Senge, P., (1994), The Fifth Discipline Fieldbook: Strategies and Tools for Building a Learning Organization
[37] Schreiber, G., (eds), (1999) Knowledge Engineering and Management, MIT Press, Mass.
[38] Sveiby, K.S., (1997) The New Organizational Wealth, Berrett-Koehler Publishers, Inc
[39] Van Der Wal, A.J.; Shao, M., (2000), Sensor fusion: the application of soft computing in monitoring and control for railroad maintenance, Proceedings of the 3rd World Congress on Intelligent Control and Automation, vol.1, pp. 341-346
[40] Yager, R.R., Zadeh L.A , (Eds.), (1991), An Introduction to Fuzzy Logic Applications in Intelligent Systems, Kluwer Academic Publishers
[41] Zadeh, L., (1973), Outline of a new approach to the analysis of complex systems and decision processes, IEEE Trans. on Systems, Man and Cybernetics SMC-3, pp. 28-44
[42] Zadeh, L., (June 2001), Applied Soft Computing – Foreword, Applied Soft Computing, Volume 1, Issue 1
[43] Zeleznikow, J., Nolan, JR, (2001), Using soft computing to build real world intelligent decision support systems in uncertain domains, Decision Support Systems (31)2 pp. 263-285

Evolving Connectionist-based Decision Support Systems

Nikola Kasabov

Knowledge Engineering and Discovery Research Institute
Auckland University of Technology
Auckland
New Zealand
E-mail: nkasabov@aut.ac.nz, http://www.kedri.info

Abstract. The chapter gives a background information on decision support systems and introduces a framework for building dynamic, adaptive decision support systems that evolve their structure and improve their knowledge-base through a continuous learning from data and through interaction with the environment. Such decision support systems use evolving connectionist systems (ECOS) – adaptive learning neural network models, thus their name – evolving connectionist based decision systems ECBDS. Case studies of ECBDS are presented from the areas of finance, economics, and bioinformatics.

Keywords: decision support systems; neural networks; adaptive learning; rule based systems; evolving connectionist systems.

1. Introduction

The complexity and the dynamics of many real-world problems, such as decision making and prediction of economics and finance, decision making in medicine, knowledge discovery in bioinformatics, predicting financial time series, adaptive intelligent control of industrial processes, on-line decision making based on large amount of continuous and dynamically changing data on the WWW, and many more, set certain requirements to the information methods and systems to be used for this purpose. Such requirements are:

- Dealing with different types of data and knowledge.
- Incremental adjustment to dynamic changes in the operating environment; accommodating new data; introducing new variables and features if needed

without the need to re-design or re-train the whole system; such adaptation may be needed in an on-line, real-time mode.

- Updating the system's structure in a dynamic way.
- Facilitating appropriate explanation of what "knowledge" is contained in the system, or what knowledge has been learned during the system's operation.

The area of information science and artificial intelligence that is concerned with the issues above is called *decision support systems*. A decision support system is an information system that helps humans to make a decision on a given problem, under given circumstances and constraints. A *decision making system* is a system that makes final decisions and takes actions, e.g. automated trading systems on the Internet, or systems that grant loans through electronic submissions.

Expert systems belong to the same subject area, as they are information systems that contain expert knowledge on a particular problem and perform inference when new data is entered that may be partial or inexact. They provide a solution to the problem when new input data is entered, which solution is expected to be similar to the solution provided by experts in the field.

An expert system usually consists of several parts (modules): (1) a knowledge base, where the expert knowledge resides; (2) a data base, where historical and new data is stored; (3) an inference machine that provides different types of inferences; (4) an efficient interface to users; (5) an explanation module that would provide an explanation of *how* and *why* a certain decision was recommended by the system; (6) a module that learns and accumulates new knowledge, based on the system's operation and on new incoming data (see Duda and Shortliffe, 1983; Kasabov,1996).

In the rest of this article we will use the collective term *decision systems* (DS) to refer to either decision support systems, or to decision making systems, or to the most sophisticated among them - expert systems.

Different methods and approaches have been used so far for building DS that deal with different types of data and problem knowledge; that work in different modes of operation – e.g. off-line or on-line mode; that use static or dynamic data; that use fixed or adaptive knowledge; etc. These methods include: logic systems; rule-based systems (Duda and Shortliffe,1983); case-based reasoning; different types of neural networks; statistical methods, e.g. clustering, principal component analysis, hidden Markov models; mathematical (analytical) modelling; finite automata; hybrid methods that combine some or all from the above (Kasabov,1996; 2002, also see relevant chapters in this book).

Most of the contemporary DS use rules to either represent human knowledge and process it, or to explain what the system has learned during its operation in a way that human beings can understand. Different types of IF-THEN rules have been used for the purpose. Some of more commonly used rule types are listed below (Kasabov, 1996):

(1) Simple propositional rules (e.g., IF x1 is A AND/OR x2 is B THEN y is C, where A,B and C are constants, variables, or symbols of true/false type) (Gallant, 1993).
(2) Propositional rules with certainty factors (e.g., IF x1 is A (CF1) AND x2 is B (CF2) THEN y is C (CFc)) (Fu,1989; Touretzky and Hinton, 1989).
(3) Zadeh-Mamdani fuzzy rules (e.g., IF x1 is A AND x2 is B THEN y is C, where A,B and C are fuzzy values represented by their membership functions) (Zadeh, 1965).
(4) Takagi-Sugeno fuzzy rules (e.g., IF x1 is A AND x2 is B THEN y is a.x1 + b.x2 +c, where A,B and C are fuzzy values and a, b and c are constants) (Takagi and Sugeno, 1985)
(5) Rules that have associated degrees of importance and certainty degrees (e.g., IF x1 is A (DI1) AND x2 is B (DI2) THEN y is C (CFc), where DI1 and DI2 represent the importance of each of the condition elements for the rule output, and the CFc represents the strength of this rule) (Kasabov,1996)
(6) Rules that represent associations of clusters of data from the problem space (e.g., Rule j: IF [an input vector x is in the input cluster defined by its center (x1 is Aj, to a membership degree of MD1j, AND x2 is Bj, to a membership degree of MD2j) and by its radius Rj-in] THEN [y is in the output cluster defined by its center (y is C, to a membership degree of MDc) and by its radius Rj-out, with Nex(j) examples represented by this rule] (Kasabov, 2001, 2002).
(7) Temporal rules (e.g., IF x1 is present at a time moment t1 (with a certainty degree and/or importance factor of DI1) AND x2 is present at a time moment t2 (with a certainty degree/importance factor DI2) THEN y is C (CFc))
(8) Temporal, recurrent rules (e.g., IF x1 is A (DI1) AND x2 is B (DI2) AND y at the time moment (t-k) is C THEN y at a time moment (t+n) is D (CFc))

There are different techniques that can be used to implement and process the above types of rules in a DS. This chapter discusses how neural networks (NN), or also called connectionist techniques, that are vague analogues of the human brain, can be employed for this purpose, i.e. how NN can be used to create *connectionist-based decision systems CBDS*.

The DS presented in this chapter are both human-oriented, and *human-like* systems. The human brain constitutes the 'ultimate' DS, so sing NN is a promising approach. The human-like implementation of DS is one aspect of the issue of the DS design. A more important aspect of DS design is their functionality. DS should help humans in their decision process, so the DS should be comprehensible by humans; DS should incorporate elements of human-like thinking and human-like information and knowledge processing, they must be *human-oriented*.

2. Connectionist-Based Decision Systems (CBDS)

2.1. Neural networks (NN) and connectionist-based decision systems (CBDS)

Neural networks (NN) and connectionist-based systems have been applied to the DS design as learning machines to learn from data, or as knowledge representation machines (knowledge-based neural networks KBNN), or as both, as it is explained below.

In some CBDS NN are used for a low-level data processing in pattern matching modules, while rule-based modules are used for a higher-level decision making (Fu, 1989; Kasabov, 1990). In this way NN can be incorporated in a production system, or in a first-order logic system for decision support.

In other CBDS a fixed set of flat rule base is incorporated in a connectionist structure with a pre-defined built-in inference mechanism (Gallant, 1993). No learning is applied.

Touretzky and Hinton (1989) suggested a production CBDS in which all elements of a production system, e.g. rules, facts, inference machine, are represented as different connectionist modules. No learning is applied on the structure.

Several CBDS use trainable knowledge-based neural networks (KBNN) that have fixed structures in terms of number of neurons and connections. KBNN are used to accommodate both knowledge (rules) and data (Fu, 1989; Towel and Shavlik, 1993; Cloete, I, and J.Zurada (eds), 2000). Such KBNN are the fuzzy neural networks (Kasabov,1996).

2.2. A general framework of a connectionist-based decision system (CBDS)

A framework of a CBDS is shown in Fig.1. It consists of the following parts:

- Pre-processing part (e.g. filtering of data, feature extraction).
- Neural network part - NN modules that are trained on data and incorporate knowledge.
- Higher-level knowledge-based part (e.g. rules for producing final recommended decisions).
- Adaptation part – it evaluates the system's performance and makes changes to the system's structure and functionality. For example, output error can be used to adjust relevant NN modules.
- Rule extraction, explanation part. This part uses both extracted rules from the NN modules and rules from the decision part to explain: (1) *what* the system 'knows' about the problem it is designed to solve; (2) *why* a particular decision for a concrete input vector has been made.

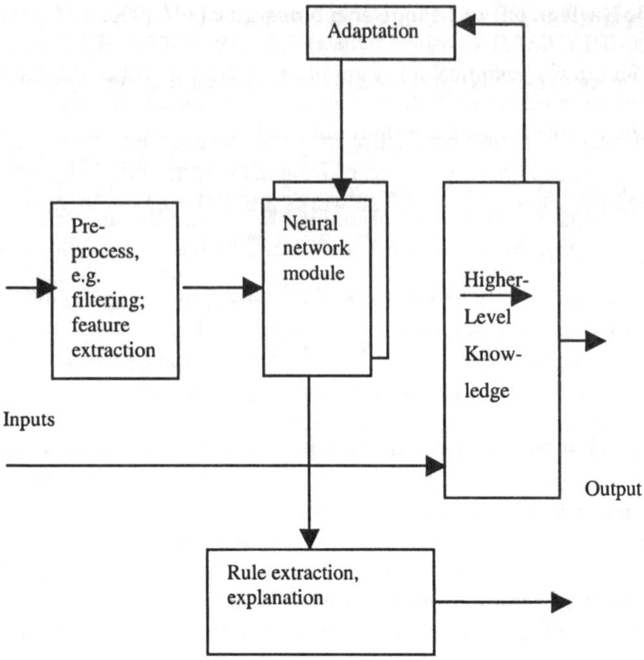

Figure 1. A framework of a connectionist-based decision system (CBDS)

Different types of NNs, e.g. multiplayer perceptrons, self-organising maps, radial-basis functions, fuzzy neural networks, etc. have been used as part of a CBDS, the most used being the KBNN as discussed in the next section.

2.3. Knowledge-based neural networks KBNN

KBNN are pre-structured neural networks that allow for data and knowledge manipulation, including learning from data, rule insertion, rule extraction, adaptation and reasoning (Towel and Shavlik, 1993; Cloete, I, and J.Zurada (eds), 2000; Mitra and Hayshi,2000). KBNN have been developed either as a combination of symbolic AI systems and NN, or as a combination of fuzzy logic systems and NN, or as other hybrid systems.

As DS deal with rules, it is important to have efficient methods for rule insertion and rule extraction in a connectionist structure. There are several methods for rule extraction from a KBNN. Three of them are explained below:

(1) Activating a trained KBNN on input data and observing the patterns of activation. The patterns are then represented as IF-THEN rules, e.g. IF (input patterns) THEN (output patterns).
(2) Rule extraction through analysis of the connections in a trained KBNN.
(3) Methods that combine (1) and (2).

In terms of applying KBNN to perform an inference on new input data, there are three types of methods that can be used:

(1) Rules, extracted from a KBNN, are interpreted in another inference machine (e.g. fuzzy inference, production system).
(2) The rule base learning and reasoning modules constitute an integrated connectionist structure, so that reasoning is performed in the connectionist structure.
(3) The two options from above are combined in one CBDS.

In terms of learning and rule extraction in a KBNN, we can differentiate the following cases: (1) off-line learning and static rule set extraction – first, learning is performed, and then rules are extracted, which is an one-off process; the rules are not updated further; (2) on-line learning – rules can be extracted at any time from the NN structure which is modified through a continuous, on-line learning process.

In terms of how much space is covered by rules when compared with the whole problem space, there are two cases: (1) global rules - they are derived from a globally optimised KBNN and partition the space globally; (2) local rules, that are derived from a locally optimised KBNN, i.e. for every new example only few connections change, and each rule covers only a "patch" of the problem space. Local optimisation in a KBNN would allow for adjusting the KBNN to new data through tuning a small number of elements, and also for extracting locally meaningful rules. The rules can be tuned as the system works ion a continuous mode.

A special type of KBNN are fuzzy neural networks (FNN) that are NN that can be interpreted in terms of fuzzy rules. Neuro fuzzy inference systems are fuzzy systems that can be structurally represented in a connectionist way (Kasabov, 1996; Cloete, I, and J.Zurada (eds), 2000). A review of neuro-fuzzy systems for rule generation is published by Mitra and Hayashi (2000).

KBNN that develop (evolve) their structure, their functionality and their knowledge in time from incoming data, starting from an initial set of knowledge if such is available, are called evolving connectionist systems ECOS (Kasabov, 2002). In the next sections ECOS are used for building adaptive, incremental, on-line learning and evolving DS called here *evolving connectionist-based decision systems ECBDS.*

3. Evolving Connectionist-based Decision Systems (ECBDS)

3.1. Evolving fuzzy neural networks

The design of ECBDS needs evolving learning models and systems. One of them used here is called evolving fuzzy neural network (EFuNN) as shown in fig.2 (Kasabov, 2001,2002). The architecture consists of five layers of neurons and four layers of connections. The first layer of neurons receives the input information. The second layer calculates the fuzzy membership degrees to which the input values belong to predefined fuzzy membership functions (MF), e.g. small, medium, or large. The MF can be kept fixed, or can change during training. The third layer of neurons represents associations between the input and the output variables, fuzzy rules. The fourth layer calculates the degrees to which output membership functions are matched by the input data, and the fifth layer does defuzzification and calculates values for the output variables.

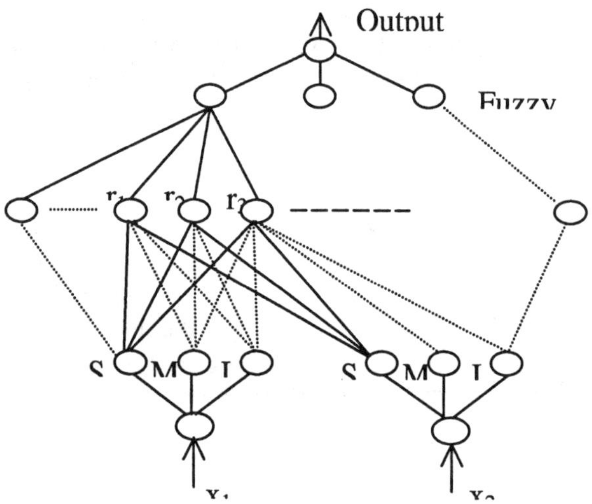

Figure 2. A simple neuro-fuzzy inference system - 2 inputs, 2 membership functions and 1 output. Rule node r1 can be represented as a rule: IF x1 is S and x2 is S THEN Output is S (certain statistical and linguistic parameters are attached)

In a simplified version, when there are no fuzzy membership functions used, it is a three layer evolving multi-layer perceptron called simple evolving connectionist system (Kasabov, 2002). This is the case with the ZISC systems that are based on evolving radial-basis function neural networks (ZISC Manual, 2001).

Papers and software of evolving connectionist systems ECOS are available from http://www/kedri.info (the entry "Book")

3.2. Building ECBDS with ECOS

Dynamic, evolving CBDS systems (ECBDS) evolve their structure, their functionality, and their knowledge from incoming data, rather than having a pre-defined knowledge and a fixed set of rules. They learn and improve continuously over time as it is graphically shown in Fig.3.

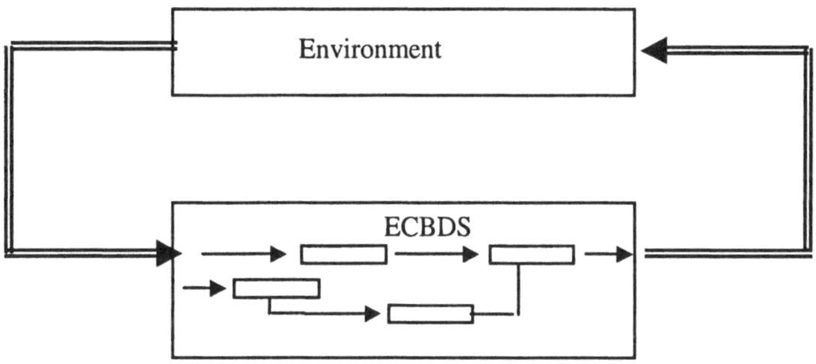

Figure 3. Adaptive, ECBDS learn their structure and functionality through interaction with the environment

ECBDS adapt to a changing environment – possibly in a real time; they can learn in a "lifelong" learning mode; they are able to explain what they have learned in terms of rules and other types of knowledge. An ECBDS would have a modular, 'open' structure evolving over time. Initially, the NN can be a mesh of nodes (neurons) with very little connections between them, pre-defined through *prior* knowledge or "genetic" information. An initial set of rules can be inserted in this structure. Gradually, through self-organisation, the system becomes more and more "wired". The network learns different patterns (exemplars, prototypes) from the training examples. For example, in an evolving connectionist system (ECOS) (Kasabov, 2001, 2002) a node is created and designated to represent individual example only if this example is significantly different from the previous ones (with a level of differentiation set through dynamically changing parameters).

Along with growing procedure, ECOS have pruning procedures defined as well. They allow for removing neurons and their corresponding connections that are not actively involved in the functioning of the ECOS thus making space for new input patterns. Different modes of learning in ECOS are possible, e.g.: (a) active learning mode - learning is performed when a stimulus (input pattern) is presented and kept active; (b) passive (e.g., sleep) learning mode - learning is performed when there is no input pattern presented at the input of the ECOS.

ECBDS operate usually as on-line DS that make decisions and adjust their knowledge through an incremental, continuous learning from incoming data. Examples are systems that learn and react to dynamic changes of a stock index, or adjust their knowledge to include new genes discovered in the due course.

4. Case Studies of ECBDS

Several case studies of ECBDS are presented in this section.

4.1. A case study of ECBDS for on-line financial decision making based on on-line prediction of the MIB30 stock index.

Figure 4 shows the use of an ECBDS for on-line learning and prediction of 5-days ahead moving average values of the MIB30 stock index (the Milan stock index). Here, as a NN, evolving fuzzy neural network (EFuNN) (Kasabov, 2001, 2002) is used. The NN is trained on-line on consecutive data vectors. Inputs to the NN are 5 day moving averages of the following variables: Dow Jones index DJ (t), DJ(t-1), MBI30(t), MIB30(t-1), exchange rate Euro/US$(t), Euro/US$(t-1).

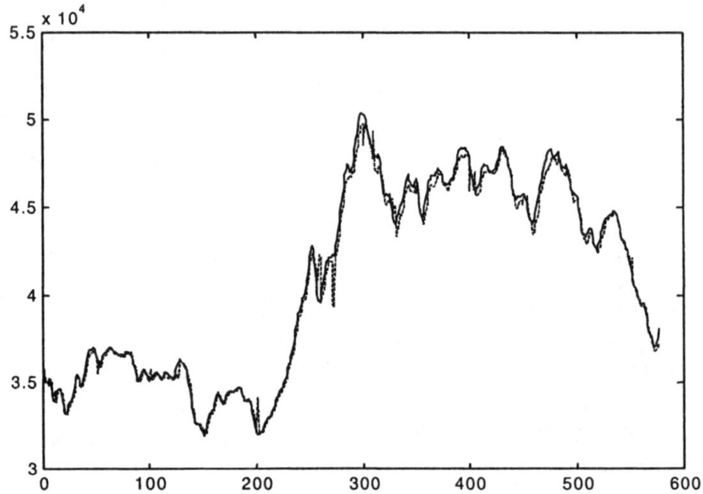

Figure 4. Adaptive ECBDS in financial decision making: The process of on-line learning and prediction five days ahead of the MIB30 index - the desired versus the predicted output value for 576 consecutive days

At any time of the functioning of the DS above rules can be extracted. The rules represent the current association between the input variables and the output variable as it is illustrated in the rule below:

> Rule i: IF [DJ(t) is (Medium 0.828) and DJ(t-1) is (Medium 0.840) and MIB30(t) is (Low 0.885) and MIB30(t-1) is (Low 0.887)] (receptive field= 0.808)
> THEN MIB30(t+5) will be (Low 0.852) (accommodated training examples 18 out of 576)

4.2. ECBDS in economics and finance

Beltraffi et al. (1996) describe a variety of CBDS in the area of finance and economics. They include CBDS for solving problems such as: stock trading, portfolio decision, exchange rate prediction, fraud detection, credit scoring, and many more. Some CBDS contain expert rules and make decisions similar to the decision made by experts. The example below illustrates this type of applications with the use of ECBDS.

Case study: An ECBDS for simulation and prediction of decisions made by the European Central Bank (ECB) on interest rate intervention (Rizzi et all, 2001)

The ECB meets regularly to make a decision on the interest rate for the European Union countries. Modelling this decision making process is extremely difficult as it requires both the comprehensive knowledge the ECB members have, and fast adaptation to new situations. The system developed by Rizzi et al (2003) consists of two parts. The first part has several NN that use 6 groups of economic indicators totaling in 17 variables and produces intermediate outputs that include the predicted values for some time series. These values, as well as other economic variables, are fed into a rule –based system that consitutes the second part of the system. The system finally suggests what ECB decisions on the interest rate might be expected at the next 3 consecutive meetings. Figure 5 shows some test results. The system learns and improves over time after each new data is entered.

4.3. ECBDS in bioinformatics

Processing large amount of data in many areas such as medicine, biochemistry, molecular biology, and making decisions based on this information, is a task where CBDS have been successfully applied. Baldi and Brunak (1998,2001) have demonstrated the application of machine learning techniques, that include NN, for solving difficult problems, such as DNA promoter recognition, RNA splice junction identification (see the example below), secondary structure protein prediction, etc.

ECB intervention	Forecasts of the expert system					
Date (t)	Date (t+1)		Date (t+2)		Date (t+3)	
3 February 2000	16 March 2000		27 April 2000		8 June 2000	
0.25	0.25	(0.25)	0	(0.25)	0	(0.50)
16 March 2000	27 April 2000		8 June 2000		31 August 2000	
0.25	0.25	(0.25)	0.25	(0.50)	0.25	(0.25)
27 April 2000	8 June 2000		31 August 2000		19 October 2000	
0.25	0.25	(0.5)	0.25	(0.25)	0.25	(0.25)
8 June 2000	31 August 2000		19 October 2000		30 November 2000	
0.5	0.25	(0.25)	0.25	(0.25)	0.25	
31 August 2000	19 October 2000		30 November 2000		18 January 2001	
0.25	0.25	(0.25)	0.25		0.25	

Figure 5. ECBDS in Economics: The interest rate decision intervention by the European central Bank (ECB) at five ECB meetings, versus the calculated ahead intervention by an expert system. After each intervention index decided by the ECB is made available, the expert system adjusts to this decision and then suggests what the ECB decision at three consecutive meetings ahead might be. The real intervention indexes are the numbers in brackets.

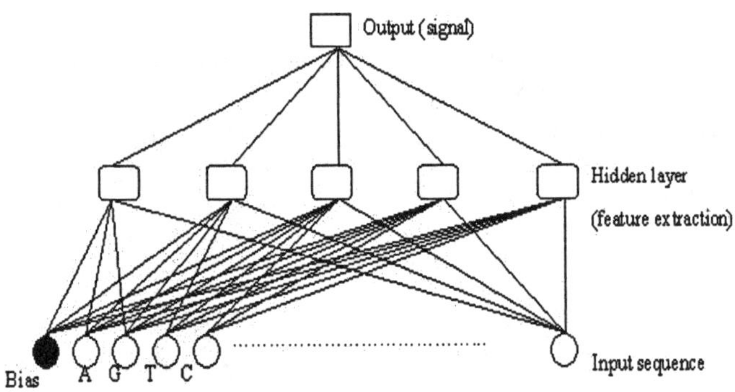

Figure 6. ECBDS in bioinformatics; A neural network that takes an input vector of 60 nucleotides from an RNA data and evaluates the probability for having exon-intron, or intron-exon splice junction in the middle of the sequence, or no junction at all.

Example: ECBDS for RNA splice junction identification

Figure 6 shows a principle diagram of using simple 3-layer-perceptron NN for identifying a biological feature, such as a splice junction from an input RNA sequence. The NN has 60 inputs, which is the length of the RNA window sequence, 5 intermediate neurons (nodes) and one output neuron to encode for the feature (1 – present, 0- not present). The data set used for training is available from: http://www.ics.uci.edu/~mlearn/MLRepository.html. The predicted by the trained NN splice junction from new input data should be further analyzed before a final decision is made.

An EFuNN-based ECBDS for on-line training and prediction on the same problem a above is available from: http://divcom.otago.ac.nz/infosci/kel/CBIIS/GenIn/. The system can also be used to extract rules, such as the rule shown below:

IF -----C--C-C-TCC-G--CTC-GT-C--GGTGAGTG--GGC---C---G-GG-C--CC- THEN [Junction Exon-Intron] Receptive field =0.216, Examples covered by the rule are 26 out of 1000

5. Future Directions for Evolving Connectionist-based Decision Systems

ECBDS is a growing area of artificial intelligence and knowledge engineering. It includes both developing new connectionist techniques for learning, data mining and knowledge discovery, and developing important practical applications in finance and economics, bioinformatics and life sciences, process control and manufacturing, medicine and social sciences.

References

Baldi, P., Brunak, S., 2001, Bioinformatics – A Machine Learning Approach, MIT Press *

Beltraffi, A., Margarita, S., Terna, P., 1996, Neural networks for economics and financial modelling, Int. Thomson Computer Press

Cloete, I, and J.Zurada (eds), 2000, Knowledge-Based Neurocomputing, The MIT Press

Duda, R.O., and Shortliffe, E.H.,1983, Expert systems research, Science, 220:261-268

Fu, L.M., 1989, Integration of neural heuristics into knowledge–based inference, Connection Science, 1;325-340 *

Gallant, S.I.,1993, Neural Network Learning and Expert Systems, MIT Press *

Kasabov, N.K., 1990, Hybrid connectionist rule-based systems, in: Jorrand, P., and Sgurev, V.(eds), Artificial Intelligence – Methodology, Systems, Applications, Amsterdam, North Holland, 227-235

Kasabov, N., 1996, Foundations of Neural Networks, Fuzzy Systems and Knowledge Engineering, CA, MA, MIT Press

Kasabov, N. (2001). Evolving fuzzy neural network for supervised/ unsupervised knowledge-based learning, IEEE Transactions on Man, Machine and Cybernetics, vol.31, No.6, December, 902-918

Kasabov, N., 2002, Evolving connectionist systems: methods and applications in bioinformatics, brain study and intelligent machines, Springer Verlag, London.

Kasabov, N. and Song, Q. (2002) DENFIS: Dynamic evolving neuro-fuzzy inference system and its application for time-series prediction, IEEE Trans. on Fuzzy Systems, vol.10, No.2, April,144-154

Mitra, S., and Hayshi, Y., May 2000, Neuro-fuzzy rule generation: survey in soft computing framework, IEEE Transactions on Neural Networks, vol.11, No.3

Rizzi, R., Bazzana, F., Kasabov, N., Fedrizzi, M. and Erzegovesi, 2002, A connectionist-based decision support system for modelling the interest rate intervention made by the European Central Bank, European Journal of Operation Research.

Sima,J., and Cervenka, J., 2000, Neural knowledge processing in expert systems, in: I.Cloete and J.Zurada (eds) Knowledge-based neurocomputing, MIT Press, 419-466

Takagi, T. and M.Sugeno (1985) Fuzzy Identification of Systems and its Applications to Modelling and Control. IEEE Trans. on Systems, Man and Cybernetics, 15, 116-132.

Towell, G.G. and Shavlik, J.W., 1993, Extracting refined rules from knowledge-based neural networks, Machine Learning, 13:71-101

Touretzky, D., and Hinton, G, 1988, A distributed connectionist production system, Cognitive Science, 12:1423-1466

Zadeh L, (1965) Fuzzy Sets. Information and Control, vol.8, 338-353

ZISC Manual (2001) Zero Instruction Set Computer, Silicon Recognition, Inc., URL www.silirec.com.

An Agent-based Soft Computing Society with Applications in Financial Investment Planning

Chengqi Zhang[1] and Zili Zhang[2]*

[1] Faculty of Information Technology
University of Technology, Sydney
PO Box 123 Broadway, NSW 2007 Australia

[2] School of Computing and Mathematics
Deakin University, Geelong, VIC 3217, Australia

Abstract. Solving problems and making decisions effectively have far-reaching influence on a nation, an organization, or a person. It is not easy work to do for complex problems such as financial investment planning. Many complex problems have many different components, each of which requires different types of processing. That is, hybrid solutions are crucial for complex problem solving and decision making. On the other hand, hybrid intelligent systems are complex because they have a large number of parts or components that have many interactions, whereas multi-agent systems (MAS) are good at complex, dynamic interactions. Thus a multi-agent perspective is suitable for modeling, design, and construction of hybrid intelligent systems. To this end, this chapter discusses the details of the analysis, design, and implementation of an agent-based hybrid intelligent system (called *agent-based soft computing society*) for complex problem solving and decision making according to the available agent-oriented software engineering methodologies. The applications of the society in financial investment planning are also addressed.

1 Introduction

There is now a growing realization in the intelligent systems community that many complex problems require hybrid solutions. Soft computing (SC) techniques such as fuzzy logic (FL), neural networks (NN), and genetic algorithms (GA) have produced encouraging results in particular applications, but there are still certain problems that cannot be solved by a single intelligent technique alone [12]. Each intelligent technique has particular computational properties that make them suitable for particular problems and not for others. They cannot be applied universally to every problem. This has been a central driving force behind the creation of hybrid intelligent systems, that is, intelligent systems which employ a combination of FL, NN, GA and related methodologies to overcome the limitations of individual techniques.

Hybrid intelligent systems are also important when considering the varied nature of application domains. Many complex domains have many different

* The author's address in China: Computer Science Department, Southwest China Normal University, Chongqing 400715.

component problems, each of which may require different types of processing. For example, in the financial investment planning application, the NN can be used as a pattern watcher for the stock market; the GA can be used to predict interest rates; the approximate reasoning based on FL can be used to evaluate clients' financial risk tolerance ability etc.

It is now widely recognized that *interaction* is probably the most important single characteristic of complex software, whereas the integration and interaction of different techniques is crucial for the construction of hybrid systems. Thus hybrid systems are complex and difficult to build.

An agent is an encapsulated computer system that is situated in some environment and that is capable of flexible, autonomous action in that environment in order to meet its design objectives. A multi-agent system can be defined as a loosely coupled network of entities that work together to make decisions or solve problems that are beyond the individual capabilities or knowledge of each entity [2]. These entities – agents – are autonomous and may be heterogeneous in nature. From a multi-agent perspective, agents in multi-agent systems are autonomous and can engage in flexible, high-level interactions. Considering the autonomous nature of agents, autonomy means that the agents have their own persistent thread of control (i.e., they are active) and that they decide for themselves which actions they should perform at what time. The fact that agents are active means they know for themselves when they should be acting and when they should update their state. The flexible nature of interactions means that agents can make decisions about the nature and scope of interactions at run-time rather than design time. MASs are good at complex, dynamic interactions. Thus a multi-agent perspective is suitable for modeling, design, and construction of hybrid intelligent systems.

Thus far, there are some hybrid intelligent systems that adopted agent concepts. One of such attempts is the MIX multi-agent platform [6][18]. The focus of MIX is the development of strategies and tools for integrating neural and symbolic technologies. Another such attempt is the PREDICTOR system [7]. In [7] Scherer and Schlageter detail the use of distributed artificial intelligence approaches for combining neural networks and expert systems. In [8], Khosla and Dillon introduce a computational architecture called IMAHDA (Intelligent Multi-Agent Hybrid Distributed Architecture). The role and knowledge content of IMAHDA consist of four layers, namely, object, software agent, intelligent agent, and the problem solving agent, respectively. A more recent attempt is the multi-agent architecture for fuzzy modeling [9]. Delgado et al. proposed a hybrid learning system that combines different fuzzy modeling techniques by means of a multi-agent architecture. In [19], Jacobsen proposed a generic architecture for hybrid intelligent systems, which is based on the conceptual learning agent architecture according to Russell and Norvig [1].

Among the above five hybrid intelligent systems, the MIX, PREDICTOR, and the architecture for fuzzy modeling only integrated very limited soft computing techniques. Both the MIX and PREDICTOR systems are focused on the integration of neural networks and symbolic technologies such as expert systems. The multi-agent architecture of Delgado et al. concentrated on the integration of different fuzzy modeling techniques such as fuzzy clustering, fuzzy rule generation, and fuzzy rule tuning techniques. In MIX and PREDICTOR systems, the way for integrating soft computing technologies into multi-agent systems is to embed the soft computing technologies in each individual agent. The MIX and IMAHDA architectures are inflexible as no middle agent [10] was used. The work in [19] is focused on the micro (intra-agent) level of agents, i.e., the integration and interaction of different components within one agent. The macro (inter-agent) level integration and interaction are ignored.

In summarizing, the approaches used in the above systems have the following limitations:

- It is impossible to embed many SC technologies within a single agent. Otherwise, the agents will be overloaded. In many applications, the agents in multi-agent systems should be kept simple for ease of maintenance, initialization, and customization;
- It is not flexible to add more SC technologies to or delete some unwanted one from the multi-agent systems. For example, one software agent may be equipped with fuzzy logic, the other with neural network etc. In such a way, one agent can only have one SC capability. If one wants the agent to possess two or more SC capabilities, the implementations must be modified;
- The agents in these systems are difficult to interoperate as they did not use some type of common or standard agent communication language.

In addition to the above limitations, all these systems did not follow any agent-oriented analysis and design methodology. (In IMAHDA, object-oriented analysis and design approaches were adopted).

With all these observations in mind, an agent-based framework for constructing hybrid intelligent system is proposed. This framework is called *agent-based soft computing society*. In this chapter, the analysis, design, and implementation of the society are detailed by following an agent-oriented methodology. The proposed framework has some crucial characteristics that differentiate this work from other hybrid intelligent systems mentioned above:

- The ability to exchange comprehensible communications (interactions at knowledge level);
- Each service requester agent (decision making agent) can easily access all the techniques provided by service provider agents (e.g., SC agents) in the system;
- The presence of the serving agent in the framework allows adaptive system organization. For example, if one is sure that one hybrid SC agent

can do a better job than a single technology SC agent, one can delete the single SC technology agent and add the hybrid agent to the society simply by adding or deleting a record in the database of the serving agent;
- Overall system robustness is facilitated through the use of the serving agent. For example, if a particular service provider (e.g., an SC agent) disappears, a requester agent (decision making agent) can find another one with the same or similar capabilities by interrogating the serving agent; and
- Agent-based hybrid intelligent systems based on the framework can make decisions about the nature and scope of interactions at run time.

With the support of the soft computing agent society, the multi-agent system developers need only to build the domain-specific parts and construct the ontologies used in the specific application field – rather than re-inventing the wheel as often happens at the moment.

The rest of this chapter is organized as follows. In Section 2, an agent-oriented methodology used to analyze and design the society is outlined. Section 3 presents the details of the analysis, design, and implementation of the society according to the methodology discussed in Section 2. A case study is provided in Section 4 to demonstrate how to apply this society to construct agent-based hybrid systems for real applications. The application field used here is financial investment planning. The evaluation of the society is discussed in Section 5. Section 6 contains conclusions.

2 The Agent-Oriented Methodology

Existing software development techniques (for example, object-oriented analysis and design) are inadequate for multi-agent system analysis and design [17]. There is a fundamental mismatch between the concepts used by object-oriented developers (and indeed, by other mainstream software engineering paradigms) and the agent-oriented perspective. In particular, extant approaches fail to adequately capture an agent's flexible, autonomous problem-solving behavior, the richness of an agent's interactions, and the complexity of an agent system's organizational structure. For these reasons, this section outlines a methodology that has been specifically tailored to the analysis and design of agent-based hybrid intelligent systems. The main models used in this methodology are summarized in Figure 1, which are mainly based on the Gaia methodology [17].

The analysis phase aims to identify what the actual organization of the multiple agents should look like. It does this by decomposing the system into abstract "loci of control"; i.e., the *roles* to be played in the organization, and the way in which they interact accordingly to specific *protocols*. This defines the *role model* and *interaction model*, respectively.

The role model identifies the key roles in the system. Here a role can be viewed as an abstract description of an entity's expected function. A role is

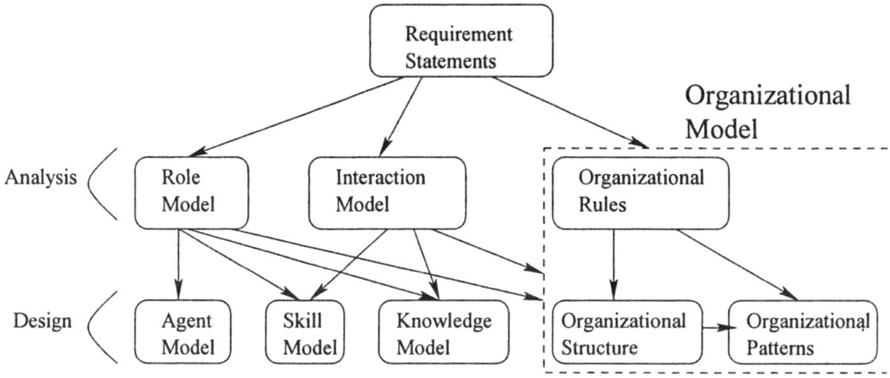

Fig. 1. Relationships between Models

defined by four attributes: *responsibilities, permissions, activities*, and *protocols.*

Responsibilities determine functionality and, as such, are perhaps the key attribute associated with a role. Responsibilities are divided into two types: *liveness properties* and *safety properties.* Liveness properties intuitively state that "something good happens". They describe those states of affairs that an agent must bring about, given certain environmental conditions. Following the Gaia notation, liveness properties are specified via a *liveness expression*, which defines the "life-cycle" of the role and is a regular expression. The general form of a liveness expression is:

$$\text{RoleName} = \textit{expression}$$

where RoleName is the name of the role whose liveness properties are being defined, and *expression* is the liveness expression defining the liveness properties of RoleName. The atomic components of a liveness expression are either *activities* or *protocols.* The operators for liveness expressions are shown in Table 1.

In contrast, safety properties are *invariants.* Intuitively, a safety property states that "nothing bad happens" (i.e., that an acceptable state of affairs is maintained across all states of execution). Safety requirements are specified by means of a list of predicates. These predicates are typically expressed over the variables listed in a role's permission attribute. By convention, safety expressions are listed as a bulleted list, each item in the list expressing an individual safety responsibility.

In order to realize responsibilities, a role has a set of *permissions.* Permissions are the "rights" associated with a role. The permissions of a role thus identify the resources that are available to that role in order to realize its responsibilities. Permissions tend to be *information resources.* For example, a role might have associated with it the ability to read a particular item

Table 1. Operators for Liveness Expressions

Operator	Interpretation
$x.y$	x followed by y
$x \mid y$	x or y occurs
$x*$	x occurs 0 or more times
$x+$	x occurs 1 or more times
x^{ω}	x occurs infinitely often
$[x]$	x is optional
$x \parallel y$	x and y interleaved

RoleSchema:	*name of role*
Description	*short English description of the role*
ProtocolsandActivities	*protocols and activities in which the role plays a part*
Permissions	*"rights" associated with the role*
Responsibilities	
Liveness	*liveness responsibilities*
Safety	*safety responsibilities*

Fig. 2. Template for Role Schemata

of information, or to modify another piece of information. A role can also have the ability to generate information. The *activities* of a role are computations associated with the role that may be carried out by the agent without interacting with other agents.

Finally, a role is also identified with a number of *protocols*, which define the way that it can interact with other roles. A role model is comprised of a set of *role schemata*, one for each role in the system. A role schema draws together the various attributes discussed above into a single place (see Figure 2).

In summary, the analysis phase is tasked with collecting all the specifications from which the design of the computational organization can start (refer to Figure 1). The output of the analysis phase should be a triple: $\langle PR, PP, OL \rangle$, where PR are the preliminary roles of the system, PP are the preliminary protocols (which have already been discovered to be necessary for the preliminary roles), and OL are the organizational rules.

The design phase starts from the models defined during the analysis phase and aims to define the actual agent system in such a way that it can easily be implemented. To this end, the design phase has to decide which classes of agents (and how many) have to play the roles identified during the analysis phase, which services agents must provide to fulfill their role, and what is the actual topology of the interactions that flows from the interaction and agent models.

The purpose of the agent model is to document the various *agent types* that will be used in the system under development, and the *agent instances* that will realize these agent types at run-time.

An agent type is best thought of as a set of agent roles. There may, in fact, be a one-to-one correspondence between roles (as identified in the role model) and agent types. However, this need not be the case. A designer can choose to package a number of closely related roles in the same agent type for the purpose of convenience. Efficiency will also be a major concern at this stage – a designer will almost certainly want to optimize the design, and one way of doing this is to aggregate a number of agent roles into a single type.

The agent model is defined using a simple *agent type tree*, in which leaf nodes correspond to roles, and other nodes correspond to agent types. If an agent type t_1 has children t_2 and t_3, then this means that t_1 is composed of the roles that make up t_2 and t_3.

The aim of the skill model is to identify the main skills with each agent role. Skills mainly consist of the basic *services* required to be able to perform a role.

A service is defined as a *function* of the agent. For each service that may be performed by an agent, it is necessary to document its properties. Specifically, one must identify the *inputs, outputs, pre-conditions*, and *post-conditions* of each service. Inputs and outputs to services will be derived in an obvious way from the interaction model. Pre- and post-conditions represent constraints on services. These are derived from the safety properties of a role. Note that by definition, each role will be associated with at least one service.

The services that an agent will perform are derived from the list of protocols, activities, responsibilities and the liveness properties of a role. The inference mechanisms used by roles are also needed to be identified in this model.

The design phase builds on the output of the analysis phase and produces a complete specification of the multi-agent system. The design stage can now be summarized as the following:

- Create an agent model: (1) aggregate roles into agent types, and refine to form an agent type hierarchy; (2) document the instances of each agent type using instance annotations.
- Develop a skill model, by examining activities, protocols, and safety and liveness properties of roles.
- Develop a knowledge model from the interaction model and agent model.
- Identify organizational structures and organizational patterns that respect the organizational rules.

3 Agent-Based Soft Computing Society for Decision Making

In order to identify which components should be contained in a typical decision making system, without loss of generality, consider a financial house providing investment advice for clients. In such a house, there are: a front counter or reception desk clerk, one or more personnel officer(s), and many financial investment experts (decision makers). The advice giving (decision making) process is initiated by a user contacting the front desk clerk with a set of requirements. The clerk asks the personnel officer to provide the experts' profile, and then delegates the task to one or more experts based on experts' profiles. The experts then work on the task and try to give their recommendations with or without external help. After the experts finish preparing a recommendation (if the task was assigned to more than one expert, the recommendations from different experts must be combined to form a final one), they pass it to the front desk clerk. Finally, the clerk sends the advice to the user. Such a typical process can help us analyze and design an agent-based hybrid system for decision making.

3.1 Analysis of Agent-Based Soft Computing Society

Based on the above description and the methodology proposed in the previous section, it is comparatively straightforward to identify the roles in the hybrid framework for decision making.

Role Schema: USERHANDLER
Description: Receives request/inquiry from the user and oversees process to ensure appropriate decision is returned
Protocols and Activities: AwaitCall, InformUser, ProduceAssignments, ReceiveDecision
Permissions: **reads supplied** *UserDetails* // personal information of the user **supplied** *UserRequirements* // what user wants *Decision* // final decision or nil
Responsibilities Liveness: USERHANDLER = AwaitCall. ProduceAssignments. ReceiveDecision. InformUser Safety: • True

Fig. 3. Schema for Role USERHANDLER

Role Schema: WORKPLANNER
Description: This role elaborates a work plan for decision making and is in charge of ensuring that such a work plan is fulfilled.
Protocols and Activities: GetUserRequirements, <u>ProducePlan</u>, GetCapabilities, DelegateTasks
Permissions: **reads supplied** *UserDetails* // personal information of the user **supplied** *UserRequirements* // detailed service requirements *Capabilities* // capabilities of other roles **generates** plan // work plan
Responsibilities Liveness: WORKPLANNER = GetUserRequirements.GetCapabilities.<u>ProducePlan</u>.DelegateTasks Safety: • *UserDetails* and *UserRequirements* are available

Fig. 4. Schema for Role WORKPLANNER

Role Schema: CAPABILITYRECORDER
Description: Add capabilities advertised by roles to CapabilityDatabase, or delete capabilities unadvertised by roles from the database
Protocols and Activities: ReceiveAdvertisement, <u>AddCapability, DeleteCapability</u>
Permissions: **reads supplied** *CapabilityInfo* // Advertised or unadvertised capability information **changes** *CapabilityDatabase* // add or delete capability
Responsibilities Liveness: CAPABILITYRECORDER = ReceiveAdvertisement.(<u>AddCapability</u>|<u>DeleteCapability</u>) Safety: • CapabilityDatabase exists

Fig. 5. Schema for Role CAPABILITYRECORDER

Role Schema: CAPABILITYMATCHER
Description: Matches capabilities requested by a role with those capabilities in CapabilityDatabase. Informs requester role with the ROLENAME whose capability matched with the requested one.
Protocols and Activities: GetRequestedCapability, AccessCapabilityDatabase, InformRequester, CompareCapability
Permissions: **reads supplied** *RequestedCapability* // capability requested by a role **supplied** *CapabilityDatabase*
Responsibilities Liveness: CAPABILITYMATCHER = GetRequestedCapability. (AccessCapabilityDatabase.CompareCapability)+.InformRequester Safety: • True

Fig. 6. Schema for Role CAPABILITYMATCHER

The front desk clerk's behavior falls into two distinct roles: one acting as an interface to the user (USERHANDLER, Figure 3) and one overseeing the process inside the organization (WORKPLANNER, Figure 4). The personnel officer's behavior falls into another two roles: one keeping track of the profiles (CAPABILITYRECORDER, Figure 5) and one checking the profiles (CAPABILITYMATCHER, Figure 6). The experts' behaviors are covered by DECISIONMAKER (Figure 7), HELPPROVIDER (Figure 8), and DECISIONAGGREGATOR (Figure 9) roles. The final role is that of the USER (Figure 10) who requires the decision.

With the respective role definitions in place, the next stage is to define the associated interaction models for these roles. Here we focus on the interactions associated with the DECISIONMAKER role.

This role interacts with the WORKPLANNER role to obtain the task this role will accomplish (ReceiveTask protocol, Figure 11a). It interacts with the INFOGATHER role to get some relevant information (known facts) for the task (GetInformation protocol, Figure 11b). It also interacts with the CAPABILITYMATCHER role to provide some roles for data pre- and/or post-processing etc. when accomplishing the task (AskforHelp protocol, Figure 11c). When the DECISIONMAKER role finishes making decision for the task, it informs the DECISIONAGGREGATOR role its alternative decision for the task (InformDecisionAggregator protocol, Figure 11d).

In the society, the most important organizational rule in the organizational model is that *if a role says it has a capability then it can perform the tasks corresponding to the capability and will do so when asked.*

Role Schema: DECISIONMAKER
Description: Reaches a conclusion for the delegated task based on the information gathered and knowledge the role has.
Protocols and Activities: ReceiveTask, GetInformation, AskforHelp, InformDecisionAggregator, <u>Reasoning</u>
Permissions: **reads supplied** *Task* // Task delegated by other roles **supplied** *Knownfacts* // Gathered information `generates` *Decision* // one alternative decision
Responsibilities Liveness: DECISIONMAKER = ReceiveTask.GetInformation.(<u>Reasoning</u>.AskforHelp)+.InformDecisionAggregator Safety: • True

Fig. 7. Schema for Role DECISIONMAKER

Role Schema: HELPPROVIDER
Description: Provides appropriate help for other roles (e.g., DECISIONMAKER) when asked.
Protocols and Activities: ReceiveHelpRequirements, ReturnProcessedResults, AskforHelp, <u>Processing</u>
Permissions: **reads supplied** *HelpRequirements* // one role asks another role for help `generates` *ProcessedResults* // do the requested processing and // generate results
Responsibilities Liveness: HELPPROVIDER = ReceiveHelpRequirements.(<u>Processing</u>.AskforHelp)$*$.ReturnProcessedResults Safety: • HELPPROVIDER has the requested capability

Fig. 8. Schema for Role HELPPROVIDER

3.2 Design of the Society

Having completed the analysis of the society, the design phase follows. The first model to be generated is the agent model (Figure 12). This shows, for most cases, a one-to-one correspondence between roles and agent types. The exception is for the CAPABILITYRECORDER and CAPABILITYMATCHER roles which, because of their high degree of interdependence, are grouped into a single agent type.

Role Schema: DECISIONAGGREGATOR
Description: Produces the final decision based on currently received alternative decisions for a task.
Protocols and Activities: ReceiveDecisions, InformUserHandler, ChooseDecision, AggregateDecisions
Permissions: **reads supplied** *Decision* // alternative decision **generates** *FinalDecision* // The "best" one of all alternative decisions or aggregated decision
Responsibilities Liveness: DECISIONAGGREGATOR = ReceiveDecisions.(ChooseDecision| AggregateDecisions).InformUserHandler Safety: • alternative decisions are available

Fig. 9. Schema for Role DECISIONAGGREGATOR

Role Schema: USER
Description: Organization or individual requesting a decision.
Protocols and Activities: MakeCall, GiveRequirements
Permissions: **generates** *UserDetails* // Owner of user information *UserRequirements* // Owner of user requirements
Responsibilities Liveness: USER = (MakeCall.GiveRequirements)+ Safety: • True

Fig. 10. Schema for Role USER

The second model is the skill model. Again to avoid redundancies the focus is on the DECISIONMAKER role and the Decision Making Agent. Based on the DECISIONMAKER role, six distinct services can be identified (Table 2).

From the ReceiveTask protocol, the service "obtain task" is derived. This service returns the *TaskRequirements* as output. It has a pre-condition that the agent or role has the corresponding capability to perform the task, but has no post-condition.

The service associated with the GetInformation protocol is "get information". Its inputs, derived from the protocol definition (Figure 11b), are the

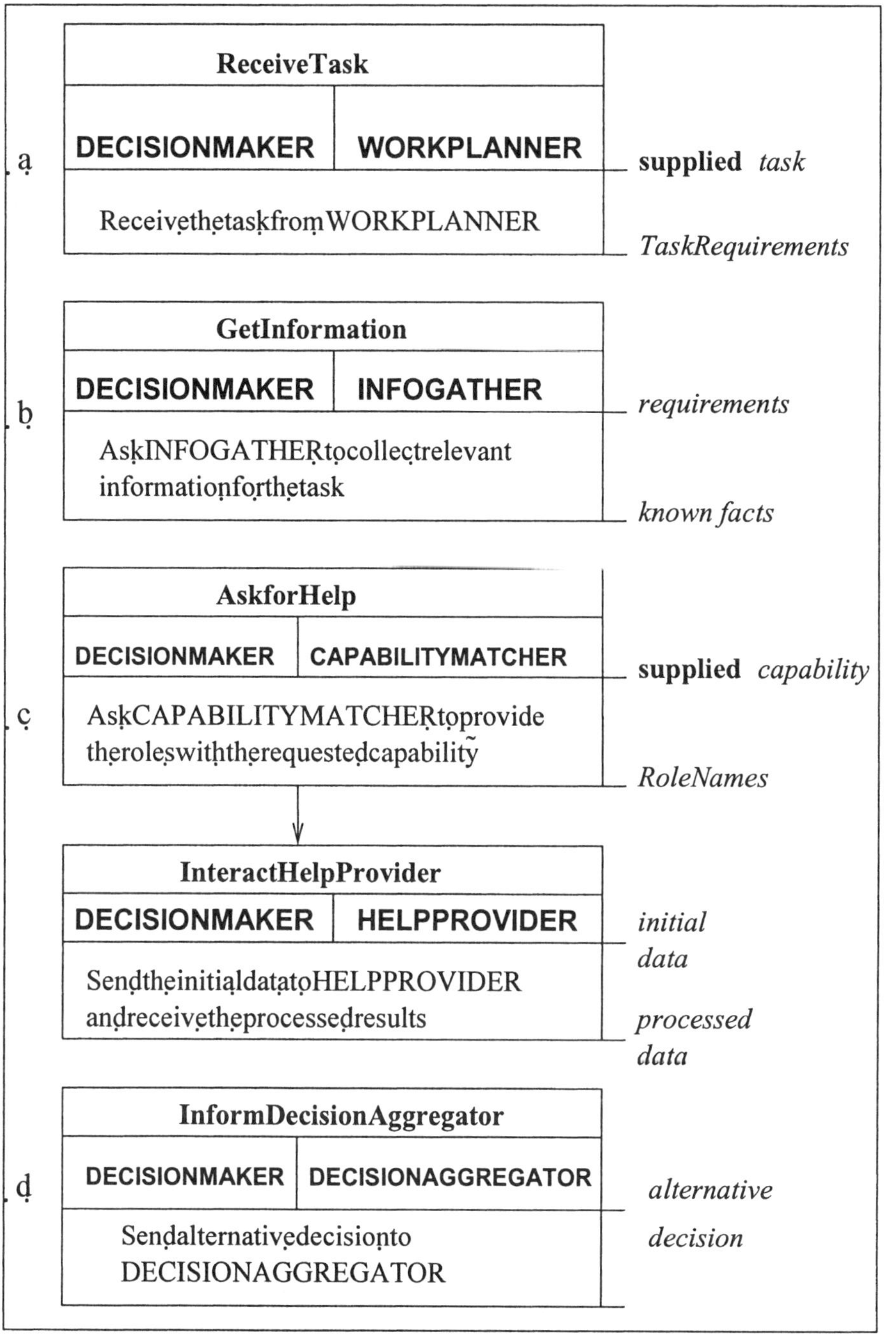

Fig. 11. Definition of Protocols Associated with the DECISIONMAKER Role: (a) ReceiveTask, (b) GetInformation, (c) AskforHelp, and (d) InformDecisionAggregator.

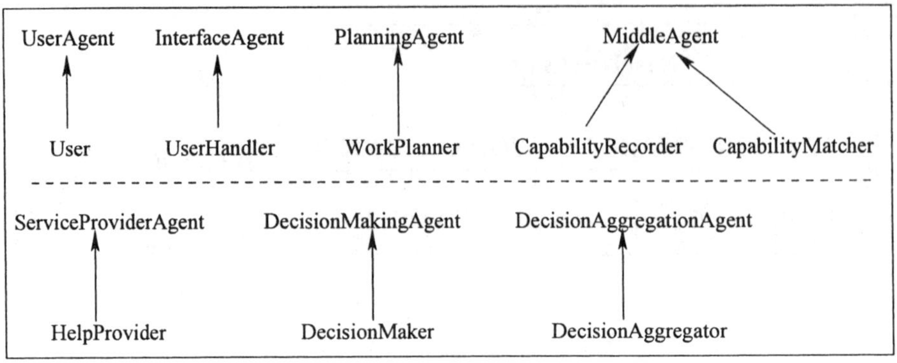

Fig. 12. The Agent Model of the Society

Table 2. Services in Skill Model

Service	Inputs	Outputs	Pre-condition	Post-condition
obtain task		task requirements	agent or role has corresponding capability	true
get information	requirements for gathering	gathered information	information sources are available	relevant information returned
accomplish reasoning	known facts	conclusions (decisions)	known facts are enough	conclusions reached
call for help	capability description	agent names or nil	true	agent name with matched capability returned
provide initial data	agent name and initial data	processed results	initial data are ready	data processed
inform aggregator	alternative decision		true	aggregator knew decision

requirements for information gathering and its outputs are known facts (gathered information). The pre-condition for this service is that some information sources are available, and the post-condition is that relevant information is returned.

The third service, "accomplish reasoning", is related to the Reasoning activity. The inputs for this service are known facts, and outputs are some conclusions. This service has a pre-condition that the known facts are enough to reach conclusions (post-condition).

The next two services, "call for help" and "provide initial data", are derived from the AskforHelp protocol. The "call for help" service takes a specific capability description as input, and returns the agent name with matched capability or nil (if no matched) as outputs. The "provide initial data" service takes the agent name and initial data as inputs and returns the processed data as outputs.

The final service involves informing the decision aggregator of the alternative decisions. Then the inference mechanisms in the skill model are checked. To make decisions, the agents must accomplish some reasoning based on their knowledge and other available information. The final model is the knowledge model, which indicates different levels of knowledge that agents should have. In the society, the agents should have three levels of knowledge. The first level of knowledge is for agent interaction and communication. This involves in domain-specific and domain-independent terminologies and their relationships etc. The identified domain-specific terms and their relationships will result in the construction of a domain-dependent ontology for a specific application. The second level of knowledge is some domain knowledge related to specific problem solving techniques. The third level of knowledge is meta knowledge that directs the activities of an agent.

3.3 Architecture of the Society

From the above analysis and design phases, it is clear that there are seven types of agents in the soft computing society – user agent, interface agent, planning agent, middle agent, service provider agent, decision making agent, and decision aggregation agent. Now it is time to find an appropriate multi-agent architecture to put all these agents together so as to form the agent-based hybrid decision making systems.

When incorporating SC technologies into multi-agent systems to form agent-based hybrid intelligent systems, one criterion that must be met, is that any agent in the system can access any of the SC and other techniques available in the system when needed. To this end, the architecture of the *agent-based soft computing society* is shown in Figure 13. The behaviors of each kind of agent in the society (except the user agent) are briefly described below:

- **Interface Agent** This agent interacts with the user (or user agent). It asks the user to provide his personal information and requirements, and provides the user with a final decision or advice that best meets the user's requirements.
- **Planning Agent** The planning agent is in charge of the activation and synchronization of different agents. It elaborates a work plan and is in charge of ensuring that such a work plan is fulfilled. It receives the assignments from the interface agent.

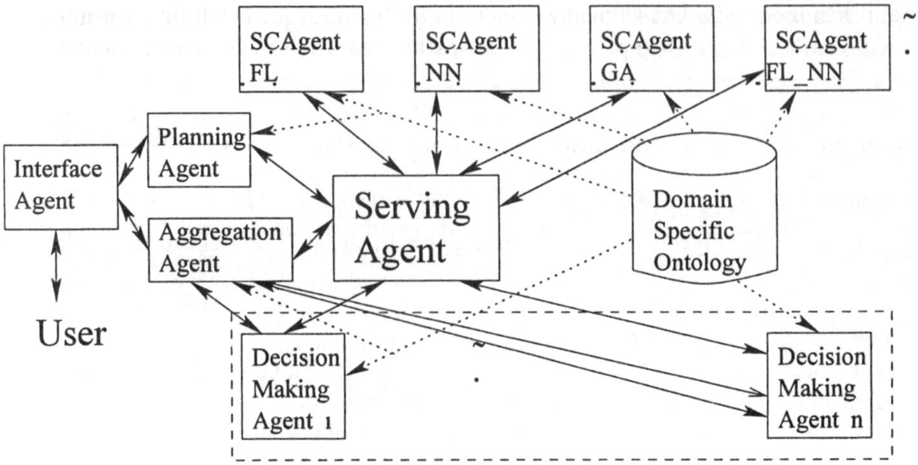

Fig. 13. Architecture of Agent-Based Soft Computing Society

- **Decision Making Agent** It is application-specific, i.e., it has its own knowledge base; it must have some meta-knowledge about when it needs the help of soft computing agents (e.g., pre or post processing some data); it can ask soft computing agents to accomplish some subtasks.
- **Serving Agent** The serving agent is a matchmaker – one kind of middle agent. It keeps track of the names, ontologies, and capabilities of all registered soft computing agents in the system; it can reply to the query of a decision making agent with appropriate soft computing agent's name and ontology.
- **Service Provider Agent** Most of the service provider agents in the society are SC agents. Each SC agent can provide services for decision making agents with one or some kinds of combined SC technologies; it can send back the processed results to decision making agents; it must advertise its capabilities to the serving agent.
- **Decision Aggregation Agent** When decision making agents finish the assigned tasks they return the results to the decision aggregation agent. The aggregation agent chooses one of the alternative decisions, or performs an aggregation of the different results into a final one.

The ontology is the foundation for agent communication. All agents in the society interpret the content of received messages based on the ontology.

4 Case Study

We have implemented a prototype system of agent-based soft computing society using Java and under the support of JATLite (*java.stanford.edu*). The application background of the prototype is financial investment planning.

Following the analysis and design phases of the proposed methodology, we worked out that the prototype consists of the following agents: one middle (serving) agent, one interface agent, one planning agent, one aggregation agent, four decision making agents (investment policy agent, portfolio selection agent based on the Markowitz model, portfolio selection agent based on the fuzzy model, and portfolio selection agent based on the possibility model), and four service provider agents (financial risk tolerance ability evaluation agent, interest rate prediction agent based on neural network, interest rate prediction agent based on fuzzy logic and genetic algorithm, and an approximate reasoning agent).

To demonstrate how the prototype works, we first examine the typical scenario for investment. We then present an example to show how the multi-agent financial investment planning prototype system gives advice to the investor.

4.1 A Typical Scenario for Investment

When a person wants to invest some money somewhere, he usually comes to the financial investment adviser for advice. The first thing the adviser needs to do is to understand the client's individual circumstances. The adviser may ask the client to provide the following information about himself: his financial position (annual income, total net-worth etc.), age, tax effectiveness etc. Based on the information, the adviser should evaluate the financial risk tolerance ability as well as the client's investment goal.

If the client's primary goal is income, then investments that provide interest or dividend payments regularly and dependably are required. If the primary goal is growth, then investments that are likely to increase in value are wanted so that they may be resold for more than their initial cost. If, however, the primary goal is to avoid risk, then investments that offer the greatest safety of principal and protection from inflation are required. Unfortunately, there is no single investment that simultaneously offers maximum income, maximum growth and minimum risk.

Suppose the adviser suggests the client invest some money in the stock market after evaluating his financial risk tolerance ability. How can one select a portfolio for the client under his constraints (risk tolerance level, return rate etc.)? The adviser should gather some information about the stock market. The information includes market data, financial report data, technical models, analysts' reports, breaking news etc. After gathering the information, the adviser then makes a portfolio selection decision based on some models (e.g., the Markowitz model, the fuzzy probability model etc.)

In short, the overall investment advising task has several component tasks: eliciting (or learning) user profile information, collecting information on the user's initial portfolio position, and suggesting and monitoring a reallocation to meet the user's current profile and investment goals.

4.2 Example

Suppose the investor provides the interface agent with the following personal information:

Amount of money to invest: \$ 150,000

Annual income: \$ 70,000

Total net-worth: \$ 200,000

Age: 40

Investment goal: *growth* (can be *income, growth*, and *avoid risk* etc. This reflects investor's attitude towards risk.)

The interface agent passes the investor's information to the planning agent (see Figure 13). The planning agent will delegate subtasks to different decision making agents. During the decision making process, decision making agents may ask different service provider agents (most of them are SC agents) for help. Finally, the prototype will give its advice to the investor through the interface agent. The whole process (from receiving inputs to giving advice) consists of the following five steps.

Step 1: Determining the investor's investment policy (aggressive or conservative).

The investment policy agent will use fuzzy rules in its knowledge base to determine the investor's investment policy. One example fuzzy rule is as follows:

If the investor's risk tolerance is *high* and the interest rate is *falling* and the investor's investment goal is *growth* **then** the investor can take an *aggressive investment policy.*

The investment policy agent will then ask for help based on its meta-knowledge. Thus, the investment policy agent sends KQML messages using *recommend-one* performative to the serving agent:

```
(recommend-one
 :sender investment_policy_agent
 :receiver serving_agent
 :language KQML
 :content (ask
            :ability risk_tolerance_evaluation
            :name ?
            :ontology ?
           )
 )
```

```
(recommend-one
 :sender investment_policy_agent
 :receiver serving_agent
 :language KQML
 :content (ask
            :ability interest_rate_prediction
            :name ?
            :ontology ?
           )
 )
```

The serving agent then retrieves its database and replies with an appropriate service provider agent's name and ontology which has the capability asked for using *reply* performative. In the prototype, there are a risk tolerance ability evaluation agent based on fuzzy logic and two interest rate prediction agents based on feed-forward neural network and fuzzy-genetic algorithm, respectively. The serving agent will choose the interest rate prediction agent based on fuzzy genetic model according to the initial values of these agents' track records [20].

```
(reply
 :sender serving_agent
 :receiver investment_policy_agent
 :content (:name SC_agent_FL
           :ontology Financial_investment
          )
)

(reply
 :sender serving_agent
 :receiver investment_policy_agent
 :content (:name SC_agent_FLGA
           :ontology Financial_investment
          )
)
```

The investment policy agent then communicates with SC_agent_FL (for risk tolerance evaluation) and SC_agent_FLGA (for interest rate prediction) directly.

The risk tolerance evaluation agent (based on fuzzy logic) uses rules such as

If the investor's annual income is *high* and total net-worth is *high* and the investor is *young* **then** the investor's risk tolerance ability is *high*;

If annual income $> \$50,000$ **then** annual income is *high*;

If age ≤ 35 **then** the investor is *young* etc.

Based on the investor's information and the financial risk tolerance model [21], the risk tolerance evaluation agent obtains the result that the investor's risk tolerance ability is *high*. During this process, the risk tolerance evaluation agent needs the help of the approximate reasoning agent.

The fuzzy-genetic based interest rate prediction agent will ask the following parameters as input: the change of real gross national product ΔGNP, the change of consumer price index ΔCPI, the change of M_2 nominal money supply ΔM_2, the change of personal wealth ΔW, and the change of previous T-bill rates Δr. All these data need to be gathered by information gathering agents.

Suppose the gathered data are as follows: $\Delta GNP = 0.50$, $\Delta CPI = -0.10$, $\Delta M_2 = 47.80$, $\Delta W = 456.70$, and $\Delta r = 0.01$. The predicted result is -0.76. Thus the interest rate prediction agent reaches the conclusion that the interest rate will *fall*.

Combining the results of the two SC agents and the investor's investment goal, the investment policy agent reaches the conclusion that the investor's investment policy can be *aggressive*.

Step 2: Determining the investment category. The investment category can be the stock market, real estate, fixed term deposit etc. One of the decision making agents based on a fuzzy model is used to accomplish this task. Example rules used are:

If the investor's risk tolerance ability is *high* and the investment policy is *aggressive* **then** the suggested investment category is the *stock market*;

If the investor's risk tolerance ability is *high* and the investment policy is *conservative* **then** the suggested investment category is *real estate*, etc.

Based on the model, the agent gives the suggestion that the investor can invest in the *stock market*.

Step 3: Delegating information gathering agents to gather stock market information (this step has not been implemented in the prototype). Here we suppose information gathering agents gathered the returns of nine securities in the past 18 years (see Table 3).

Step 4: Selecting portfolios from these nine securities that satisfies the investor's risk tolerance level and maximizes the return. The selected portfolios (with the expected average return $r_s = 0.17$) based on the three portfolio selection models – the Markowitz model, the fuzzy probability model, and the possibility distribution model are shown in Figure 17 (denoted by P_{mar}, P_{fuz}, and P_{pos}, respectively).

Step 5: Aggregating the portfolios based on different models and giving a final portfolio to the investor.

How can one aggregate the portfolios from different models and produce a final one that is the best among those? This is a very tough question to answer. Currently, we aggregated these portfolios based on the ordered weighted averaging (OWA) aggregation algorithm based on fuzzy logic [23][24]. Based

Table 3. Returns on Nine Securities

Year	S_1	S_2	S_3	S_4	S_5	S_6	S_7	S_8	S_9
1982	-.305	-.173	-.318	-.477	-.457	-.065	-.319	-.400	-.435
1983	.513	.098	.285	.714	.107	.238	.076	.336	.238
1984	.055	.200	-.047	.165	-.424	-.078	.381	-.093	-.295
1985	-.126	.030	.104	-.043	-.189	-.077	-.051	-.090	-.036
1986	-.280	-.183	-.171	-.277	.637	-.187	.087	-.194	-.240
1987	-.003	.067	-.039	.476	.865	.156	.262	1.113	.126
1988	.428	.300	.149	.225	.313	.351	.341	.580	.639
1989	.192	.103	.260	.290	.637	.233	.227	.473	.282
1990	.446	.216	.419	.216	.373	.349	.352	.229	.578
1991	-.088	-.046	-.078	-.272	-.037	-.209	.153	-.126	.289
1992	-.127	-.071	.169	.144	.026	.355	-.099	.009	.184
1993	-.015	.056	-.035	.107	.153	-.231	.038	.000	.114
1994	.305	.038	.133	.321	.067	.246	.273	.223	-.222
1995	-.096	.089	.732	.305	.579	-.248	.091	.650	.327
1996	.016	.090	.021	.195	.040	-.064	.054	-.131	.333
1997	.128	.083	.131	.390	.434	.079	.109	.175	.062
1998	-.010	.035	.006	-.072	-.027	.067	.210	-.084	-.048
1999	.154	.176	.908	.715	.469	.077	.112	.756	.185

on the OWA aggregation ($\alpha = 0.7$), the final portfolio is obtained (P_{owa} in Figure 17).

4.3 Running the System

Figure 14 shows the user interface of the system, which can start from any Internet Browser or appletviewer.

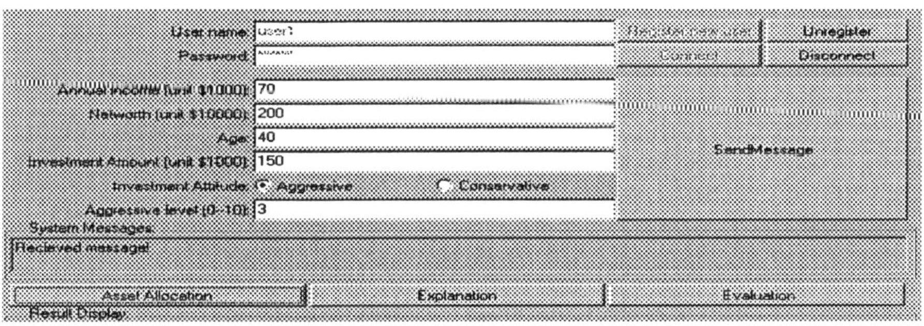

Fig. 14. User Interface of the System

To use the system, the user needs to type the user name and the password he likes in the corresponding fields and click "register new user" to register for the first time. Thereafter, just input the registered user name and password and click "connect". The user then can input his annual income, networth etc. information to the system and click "sendmessage" button. The system provides asset allocation information, explanations of how to get the results, and the evaluation of different portfolios in the "result display" window. If the user wants to leave the system, click "unregister" and "disconnect".

The system can provides reasonable financial investment planning information based on the user provided data and some relevant models. Figure 15 shows the asset allocation results when the annual income is $70,000, networth $200,000, age 40, investment amount $150,000, and investment attitude is aggressive (level 3).

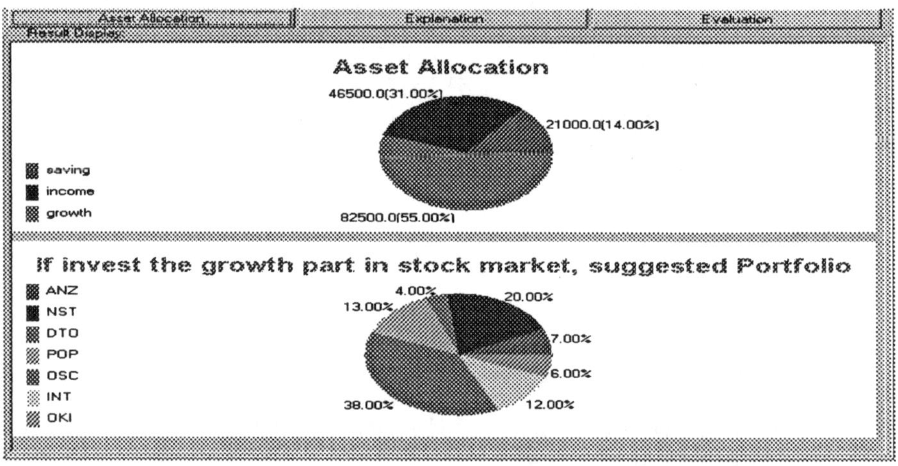

Fig. 15. Example Asset Allocation Results

By clicking the "explanation" button, the corresponding explanation of how to get the results is displayed in the "result display" window (Figure 16). This indicates that the explanation mechanisms of expert systems are integrated into the system.

If the growth part is invested in stock market, the system can provide a portfolio for the user (Figure 15). The portfolio is the aggregated result of three portfolios based on Markowitz's portfolio selection model, the fuzzy probability portfolio selection model, and the possibility distribution portfolio selection model, respectively. The aggregation algorithm used is ordered weighted averaging (OWA) aggregation algorithm [24]. By clicking the "evaluation" button, the system will provide the comparisons of the four portfolios (Figure 17).

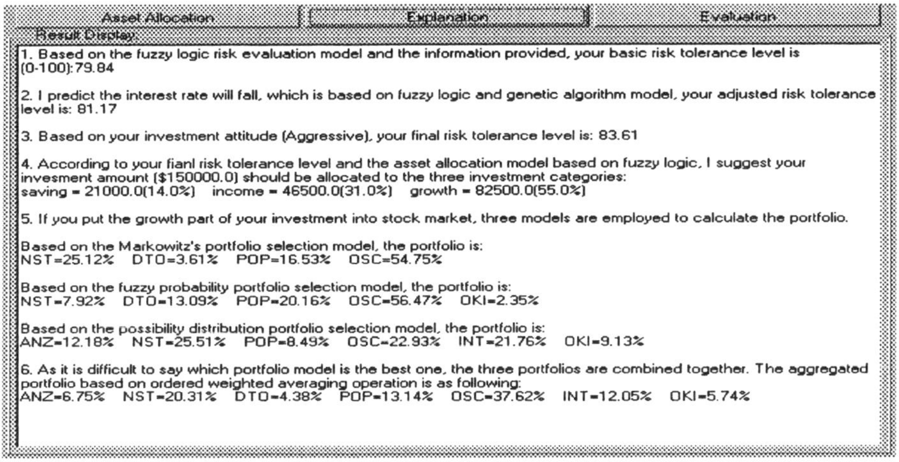

Fig. 16. Example Explanations

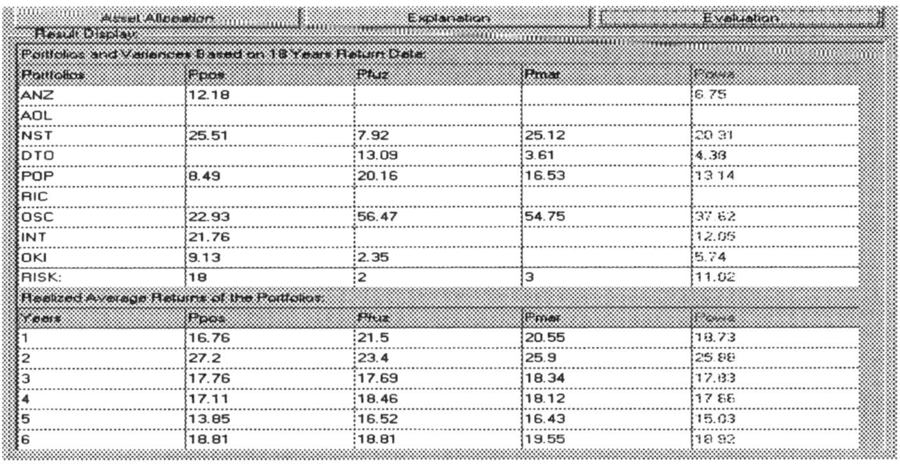

Portfolios and Variances Based on 18 Years Return Data:

Portfolios	Ppos	Pfuz	Pmar	Powa
ANZ	12.18			6.75
AOL				
NST	25.51	7.92	25.12	20.31
DTO		13.09	3.61	4.38
POP	8.49	20.16	16.53	13.14
RIC				
OSC	22.93	56.47	54.75	37.62
INT	21.76			12.05
OKI	9.13	2.35		5.74
RISK:	18	2	3	11.02

Realized Average Returns of the Portfolios:

Years	Ppos	Pfuz	Pmar	Powa
1	16.76	21.5	20.55	18.73
2	27.2	23.4	25.9	25.88
3	17.76	17.69	18.34	17.83
4	17.11	18.46	18.12	17.86
5	13.85	16.52	16.43	15.03
6	18.81	18.81	19.55	18.92

Fig. 17. Comparison of the Portfolios

4.4 Integrating different Techniques

In the agent-based financial investment planning system, the following models/techniques have been integrated together: a client financial risk tolerance model and a client asset allocation model, both are based on fuzzy logic [21]; two interest rate prediction models, one based on neural networks, the other based on fuzzy logic and genetic algorithms [25]; three portfolio selection models–Markowitz's model [26], the fuzzy probability model, and the possibility distribution model [27]; and expert systems with explanation mechanisms. In addition to these models/techniques, an operations research software package called *LINDO* for solving quadratic programming (*http://www.lindo.com/*) and a matrix software package called *MatrixLib* for solving eigenvalues of matrices (it http://www.mathtools.com/) have also been integrated.

The agent-based soft computing society proposed in this chapter is not aimed to improve the performance of individual soft computing techniques but to facilitate the construction of hybrid intelligent systems. By integrating different techniques, many complex problems can be solved by using individual techniques alone or better results can be obtained. Take the portfolio selection as an example. There are many portfolio selection models available. It is too difficult to choose the "best" model to use in real-world financial investment applications. Each model has its own strengths and weaknesses. One model may put more emphasis on some factors or attributes in the portfolio selection process but ignores others. No single model can take all the factors into consideration. The results in Figure 17 show that the Markowitz portfolio has four securities, the fuzzy portfolio has five, while the possibility portfolio has six. It means that the possibility portfolio tends to take a more distributive investment than the fuzzy probability and Markowitz ones. Based on this one still cannot simply say that the possibility model is the best. It is more reasonable to combine these portfolios together in some way in real investment problems. The lower part of Figure 17 is the realized average returns of the four portfolios. From this one can see that in some year the average returns of P_{pos} is better than those of P_{fuz} or P_{mar}, or vice versa. Considering the overall results, the average returns of the aggregated portfolio P_{owa} is better.

5 Evaluation

Complex problems such as financial investment planning are *uncertain*, *dynamic*, *distributed*, and *heterogeneous* in nature. Decision making on such problems needs the combination of *adequate information*, *rich knowledge*, and *great skills* to use knowledge and information. The framework proposed here is suitable for such tasks and can overcome the shortcomings of traditional hybrid intelligent system frameworks. Here, the evaluation of the framework is given by considering the following:

- *Flexibility* In the framework, we introduced a special kind of middle agents – serving agent. The presence of the serving agent in the framework allows adaptive system organization. The SC society proposed has the ability to add and delete SC agents (or other service provider agents) dynamically as needed.
- *Robustness* In the framework, if a particular service provider disappears, a requester agent can find another one with the same or similar capabilities by interrogating the serving agent. Thus overall system robustness is facilitated through the use of the serving agent.

- *Reliability* Compared with most currently available MAS frameworks, the matchmaking results of the serving agent (matchmaker) in this framework is more reliable due to the consideration of the agents' track records. The serving agent has less chance of a single point of failure compared with the OAA Facilitator [3][4] and *InfoSleuth's* Broker [5], since after a requester has been given a list of providers, it can continue its transactions directly even when no serving agent is present. In addition, a requester can cache providers' contact information and reuse them without resorting to a serving agent every time.
- *Availability of agent capabilities* As any agent under this framework should register and advertise its capabilities to the serving agent, any other agent can easily access all the capabilities available in the system whenever needed.
- *Interoperability* One main obstacle of a meaningful interoperation and mediation of services is the syntactic and semantic heterogeneity of data and knowledge the middle agent does access and receive from multiple heterogeneous agents, information systems and sources. By using domain-specific ontologies, the semantic heterogeneities are resolved among agents in the framework.
- *Inter-agent interactions* The interactions among agents are easy as all agents use the standardized agent communication language – KQML.
- *Platform independent* The prototype was implemented by using Java and under the support of JATLite. It can execute on any platform with a Java Virtual Machine. Thus it is platform independent.

All the agents in the prototype, which were derived from the methodology, exhibited their behaviors correctly as specified. However, we are not clear whether and how the adequacy and completeness of the outputs of the methodology can be assessed at this stage .

6 Conclusions

To make good decisions for complex problems, agent-based hybrid intelligent systems in decision making are needed. This is because:

- multi-agent approaches are good at complex problems;
- employing a combination of FL, NNs, and GAs etc. SC technologies can increase agents' "intelligence"; and
- hybrid systems combining FL, NN, and GA etc. allow the representation and manipulation of different types and forms of data and knowledge which may come from various sources. Refined system knowledge is used during reasoning and decision making processes producing more effective results.

A flexible framework – the agent-based SC society is presented – to construct multi-agent application systems that need to incorporate different kinds of SC technologies into them. Based on the proposed agent-oriented methodology, the analysis, design, and implementation of the agent-based hybrid framework for decision making on complex problems was systematically discussed . The framework has the following crucial characteristics that differentiate this work from others:

- The ability to exchange comprehensible communications (interactions at knowledge level [11]);
- Each service requester agent (decision making agent) can easily access all the techniques provided by service provider agents (e.g., SC agents) in the system;
- The presence of the serving agent in this framework allows adaptive system organization. For example, if one is sure that one hybrid SC agent can do a better job than a single technology SC agent, one can delete the single SC technology agent and add the hybrid agent to the society simply by adding or deleting a record in the database of the serving agent;
- Overall system robustness is facilitated through the use of the serving agent. For example, if a particular service provider (e.g., an SC agent) disappears, a requester agent (decision making agent) can find another one with the same or similar capabilities by interrogating the serving agent; and
- Agent-based hybrid intelligent systems based on the framework can make decisions about the nature and scope of interactions at run time.

A case study was also presented for providing investment advice for investors using the proposed SC society, and some experiments were conducted to test the society under the support of JATLite. The tests indicated that the framework is fully functional. Based on the experiments conducted, it is evident that the proposed framework is applicable to the situations where there are a great number of heterogeneous computing techniques/packages.

References

1. S. Russell and P. Norvig, *A Modern Approach to Artificial Intelligence*, Pretice-Hall, 1995.
2. E. H. Durfee and V. Lesser, Negotiating Task Decomposition and Allocation Using Partial Global Planning, in: L. Gasser and M. Huhns (Eds.), *Distributed Artificial Intelligence Volume II*, Pitman Publishing and Morgan Kaufmann, 1989, 229-244.
3. A. Cheyer and D. Martin, The Open Agent Architecture, *Autonomous Agents and Multi-Agent Systems*, Vol. 4, No. 1-2, 2001, 143-148.
4. V. Subrahmanian, P. Bonatti, J. Dix, et al., *Heterogeneous Agent Systems*, The MIT Press, 2000.

5. R. J. Bayardo, W. Bohrer, R. Brice et al., InfoSleuth: Agent-Based Semantic Integration of Information in Open and Dynamic Environments, in: M. N. Huhns and M. P. Singh (Eds.), *Readings in Agents*, Morgan Kaufmann, CA, 1998, 205-216.
6. M. Hilario, C. Pellegrini, and F. Alexandre, Modular Integration of Connectionist and Symbolic Processing in Knowledge-based Systems, in: *Int. Symposium on Integrating Knowledge and Neural Heuristics*, Pensacola, Florida, 1994, 123-132.
7. A. Scherer and G. Schlageter, A Multi-Agent Approach for the Integration of Neural Networks and Expert Systems, in: S. Goonatilake and S. Khebbal (Eds.), *Intelligent Hybrid Systems,* Wiley, 1995, 153-173.
8. R. Khosla and T. Dillon, *Engineering Intelligent Hybrid Multi-Agent Systems*, Kluwer Academic Publishers, Boston, 1997.
9. M. Delgado, A. F. Gómez-Skarmeta et al., A Multi-Agent Architecture for Fuzzy Modeling, *International Journal of Intelligent Systems*, Vol. 14, 1999, 305-329.
10. K. Decker, K. Sycara, and M. Williamson, Middle Agents for the Internet, *Proceedings of 15th International Joint Conference on Artificial Intelligence*, Nogoya, Japan, 1997, 578-583.
11. A. Newell, The Knowledge Level, *Artificial Intelligence,* Vol. 18, 1982, 87-127.
12. S. Goonatilake and S. Khebbal (Eds.), *Intelligent Hybrid Systems,* Wiley, 1995.
13. L. R. Medsker, *Hybrid Intelligent Systems*, Kluwer Academic Publisher, 1995.
14. L. C. Jain and R. K. Jain (Eds.), Hybrid Intelligent Engineering Systems, World Scientific, Singapore, 1997.
15. N. Lertpalangsunti and C. W. Chan, An Architecture Framework for the Construction of Hybrid Intelligent Forecasting Systems: Application for Electricity Demand prediction, *Engineering Applications of Artificial Intelligence*, Vol. 11, No. 4, 1998, 549-565.
16. S. Li, The Development of a Hybrid Intelligent Systems for Developing Marketing Strategy, *Decision Support Systems*, Vol. 27, No. 4, 2000, 395-409.
17. M. Wooldrige, N. Jennings, and D. Kinny, The Gaia Methodology for Agent-Oriented Analysis and Design, *Journal of Autonomous Agents and Multi-Agent Systems*, Vol. 3, No. 3, 2000, 285-312.
18. C. Iglesias, J. Gonzales, and J. Velasco, MIX: A General Purpose Multiagent Architecture, in: M. Wooldridge, J. Muller, and M. Tambe (Eds.), *Intelligent Agents II (ATAL'95)*, LNCS 1037, Springer, 1996, 251-266.
19. H.-A. Jacobsen, A Generic Architecture for Hybrid Intelligent Systems, in: T. Furuhashi et al. (Eds.), *Deep Fusion of Computational and Symbolic Processing*, Physica-Verlag, 2001, 145-173.
20. Z. Zhang and C. Zhang, Considering Agents' Track Records in Matchmaking of Middle Agents, *Proceedings of 4th Pacific Rim International Workshop on Multi-Agents*, Taipei, 2001, 281-292.
21. G. Bojadziev and M. Bojadziev, *Fuzzy Logic For Business, Finance, and Management*, World Scientific, Singapore, 1997.
22. C. Zhang, Z. Zhang, and S. Ong, An Agent-Based Soft Computing Society, *Proceedings of 2nd International Conference on Rough Sets and Current Trends in Computing*, Banff, Canada, 2000, 621-628.
23. R. R. Yager, On Ordered Weighted Averaging Aggregation Operators in Multi-criteria Decision Making, *IEEE Transactions on Systems, Man, Cy bernetics*, Vol. 18, Jan/Feb, 1988, 183-190.

24. Z. Zhang and C. Zhang, Result Fusion in Multi-Agent Systems Based on OWA Operator, *Proceedings of 23rd Australasian Computer Science Conference*, IEEE Computer Society Press, NJ, 2000, 234-240.
25. Stephen T. Welstead, *Neural Network and Fuzzy Logic Applications in C/C++*, Wiley, New York, 1994, 395-421.
26. H. Markowitz, *Portfolio Selection: Efficient Diversification of Investments (2nd ed.)*, Blackwell, 1991.
27. H. Tanaka, P. Guo, and I. Turksen, Portfolio Selection Based on Fuzzy Probabilities and Possibility Distributions, *Fuzzy Sets and Systems*, Vol. 111, 2000, 387-397.

A Rough Sets/Neural Networks Approach to Knowledge Discovery for the Development of Decision Support Systems

Ilona Jagielska

Decision Support Systems Laboratory
Faculty of Information Technology
Monash University
26 Sir John Monash Drive
Caulfield East, VIC 3145
Australia
E-mail: ilona.jagielska@sims.monash.edu.au

Abstract. The problem domain of decision support systems (DSS) has been characterised as ill-specified, influenced by many different sources, and subject to change. The complex and changing environment in which decision support systems are developed makes specification of the final system requirements difficult. Decision support systems need to be designed using an adaptive process of learning and experimentation. The emphasis in DSS development is on supporting the decision making process by improving the decision maker's understanding of the decision domain. This paper proposes a conceptual and a technological framework for adaptive development of decision support systems (DSS) through knowledge discovery from historical data. In this approach the users and developers gradually learn about their decision domain by an iterative process of extracting patterns from the database and building and refining predictive models using these patterns. A hybrid rough sets/neural networks framework that can facilitate this development is proposed. Finally a case study where the approach was applied in the development of a real-world decision support system is described and discussed. The framework can be applied to a variety of industrial and business decision problems where historical data exists.

Keywords: decision support systems, knowledge discovery, soft computing, neural networks, rough sets

1. Introduction

Decision support systems (DSS) are computer-based systems that are designed with the purpose of improving the process and outcome of decision making. DSS predominantly address management decisions although they are also applicable to other types of decision making. The problem domain of DSS has been characterised as ill-specified, influenced by many different sources, and subject to change (Sprague and Carlson 1982, Keen 1980). Decision domains are "poorly structured" in the sense that the domain knowledge (the facts and relationships pertinent to the decision situation) is not readily accessible and the procedures used to come to a decision are poorly understood. The emphasis in DSS development is on supporting the decision making process by improving the decision maker's understanding of the decision domain. It is essential therefore that the user and analyst fully explore and understand all possible aspects of the decision or problem domain.

Knowledge discovery has been described as the process of identifying and extracting useful and understandable information from large data sets (Fayyad *et al.* 1996a). Data that is extracted from organisational databases may contain valuable information for users and decision makers about their application or problem domain.

This paper presents an approach in which knowledge discovery is used as a mechanism for simultaneously learning about a decision domain from historical data and for building a decision support system based on this understanding. The approach is described in terms of a conceptual model for the development of a DSS using knowledge discovery, and a technological framework for the implementation of the approach.

The technological framework consists of a hybrid of two soft computing paradigms: neural networks and rough sets. Current research in soft computing has focused on integrating various soft computing methods in order to enhance the capabilities of systems developed using these methods. A common characteristic of rough sets and neural networks is that both approaches have the ability to learn decision models from examples. Learning is an important component of knowledge discovery and of the decision-making process (Spangler 1991). Rough sets and neural networks can produce good results even when data is noisy, imprecise or contradictory. The ability to learn patterns from noisy and inconsistent data (that are often contained in organisational databases) is an essential component of any knowledge discovery system and is crucial to its performance. The technological framework described in this paper enables users and analysts gradually to gradually learn about the problem domain through an iterative process of extracting patterns from the relevant database and building and refining decision models using these patterns. The paper reports a case study of the use of the approach in practice which has demonstrated its effectiveness.

The focus of the research described in this paper is thus on the human-centred, process-oriented application of knowledge discovery to decision support rather than on the development of efficient techniques for predictive data mining. It emphasises the role of humans (the users and analysts) in the knowledge discovery process.

2. Research Approach

The research described in this paper consisted of two phases. The first phase involved the development of the approach to incorporating knowledge discovery into the DSS development process and the specification of the technological framework which implements it. In the second phase, a case study was used to examine the use of the approach and its associated technological framework in practice. The case study provides empirical evidence of the effectiveness of the approach.

A computerised decision support system for issuing smog alerts was required by the Environment Protection Authority (EPA) in Melbourne, Australia. There was twelve years of historical data relating to environmental conditions and smog alerts held in the EPA's smog events database. The significance of the historical data and the inability of existing applications and manual methods to extract from and interpret patterns in the existing data provided the motivation for the EPA to participate in the study of the use of the proposed approach. The participants in the case study included senior meteorologists and a systems analyst and information technology support staff member from the EPA. Data collection sources for the case study included interviews with the participants and documentation about the smog alerts database and the existing manual system.

3. Using Knowledge Discovery in the Development of DSS

Theories in the area of decision support systems view the development of DSS as an evolutionary or adaptive process (Arnott 1998, Keen 1980). The environment in which decision support systems are developed is subject to change so that the system requirements are likely to change over time. The unstructured nature of decision problems and the complex and changing environment in which decision support systems are developed makes it difficult to specify the final system requirements. DSS are developed using an adaptive process of learning and experimentation in which the requirements evolve during a series of development cycles. The system emerges as a result of an iterative process of learning and design rather than being built according to a predetermined set of specifications.

Both the user (or decision maker) and the systems analyst contribute to the shape and design of the system during this process as the understanding of their decision situation increases.

The key idea of evolutionary or adaptive systems design is that a system is developed in a series of cycles during which the user and the systems analyst gain a better understanding of their decision task through the use of the system they are developing (Keen 1980). The user's depth of understanding of the decision domain increases with each iteration, and the system is revised and changed in order to adapt it better to the decision task. The process of gradually accumulating improved knowledge and understanding of the nature of the decision domain is therefore central to the development of DSS. Knowledge discovery can facilitate this process in situations where historical data related to the problem domain exists.

This paper proposes an approach in which knowledge discovery is used to assist the evolutionary development of a DSS as knowledge and understanding of the problem domain emerges. The objective of using knowledge discovery in this approach is to uncover hidden patterns in existing data which can be used to assist users in gradually improving their understanding of the problem domain and consequently in refining their system requirements and building decision models. The user and analyst gain valuable insights into the problem domain by developing the system through a process of extracting patterns from databases. The approach uses knowledge discovery as a means for simultaneously learning about a decision domain from historical data and building a system based on this understanding. The knowledge discovery process consists of a series of cycles involving feature and data selection, pattern extraction, pattern interpretation, and testing. Knowledge discovery includes many other tasks such as data cleansing, preprocessing and sampling. However, only the tasks relevant to understanding the proposed approach are considered in this paper.

The basic steps of the proposed approach are as follows. At the start of the development process, the user has some preliminary knowledge about the application domain. This is usually incomplete and inadequate for specification of their detailed system requirements. In the first step of the proposed approach, this knowledge is applied to the selection of the data and features that will be used in the knowledge discovery process. The first step identifies the most general patterns or decision rules in the existing data. In the next step, the users and analysts interpret and test these patterns. This will result in some initial insights into the application domain. In the third step, the users and analysts are able to use the knowledge they have acquired to specify more precise goals for further knowledge discovery and learning about the domain. This may, for example, require changes to the parameters used in the knowledge discovery methods in order to extract more detailed or more specific types of patterns in the data. The users and analysts may also decide to remove or to add some features from the selected sample. If the results of the tests of the patterns are unsatisfactory, it may be necessary to search for other sources of data in the environment of the application domain. The steps of extracting patterns from the data and building and refining predictive models using these patterns are repeated until the point is reached where it is believed that no further useful information can be extracted. In every iteration of the steps in the process, the users and analysts acquire deeper understanding of the application domain which allows them to further refine the patterns extracted and thus build better decision models. The process is thus interactive and iterative, involving a number of steps in which the users and analysts control the pattern extraction and testing activities.

4. The Technological Framework for the Use of Knowledge Discovery in the Development of DSS

In this section a technological framework that can be adopted to implement the proposed approach to using knowledge discovery in the development of DSS is described. In the approach the information extracted using knowledge discovery is used to assist in clarifying and refining understanding of the domain and of the users' requirements. It is critical, therefore, that the knowledge extracted is in a form that is comprehensible to the users and systems analysts involved. It should be represented using a high-level language such as If-then rules or decision trees. If-then rules are popular because they are easy to understand and can subsequently be used for building a rule-based expert system. The number of rules used and the complexity of individual rules (number of clauses) also contribute significantly to the comprehensibility of extracted knowledge. Knowledge discovery systems used in the development of DSS should also provide mechanisms that enable their users to build decision models using the patterns discovered and to analyse the performance of the models. The user has to be confident that the knowledge extracted from a database is accurate and will generalise over new cases. Many rule extraction techniques generate all rules at once, and do not provide mechanisms for controlling the rule extraction process. Since it is difficult to understand patterns expressed by a large number of rules, it is important that the knowledge discovery system allows gradual extraction of patterns, enabling the user to control pattern extraction and refinement.

The proposed technological framework includes techniques that support high-level representation of knowledge extracted as well as the gradual extraction of knowledge. Neural networks and rough sets are used co-operatively in a series of information extraction cycles under the guidance of the user. Rough sets and neural networks have the ability to learn decision models from examples even when data is noisy, imprecise or contradictory. Rough sets in particular have an inherent mechanism for handling contradictions in data (Pawlak *et al* 1995), and neural networks often provide better predictive models when some amount of noise is present. The structure of the hybrid rough sets/neural network framework is shown in Figure 2.

Rough sets based methods are used for attribute analysis, for deriving decision rules from the data, and for feature selection for neural network models. The rules are used to understand the relationship between the input features and the decisions. The predictive ability of the rules is also checked. The results of the rough sets input feature analysis are used to guide the choice of input features to the neural network models. The predictive models generated by the rough sets and the neural networks are compared and the best model is selected.

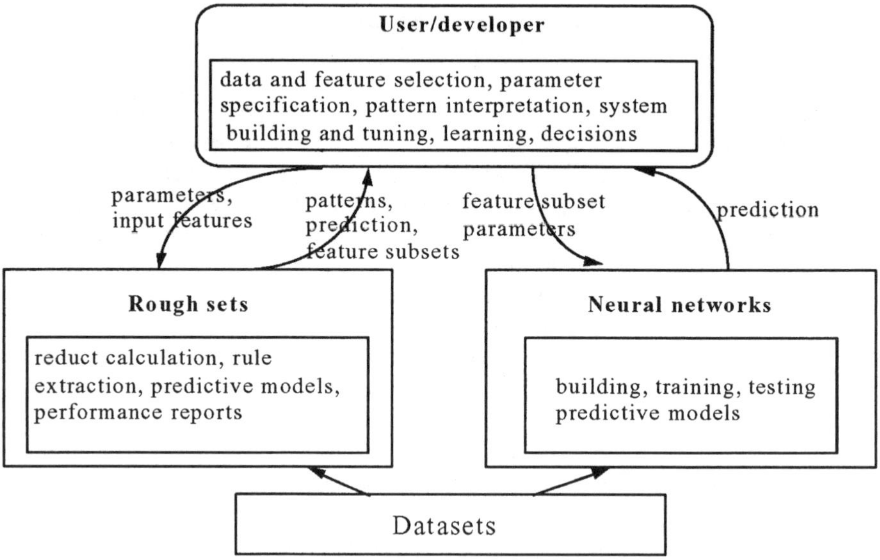

Figure 2. The structure of the hybrid rough sets/neural network framework

4.1 Rough Sets Overview

Basic concepts

The rough sets approach proposed by Pawlak (Pawlak 1991) provides rigorous mathematical techniques for discovering patterns in data and is particularly useful for dealing with imprecise and inconsistent data (for an introduction to rough sets see Pawlak *et al.* 1995). At the basis of rough sets theory is the analysis of the limits of discernibility of subset X of objects from the universe of discourse *U*.

Let U be a set of objects (universe of discourse), A be a finite set of attributes, and $f(x,a)$ be a function assigning a unique value of attribute "a" to each object x for every $a \in A$, $x \in U$ (a decision table where the rows are objects and the columns are attributes is an example of such system).

Let $x,y \in U$, and $P \subseteq A$. Then x and y are *indiscernible* by the set of attributes P

$$\text{iff} \quad f(x,a)= f(y,a) \text{ for every } a \in P$$

Each such subset P of attributes generates an equivalence relation on U which is called the *indiscernibility relation* denoted as $IND(P)$.

$IND(P)$ classifies objects into *elementary sets.* Each elementary set consists of groups of objects with the same value of attributes from P. With each set $X \subseteq U$ and equivalence relation $IND(P)$ the following subsets are associated:

$$\underline{PX} = \bigcup\{Y \in U \,/\, IND(P): Y \subseteq X ,$$

called the *P-lower approximation* of X, that contains the union of those elementary sets that are completely included in X,

$$\overline{PX} = \bigcup\{Y \in U \,/\, IND(P): Y \cap X \neq 0\},$$

called the *P-upper approximation* of X, that contains the union of those elementary sets that have non-empty intersection with X.

$$BN(X)= \overline{PX} - \underline{PX}$$

is called the *boundary region.*

If the set $BN(X)$ *is* not empty the set X is called *rough.* The set approximation can become more detailed if more attributes are added to A. Then the boundary region can get smaller, and the set is "less rough". Classification rules can be induced from the lower and the upper approximation of a set (Pawlak *et al* 1995).

Figure 3 illustrates the approximation regions in rough sets. Each cell in Figure 3 depicts an elementary set described by some combination of attribute values. When the amount of detail (attributes) in a set approximation is small, the set "roughness" is high (large boundary region), and only strong rules (patterns) can be identified. If we add more detail to set approximation (smaller elementary sets), the boundary region gets reduced and more detailed rules can be extracted. Changing the detail of set approximation is particularly useful in knowledge discovery as it enables the user to discover and analyse strong patterns (rules) gradually (REDUCT Systems Inc. 1994)

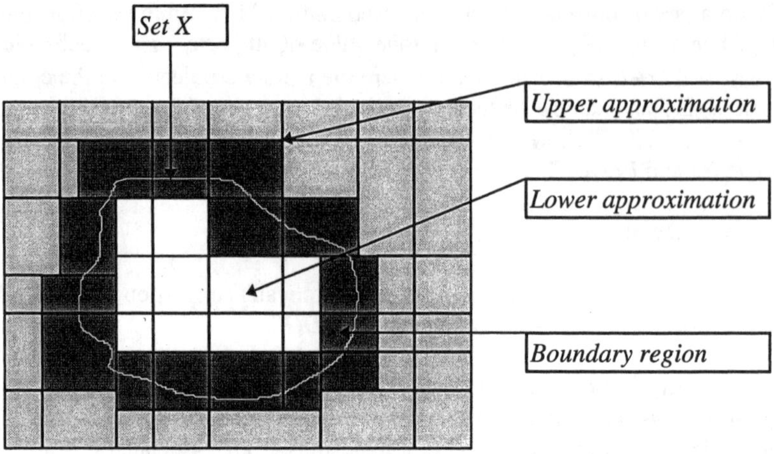

Figure 3. Approximation regions in rough sets

Another important concept in rough sets is that of a *reduct* which is a minimal set of attributes that can classify objects with the same accuracy as the original set of attributes. Finding such a subset of attributes allows elimination of irrelevant attributes without any loss of essential information. Elimination of redundant attributes can help in identification of strong, non-redundant classification rules.

Applications and implementations

Systems based on the theory of rough sets have been developed and successfully applied to a broad range of applications, especially in the areas of automated knowledge acquisition and knowledge discovery (Slowinski 1992, Ziarko *et al.* 1993, Pawlak *et al.* 1995, Ziarko 1996). The approach used typically involves approximation of classes, reduction of non-significant attributes, derivation of decision rules from the reduced system, and calculation of the accuracy of classification.

For the implementation of the proposed framework, rule induction software from Reduct Systems called Datalogic/R was chosen. Datalogic/R has the ability to extract knowledge at different levels of representation. It enables the user to control the process of pattern extraction so that patterns can be discovered and tuned gradually. This capability makes Datalogic/R a particularly useful tool for the gradual elaboration of knowledge which is at the core of the proposed approach to the evolutionary development of DSS. The software uses two parameters to control the rule extraction process: rule roughness (values from 0 to 1) and rule precision threshold (values from 0 to 1). When the level of roughness is close to 1, only the most general patterns are extracted. As roughness is decreased, the set approximation becomes more detailed and the model becomes more exact. The rule precision threshold is used to control the minimum rule probability where the rule probability is defined as:

> *No. of cases matching rule conditions & decision / No. of cases matching rule conditions.*

Changing the detail of set approximation is particularly useful for the incremental learning about a domain from data that can occur during DSS development. For example, the user or analyst may want to identify only the strongest patterns first. This can be achieved by setting a high set roughness. If more precise knowledge is required, the user can decrease the roughness and weaker patterns can be identified. It is difficult to determine the exact effect of the parameters on the derived rules without experimentation. Different data sets may require different combinations of the parameters in order to produce high quality decision rules. The software also generates an attribute and rule strength report. The rule strength is measured by the number of cases supporting the rule. The attribute strength (values from 0 to 1) is a measure of the relative importance of that attribute in distinguishing between decision classes.

4.2 Neural Networks Overview

Neural networks (also known as connectionism or neurocomputing) are the basis of a well known soft computing paradigm which has already found its way into real-world applications. Neural networks are parallel information processing systems with their architecture inspired by the structure and functioning of the brain. For an introduction to neural networks see, for example, Caudill and Butler (1994) or Zurada (1992). Neural networks consist of layers of connected processing units (artificial neurons). The strength of the connections between the neurons is expressed by their connection weights. Neural networks have the ability to "learn" patterns in datasets through the process of training on examples. Training neural networks is the process of adapting or modifying the connection weights using a learning algorithm. A simple neural network with one hidden layer is illustrated in Figure 4.

Trained neural networks have the ability to correctly predict the output when presented with previously unseen input. This makes them particularly useful for solving prediction and classification problems. Neural networks have been used for bankruptcy prediction, credit card risk assessment and financial modelling (Beltratti *et al.* 1996, Jagielska and Jaworski 1996). Neural networks are also well known for their ability to learn patterns even when data is noisy, imprecise and incomplete. Their capabilities make them a popular tool for data mining (Bigus 1996).

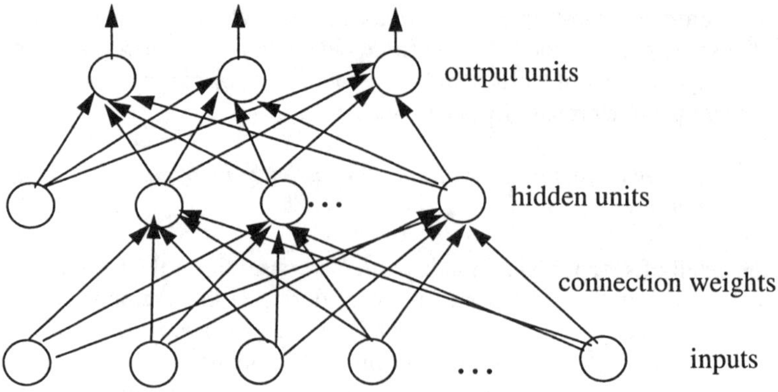

Figure 4. Fully connected, feed-forward neural network with one hidden layer

4.3 Combining Rough Sets and Neural Networks

Recently, hybrid systems have become a very promising area of research. Neural networks, fuzzy sets, genetic algorithms, and rough sets are being combined in various ways in order to increase the effectiveness of systems (Zadeh 1994, Goonatilake and Khebbal 1995, Polkowski and Skowron 1998). Rough sets are used in the framework proposed in this paper because the patterns discovered by rough sets based systems can be represented using simple rule statements, and because they enable the user to extract knowledge from data at different levels of generalisation. This facilitates gradual learning and evolutionary development of a decision support system through a series of revisions, each further refining the model of the problem or situation. Neural networks do not represent discovered patterns in a form understandable by humans but they can be used to develop models of high predictive capability (Weiss 1998). Integration of rough sets with neural networks enhances a system's explanatory capability.

Combining both techniques in the framework enables the user to compare and select the best predictive model, as it is difficult in general to determine which technique will give better results for a particular application (Fayyad *et al.* 1996a, Weiss 1998). Another reason for combining rough sets with neural networks in this framework is to remove any redundant features used as input for neural network models. The presence of irrelevant features often causes a degradation of performance of predictive systems (Kohavi and Frasca 1994). In addition, the reduction of irrelevant features reduces the dimensions of the network, thus reducing the training time. Rough sets-based systems identify a minimal set of attributes before they generate patterns. In the framework, this capability is used to select a subset of input features to the neural network models. This aspect of the framework was investigated in Jagielska (1999) where it was demonstrated that in the framework the application of rough sets for feature selection for neural network models can reduce the size of neural network models and has the potential to enhance the systems predictive capability.

4.4 Applying the Framework in the Evolutionary Development of DSS

The main steps in the process of DSS development using the framework described above are as follows:

1. The users and analysts select subsets of data from the available databases based on their knowledge of the domain. This step includes feature selection and the choice of the size of the data sample and the sampling method.
2. The users and analysts carry out iterative extraction of patterns (IF-then rules) using rough sets by changing the rule roughness and the rule precision threshold. At each iteration, the extracted rules are tested for their predictive capability.
3. The rules extracted, attribute strengths, rule strengths, rule coverage, classification rate and other information generated by the system are then analysed. Through this process, the users and analysts acquire improved understanding of the application domain and can then decide whether to continue pattern extraction, to add or remove input features, or to use different samples from the database.
4. Based on the list of attributes generated by the rough sets, the users and analysts select the input features for the neural network models.
5. Predictive models are then generated using neural networks.
6. The users and analysts compare the neural network and rough sets predictive models.
7. The models are finetuned as necessary in order to achieve the best predictive results.

At any stage the users and analysts may decide to repeat any of the previous steps. Application of the framework involves a collaborative learning process in which users and analysts together explore the problem domain and improve their understanding of it using the information extracted via the knowledge discovery process.

5. A Case Study of the Use of the Rough Sets/Neural Network Knowledge Discovery Framework in the Development of a DSS

In order to examine the use of the proposed knowledge discovery approach in the development of a DSS in practice, a real-world case study was undertaken. The study involved the application of the hybrid rough sets and neural network framework in the development of a decision support system for issuing smog alerts for the Environment Protection Authority (EPA) in Melbourne. In this section, the use of the knowledge discovery approach in the study is described and analysed. Important findings from the case study are then discussed.

5.1 Background to the Case Study

In 1984 the EPA developed a smog alert forecasting scheme which is used to issue smog alerts when a "smog event" is likely to occur. The principal component of photochemical smog is ozone. High concentrations of ozone in the atmosphere have a detrimental effect on the health of humans, animals and vegetation. The EPA's scheme is based on assessment of various meteorological conditions. When determining whether a smog event is likely, a form with ten questions relating to the meteorological conditions is completed and used by a meteorologist as input to their decision whether to issue a smog alert for the next day. The ten questions to be answered on the smog events forecast form use meteorological data which provides the basis for predicting smog alerts. The decision process is not well defined and to some extent it is intuitive, relying on the tacit knowledge of the meteorologist. There seems to be more importance allocated to some questions, for example, high pressure centre gradient, wind speed, and expected sunshine hours. The current forecasting scheme is considered inadequate as its average accuracy over 11 years has been less than 50 % correct forecasts. A performance level of 60% accuracy is considered adequate, although the desirable level of performance is 80% and above (Cook 1995).

The EPA management decided to develop a computerised DSS which would assist the EPA meteorologists in their decisions about whether to issue smog alerts. There was 12 years of data recorded in the EPA smog events forecast database. EPA senior meteorologists and the smog events forecast database were the main sources of knowledge used for the development of the requirements for the DSS. No prior specifications for the system were prepared. The requirements were to evolve through the process of users and analysts learning about the application domain from the historical data and building and finetuning predictive models.

5.2 Development of a Decision Support System for Issuing Smog Alerts Using the Proposed Approach.

5.2.1 The data

For this study, data for 259 forecast days recorded between the years 1984 and 1996 were used. The data were divided into two groups:

Set 1 - this consisted of a senior meteorologist's assessment of the relevant meteorological conditions. The set had ten attributes corresponding to ten questions about the meteorological conditions. The attribute values are yes/no answers to the questions. The questions were those used by meteorologists to

predict smog events. Three example questions are shown below:
Is gradient < 0.8hPa/100km ?
Are winds below subsidence decreasing to <15 knots?
Is tomorrow a working day?

Set 2 – this consisted of data about meteorological conditions such as the location of high pressure systems, the pressure gradient across Melbourne, geostrophic wind direction over Melbourne, thickness at Melbourne, temperature inversions, wind speed below subsidence, upper level winds and cloud at Laverton and Mt Gambier, rain, and the maximum temperature forecast. The database also contained the EPA predictions of smog events (yes/no) and the actual occurrences of smog events (yes/no).

5.2.2 Applying the Technological Framework

Rough sets and neural networks were used co-operatively in a series of knowledge discovery cycles under the guidance of the users (meteorologists) and analysts. The rough sets software used in the study was Datalogic/R. The level of roughness was gradually changed from very high to low values (between 1 and 0.1). This allowed identification and analysis of the strongest attributes and rules first, and subsequently when the roughness was reduced, more detailed rules were generated and analysed. For each level of roughness the system generated If-then rules and their associated strength, a list of attributes used in the rules, and the attribute strengths. The classification accuracy of each set of rules (or model) generated at each iteration was determined then the predictive capability (accuracy of prediction) of each model with high classification accuracy was tested The classification accuracy is the classification rate of the rules on the examples from which the rules were derived. In order to test the accuracy of prediction, the examples were divided randomly into ten training and testing sets: 10% of examples for testing and 90% for training. Rules were derived from the training sets and then tested on testing sets. Ten runs were executed and an average accuracy of classification on testing examples was calculated and used as a measure of the predictive accuracy of the models.

The same datasets were subsequently used in building neural network predictive models and the results were compared with the results obtained from rough sets models. The neural network type used in the study was a multilayered Perceptron. A fully connected, one hidden layer topology was adopted and the neural networks were trained using the backpropagation learning rule. In neural networks the number of neurons in hidden layers and the parameters controlling the learning process, such as the learning rate and the momentum term, have to be determined experimentally for each data set. In this study the number of neurons in the hidden layer and the parameters were determined for each data set in a series of preliminary trials. The neural network models were first built using the full set of attributes. As was explained in section 4.2.3, neural networks often produce better predictive results when irrelevant or weak features are not present. In order to investigate whether better results could be achieved with a smaller number of features, the input features were reduced to those selected by the users (based on the results from rough sets experiments) and the neural networks were trained and tested using the reduced data. The performance of the networks with the full set of features was then compared with the performance of the networks with the reduced set of features.

Data representation and preprocessing

Entries with missing data were not included in the sets. The total number of examples(days) after the elimination of missing values was 228.

Rough sets models

In Set 1 all attributes have categorical values Yes or No. They were represented as a string type with two string values "Y" , "N". The decision variable had two values "Y" and "N" corresponding to occurrence and non-occurrence of smog event respectively. In Set 2 many attributes have numerical values. Numerical values were preprocessed by DataLogic/R using a parameter called data precision. The data precision is used to define the scale DataLogic uses to represent numeric values. It is calculated by the formula:

scale = (maximum attribute value - minimum attribute value)/precision value

In this study the default precision value of 10 was used in all experiments.

Neural network models

Neural networks can process only numerical values, therefore non-numerical input and output data have to be converted to numerical representation. In Set 1 the ten attributes were allocated 10 input nodes and the input data were represented as two numerical values: 0 for answer *No* and 1 for *Yes.* In Set 2 offset and scaling operations were performed on numerical values so that the values would fit in the range 0 to 1. In both sets the decision variable was represented by one node and had two values, 1 and 0, corresponding to occurrence and non-occurrence of a smog event respectively.

5.2.3 The Evolutionary Development Process

This section describes the iterative steps in the development of the DSS using knowledge discovery as described in Section 4.3.

Knowledge discovery from Set 1

Initially the users decided to analyse only the data in Set 1 in order to determine whether it is possible to predict smog events with an acceptable accuracy from just the ten yes/no answers to the questions about meteorological conditions. In particular, the users wanted to determine the importance of each question for smog event forecasting, and thus understand the relationship between the questions about meteorological conditions and the actual occurrences of smog events.

Rough sets models

In the first cycle of knowledge discovery, rule roughness was set to a high value (0.95) in order to find the most general patterns. The rules extracted for *non-occurrence* of a smog event were:

> IF (thickness change > 0 and Prog. thickness > climatic average) = No
> OR
> IF (tomorrow is Saturday, Sunday or a public holiday) = Yes
> OR
> IF (gradient <0.8hPa/100km) = No.

These rules used the three strongest attributes: thickness change, pressure gradient, and weekend/holiday (see Table 2) and represent the most general pattern discovered by the use of rough sets. There were no rules extracted for the occurrence of smog events for this level of roughness.

Table 2. Attributes and their strength for roughness = 0.95

Attribute No.	Attribute description (question about)	strength
A5	thickness change	0.24
A3	pressure gradient	0.19
A1	weekend/holiday	0.16

In subsequent cycles, in order to extract more detailed knowledge from the data, the level of rule roughness was gradually reduced. After each cycle, the rules extracted were analysed by the users. Selected results are shown in Table 3 below.

The predictive capability of the models generated in the process was determined. The best model achieved an accuracy of prediction of 66.8 (see Table 6). The list of attributes generated in the model contained the nine attributes shown in Table 4 below.

At this stage of the development it was observed that temperature inversion was not included in the minimal set of attributes, indicating that this parameter had no importance for the prediction of smog events in these models. Attributes concerning Laverton winds, high pressure system, Mt. Gambier winds, and sunshine forecast all had strengths below 0.05, thus they were considered by the users as not contributing to the decision process.

Table 3. Results of selected rough sets models for Set 1

Roughness/rule prec. threshold	No. of rules	Classification accuracy	No. of attributes
0.95/0.6	3	50.0 %	3
0.93/0.6	5	49.6 %	4
0.90/0.6	13	71.5%	6
0.85.0.6	13	71.9%	7
0.80/0.6	23	74.1%	9
0.10/0.4	25	81.1%	9

Table 4. Attributes and their strength

Attribute No.	Attribute description (question about)	Strength
A5	thickness change	0.24
A3	pressure gradient	0.19
A1	weekend/holiday	0.16
A7	wind speed	0.13
A4	wind direction	0.10
A9	Laverton winds	0.04
A2	high pressure system	0.03
A8	Mt Gambier winds	0.03
A10	sunshine forecast	0.02

Neural network models

The same data were subsequently used in building neural network predictive models. The best accuracy of prediction achieved was 63.9 %. This was less than the accuracy of prediction of the rough sets models (66.8%). In order to investigate whether excluding irrelevant and weak features could improve the results of neural network models, another set was created which included only the five strongest attributes, and neural network models were developed using the reduced set. The best model generated using the reduced features achieved an accuracy of prediction of 66.7% (see Table 6) which was higher than the model that used the entire set of features.

Knowledge discovery from Set 2

The users and analysts now decided to investigate the raw meteorological data stored in the database (Set 2). Based on the understanding of the application domain they had acquired from the previous cycles of the process, a new set of features (attributes) was selected for subsequent cycles of the knowledge discovery process. This set included pressure gradient, high pressure zone, wind direction, thickness change, maximum temperature and questions about wind speed below subsidence and weekend/holiday. The steps in the process were carried out in the same way as before. Of the seven attributes, only four (shown below in Table 5) were found to be of importance to the prediction process:

Table 5. Attributes and their strength (Set 2)

Attribute	Attribute name	strength
A2	pressure gradient	0.28
A7	wind speed	0.27
A4	thickness change	0.19
A5	max. temperature	0.17

An example of the strongest rules extracted from Set 2 for decision "Yes" and "No" is shown below:

Decision - Yes
[28≤ A5 ≤37] & [A2≤0.86] & [2.51< A4 ≤6.35] & [A7=Y]
Decision - No
[A5<28 or A5>37] & [A2≤0.4 or A2>0.86]

The rules indicate that a smog event is most likely to occur when the maximum temperature forecast is between 28 and 37 deg C, the pressure gradient is lower than 0.86 Hpa/100 km, the thickness change is between 2.51 and 6.35 Dm, and the winds below subsidence are decreasing to < 15 knots. The best rough sets model had an accuracy of prediction of 68.2% and the best neural network model achieved accuracy of 68.3% (see Table 6).

Observations

At this stage of the knowledge discovery process, the users had discovered that of the meteorological factors recorded in the forecast database, the maximum temperature, pressure gradient, and thickness change factors had the highest significance. The rules derived in the process helped the users to increase their understanding of the relationship between the meteorological factors and smog events. The rules generated using high level of roughness were easy to interpret, however, when the roughness was set to low values a large number of rules was extracted and many of the weaker rules were difficult to interpret.

5.2.4 Discussion of Predictive Performance

Table 6 contains a summary of predictive results of the best decision models derived from the smog events database during the development of the smog alerts DSS. Comparison of the accuracy of the models derived from Set 1 and Set 2 indicate that the models generated from the set with meteorological data (Set 2) had slightly better prediction accuracy than the models generated from the set of answers to the questions about meteorological conditions (Set 1). Comparison of neural network results with rough sets results indicates that in terms of predictive capability, neural network and rough sets models were very similar.

Table 6. Comparison of predictive results

Model type	Accuracy of prediction	
	Set 1	Set 2
Neural net	66.7%	68.3%
Rough sets	66.8%	68.2%

In all cases the accuracy of prediction of smog events achieved by the models generated by the proposed rough sets/ neural networks framework was superior to the accuracy of prediction of the current forecasting scheme. As mentioned above the current EPA forecasting scheme is considered inadequate as its average accuracy over 11 years has been less than 50 % correct forecasts. A performance level of 60% accuracy is considered adequate, although the desirable level of performance is 80% and above (Cook 1995). In this study the accuracy of prediction of the generated models was higher than the adequate level but lower than the desirable level. It has been observed that the patterns in the smog events database are very weak. If there are strong patterns in data, the attribute strength in each class is close to 1. In this case study even the most significant attributes in the smog events database had their relative strength below 0.5. Hence, in order to further improve smog events forecasting the EPA meteorologists will have to identify additional input parameters that can better capture the input feature space of the smog event prediction problem.

5.3 Discussion of the Case Study

The case study demonstrated that the proposed approach was effective in practice. The application of the knowledge discovery framework provided the users and analysts with valuable insights into the nature of the application domain (smog event forecasting) which would not have been otherwise available to them.

Gradual extraction of decision rules allowed the users and analysts to develop an understanding of the relationships between the various factors contributing to smog events. The data concerning attribute strengths enabled them to understand the relative importance of the meteorological factors for the prediction of smog events. Some of the knowledge acquired, for example, the significance of the maximum temperature, pressure gradient, and thickness change factors, confirmed the users' beliefs. However, in contrast to their expectations, the factors relating to Mt Gambier and Laverton winds had no significance for smog event prediction. In addition, temperature inversion was found to be an irrelevant factor, and sunshine forecast and location of the high pressure system had significance close to zero. The information acquired was thus an important source of knowledge about the application domain, and helped to clarify and refine the users' understanding of the domain and of their requirements for the computerised DSS.

The case study also confirmed that the framework proposed for the implementation of the approach provides a viable and effective technological environment for the development of DSS through the knowledge discovery process. Valuable information about the application domain was extracted from the existing smog events database through the co-operative use of rough sets and neural networks. The predictive decision models generated in the case study provided superior results to the current forecasting system. The models were used as the basis for a first-cut prototype system. The benefits of using both techniques co-operatively were demonstrated, as rough sets provided users and analysts with understandable knowledge representations and important information about domain features, and enabled the gradual extraction and refinement of patterns, while neural networks provided predictive models that allowed the users to compare the results thus assuring the reliability of the results. The framework facilitated the process of gradually accumulating improved knowledge of the domain through the iterative application of the combination of techniques. In addition, when input features for neural networks were selected using the rough sets the neural networks provided better predictive models. This finding was confirmed by another study reported in Jagielska (1999) where the effect of feature selection using rough sets on the performance of neural networks was investigated on several artificial and real-world datasets.

Although the case study results confirmed the effectivness of the approach, there are further opportunities to improve the knowledge discovery process and to improve the predictive capability of the decision models. It was found in the study that when the roughness was set to low values a large number of rules was extracted and many of the weaker rules were difficult to interpret. This was particularly evident in models with continuous variables. One possible way of overcoming this problem is to further refine the models by experimenting with different representations of the continuous variables.

Irrespective of the technology used, the input features to predictive systems must adequately capture the nature of the problem. It was observed that the input features (attributes) and patterns in the smog events database are weak. If there are strong patterns in data, the attribute strength is close to 1. The rules produced from such data are often strong and have very good predictive ability. In this application even the most significant attributes had a strength below 0.5. In order to further improve the accuracy of prediction of smog events, further research needs to be directed towards the identification and inclusion in the model of additional factors that are conducive to the occurrence of smog events.

6. Conclusion

This paper proposes an approach in which knowledge discovery is used in the evolutionary development of DSS by providing enhanced knowledge of a decision domain and predictive decision models in situations where past data related to the domain exist. In the approach, users and analysts gradually learn about the problem domain through the iterative process of extracting patterns from the data and building and refining predictive models. The approach thus facilitates the gradual elaboration of requirements and decision models as knowledge and understanding of the problem domain evolves.

A technological framework which implements the approach using a hybrid of rough sets and neural networks was described in the paper. The proposed approach and its associated technological framework contribute to research in the areas of knowledge discovery and the application of soft computing methods to decision support. They demonstrate how the soft computing methods can be used to assists decision makers in dealing with complex, real-world decision problems. A case study of the use of the approach in practice has demonstrated its usefulness and confirmed the viability of the framework as a vehicle for its implementation.

Dissemination of research results demonstrating the benefits of using knowledge discovery and soft computing methods in the development of DSS may help to encourage practitioners to apply knowledge discovery techniques in appropriate situations as a means of improving DSS development practice.

References

Arnott, D. R. (1998), *A Framework for Understanding Decision Support Systems Evolution*, Proceedings of The Ninth Australasian Conference on Information Systems, Sydney, Australia.

Beltratti, A., Margarita, S. and Terna P. (1996) *Neural Networks for Economic and Financial Modelling,* Thomson Computer Press.

Bigus, J. P. (1996) *Data Mining with Neural Networks*, McGraw-Hill.

Brachman, R. J. and Anand, T. (1996) *The Process of Knowledge Discovery in Databases*, in U. Fayyad, G. Piatetsky –Shapiro and P. Smyth (eds.) *Advances in KnowledgeDiscovery and Data Mining*, MIT Press.

Caudill, M. and Butler C. (1994) *Understanding Neural Networks*, MIT Press.

Cook, B. (1995) *Smog Alert Review - Phase 1 Review of the Forecasting Scheme 1984-1994*, Internal Working Paper, EPA, Melbourne (July).

Fayyad, U., Piatetsky–Shapiro, G., and Smyth, P. (1996a) *Knowledge Discovery and Data Mining: Towards a Unifying Framework,* Proc. of the Second International Conference on Knowledge Discovery & Data Mining (KDD-96), Oregon.

Fayyad, U., Piatetsky–Shapiro, G., and Smyth, P. (1996b) *From Data Mining to KnowledgeDiscovery: An Overview* in U. Fayyad, G. Piatetsky–Shapiro and P. Smyth (eds.) *Advances in Knowledge Discovery and Data Mining*, MIT Press.

Goonatilake, S. and Khebbal, S. (1995) *Intelligent Hybrid Systems*, Wiley.

Jagielska, I., Matthews, C., and Whitfort, T. (1999) *An Investigation into the Application of Neural Networks, Fuzzy Logic, Genetic Algorithms, and Rough Sets to Automated Knowledge Acquisition,* Neurocomputing 24, Elsevier.

Jagielska I. (1999) *Using Rough Sets for Practical Feature Selection in a Rough Sets/NeuralNetwork Framework for Knowledge Discovery*, International Journal of Advanced Computational Intelligence vol. 4, no. 1, February 2000, 31-38.

Jagielska, I. and Jaworski J. (1996) *Neural Network for Predicting the Performance of Credit Card Accounts,* Computational Economics, 9, 77-82, Kluwer.

Jagielski R., Jagielska I.(1998), *Using Genetic Programming for Data Mining in the Development of a Decision Support System,* The Australian Workshop on Intelligent Decision Support and Knowledge Management AWIDS'98, Sydney, (September).

Keen, P.G.W. (1980) *Decision Support Systems: A Research Perspective* , Data Base, 12 (1/2), pp15-25.

Kohavi, R. and Frasca, B. (1994) *Useful Feature Subsets and Rough Sets Reducts*, Third International Conference on Rough Sets and Soft Computing.

Koza, J. (1992) *Genetic Programming. On the Programming of Computers by Means of Natural Selection*, MIT Press.

McFadden, F. (1996) *Data Warehouse for EIS: Some Issues and Impacts*, Proc. 29th Annual Hawaii International Conference on System Sciences, Hawaii, (January).

Oakley, E.H. (1994) *The Application of Genetic Programming to the Investigation of Short,Noisy, Chaotic Data Series* in T.C.Fogarty (ed.) Evolutionary Computing – AISB Workshop, Springer-Verlag.

Pawlak, Z. (1991) *Rough Sets: Theoretical Aspects of Reasoning About Data*, Kluwer.

Pawlak, Z., Grzymala-Busse, J., Slowinski, R., and Ziarko, W. (1995) *Rough Sets,* Communications of the ACM, 38 (11)

Polkowski, L. and Skowron A. (eds) (1998) *Rough Sets in Knowledge Discovery*, Physica-Verlag.

REDUCT Systems Inc. (1994) *DataLogic/R Reference Manual,* Regina, Canada.

Slowinski, R. (ed.) (1992), *Intelligent Decision Support*, Kluwer.

Spangler, W.E. (1991) *The Role of Artificial Intelligence in Understanding the Strategic Decision Making Process*, IEEE Transactions on Knowledge and Data Engineering Vol. 3, No. 2.

Sprague, R.H. and Carlson E.D. (1982) *Building Effective Decision Support Systems*, Prentice-Hall.

Weiss, S.M. and Indurkhya N. (1998) *Predictive Data Mining*, Morgan Kaufmann.

Zadeh, L.A (1994) *Fuzzy Logic, Neural Networks ,and Soft Computing,* Communications of ACM, March 1994/Vol 37, No3.

Ziarko, W. (1996) *Discovering Classification Knowledge in Databases Using Rough Sets*, Proc. of the Second International Conference on Knowledge Discovery & Data Mining (KDD-96), Oregon.

Ziarko, W., Golan, R. and Edwards, D. (1993) *An application of Datalogic/R Knowledge Discovery Tool to Identify Strong Predictive Rules in Stock Market Data*, Proc. AAAI Workshop on Knowledge Discovery in Databases, Washington.

Zurada, J. M. (1992) *Introduction to Artificial Neural Systems,* West Publishing Company.

Part II:

Applications and Implementations

Decision Support Systems in Healthcare: Emerging Trends and Success Factors

Tze-Yun Leong

Department of Computer Science
School of Computing
National University of Singapore
3 Science Drive 2
Singapore 117543
Email: leongty@comp.nus.edu.sg

Abstract. Emerging over 50 years ago, decision support systems have a long history in the healthcare setting. Health decision support systems incorporate a variety of soft computing techniques to derive useful information from data repositories, human knowledge, and on-line literature to support decision making across operational and clinical processes in healthcare organizations. This article summarizes the historical and state of the art decision support systems in healthcare, analyzes the success factors needed for widespread deployment, and postulates the future trends of the field in the context of a new decision management paradigm.

1. Introduction

Advancement in biomedical sciences has produced a plethora of information that healthcare providers and practitioners need to manage. Escalating healthcare expenditure and variation in practice have led to a global mandate to improve quality and reduce cost. Computer and information technology, ranging from simple information exchange platforms to advanced analytic capabilities for improving effectiveness and efficiency of care processes, is indispensable in modern day healthcare. Nevertheless, development and implementation of health information and decision support systems are still very challenging in practice.

A health information system is a compendium of systems and technologies that support the operational and clinical processes in a healthcare organization. Such systems include different applications and modules that capture, organize, and analyze the financial, administrative, and clinical data. Health decision support systems, or HDSSs, integrate and derive useful information from the different data sources in a health information system to support effective decision making. In addition to facilitating information access and usage, an HDSS extends the range and enhance the capability of cognitive and decision processes. The main objective of an HDSS is to help healthcare providers and practitioners determine the optimal strategies, in terms of both financial and clinical outcomes, for prevention, treatment, and wellness maintenance. The main success criteria for an HDSS include its efficacy, efficiency, and effectives. Efficacy measures whether the system is doing what it is supposed to do; efficiency measures whether the system is responsive and cost justified; and effectiveness measures whether the system contributes to achieving improved care (Tan 1998b).

This article summarizes the historical and state of the art HDSSs, analyzes the success factors needed for widespread deployment, and postulates the future trends of the field in the context of a new decision management paradigm.

1.1 Why Computer-Based Decision Support in Healthcare?

Computer-based decision support is essential in situations where human cognitive and reasoning capabilities cannot effectively cope with the amount and the complexity of available information. In healthcare, unlike many other domains, judgment often has to be made under incomplete or imprecise information; people sometimes make mistakes that are extremely costly. In the United States alone, for example, about 100,000 patients die because of medical errors every year (Kohn et al. 2000). Pressure has been mounting for healthcare practitioners to keep up with the ever-increasing medical knowledge and new technology. Global trends also mandate evidence-based practice both to improve quality and reduce cost of healthcare. Computer and information technology is particularly effective in supporting routine and complex decision situations by reducing variation and promoting best practice; it is also more efficient to automate repetitive decisions that may involve multiple information sources.

1.1.1 Examples

HDSSs facilitate both operational and clinical decision making in a healthcare organization. Some examples of the queries that such systems aim to answer include the following:

- Does drug A have any complications when used together with drug B and drug C in patients with risk factors profile X?
- What is the likelihood that a patient has a myocardial infarction, given data in the patient history and the ECG results?
- What is the likelihood that the hand, foot, mouth disease is caused by a certain type of virus, given the signs and symptoms and demographic patterns?

- A 60 year-old with diabetes and renal dysfunction has undergone 2 angioplasties before. An obstruction of more than 70% is seen in the left coronary artery. What is the most cost-effective therapy?
- What is the most cost-effective screening test for colorectal cancer in a group of patients with risk factors profile X?
- What is the best practice pattern, in terms of cost-effectiveness, for coronary bypass surgery for a group of doctors in the Cardiac Surgery department over the past 5 years? Why?
- A group of 45-60 year-old patients in the Medicine Department of Hospital X in Country Y with hypertension and specific risk factors and genetic predispositions are not responsive when treated in accordance with a US-based clinical practice guideline. How to improve the treatment efficacies?

1.2 Historical Perspective

First emerged in the 1950s and 1960s, early HDSSs were mainly isolated applications focused on mathematical or algorithmic routines that would enhance the clinicians' diagnostic capabilities. Ledley and Lusted's work (Ledley and Lusted 1959; Lusted 1960), for example, advocated the use of Bayesian techniques to aid physicians in quantitative analysis, differential diagnosis, and evaluation of the best course of action among different alternatives. Zadeh's work (Zadeh 1969; Bell and Zadeh 1970), on the other hand, inspired a few attempts at applying fuzzy logic in biomedical decision making, especially in complex, ill-defined, and not easily analyzed domains, e.g., psychology, image processing, etc (Zadeh 1969; Bell and Zadeh 1970; Teodorescu et al. 1999b). These pioneering efforts aimed to conceptualize and formalize the clinical diagnosis and decision making processes and to improve understanding of the nature and structure of medical knowledge.

In the 1970s and 1980s, there were three main categories of HDSSs: 1) specialized clinical decision support systems for improving diagnostic and therapeutic effectiveness in a clinical domain 2) generalized operational decision support systems for containing costs and managing resources in a healthcare organization; and 3) large-scale, organization-wide decision support systems integrating patient records and problem solving capabilities for improving patient care. Some representative specialized diagnostic and treatment consultation systems are CASNET (Kulikowski and Weiss. 1982), MYCIN (Shortliffe 1976), PIP (Pauker et al. 1976), and INTERNIST-1 (Miller et al. 1982). Various analytic methodologies, including statistical modeling, rule-based reasoning, case-based reasoning, causal and heuristic-based reasoning, and fuzzy logic, were being actively developed and increasingly applied in HDSSs. These efforts led to the knowledge-based or expert systems paradigm that has been widely accepted and continues to be deployed in many different vertical domains, including law, finance, manufacturing, and logistics. There was a paralleled development of organization wide management support capabilities such as the Casemix system (Lichtig. 1986) for utilization review and resource allocation. There were also a few attempts at integrating the two types of systems to form organization-wide HDSSs, such as the HELP system (Kuperman et al. 1991).

In the 1990s, HDSSs mainly evolved from the research environment into practical setting. A variety of systems were implemented; new theoretical foundations and problem solving paradigms were developed; standardized data representation and communication protocols were proposed and deployed. Data-mining, for example, has become one of the most promising analytic methodologies in HDSSs built on top of large databases. Neural networks, neural-fuzzy systems, and genetic algorithms have also started to emerge as alternative analytic frameworks for clinical decision support, especially in diagnosis, pattern recognition, prognosis, and control (Teodorescu et al. 1999b; Lisboa 2002). There was growing acceptance and adoption of standards including the ICD-10 (Tenth-Edition of the International Classification of Diseases) (WHO 1992), SNOMED (Systematized Nomenclature of Human and Veterinary Medicine) (Cote et al. 1993), and UMLS (Unified Medical Language System) (Humphreys 1990) for medical vocabularies and nomenclatures, the HL7 (Health Level 7)[1] communication protocol for exchange of healthcare data, the Arden Syntax (Hripcsak et al. 1994) for decision logic encoding, the GLIF (Ohno-Machado et al. 1998; Kaplan 2001b; Peleg et al. 2001) exchange format for representation of clinical practice guidelines. This era is also represented by the emergence of protocol-based consultation systems, which provide support for identifying and implementing best practices and guidelines (Beliakov and Warren 2001), and a variety of efforts to formalize the assessment and evaluation criteria for HDSSs (Kaplan 2001a; b).

A number of comprehensive and excellent surveys on HDSSs in general, and clinical decision support systems in particular, have been published over the years (Hudson and Cohen 1994; Miller 1994; 1996.; Tan 1998a; Berner 1999; Teodorescu et al. 1999a; b; Musen et al. 2001; Lisboa 2002). All the surveys, however, have come to similar conclusions about HDSSs: Despite that HDSSs are recognized to have tremendous potential, only the operational decision support systems have been commonly deployed; very few clinical decision support systems are in routine use.

In the early 2000s, under the global mandate to improve care quality and reduce cost, and the professional emphasis on practicing evidence-based medicine, a new class of evidence-adaptive HDSSs has been proposed (Sim et al. 2001). Evidence-adaptive HDSSs support analysis of past data and information from multiple sources to derive best-practice guidelines that would reduce variations and errors in various operational and clinical processes. While many challenges still remain, this framework seems most promising to lead to large-scale deployment and adoption of HDSSs in the near future. We propose a methodology in the second half of this article on how such a framework could be implemented.

---

[1] www.hl7.org

2. Characterizing Health Decision Support Systems

Decision making in healthcare is an iterative information-seeking and problem-solving process that involves identifying the decision situation, designing an action plan, analyzing the possible outcomes and consequences of each decision alternative, and reviewing the process after the decision has been made and the outcomes known. The process depends on effective and efficient management of information, possibly from multiple sources, and the ability to focus attention on the most relevant information. The process also facilitates consultation and communication among a group of stakeholders or co-decision makers when necessary. The knowledge involved in healthcare decision making includes both scientific or formal knowledge and experimental knowledge, and both subjective and evidence-based knowledge.

2.1 Major Types of Decisions

Decision support in healthcare can be characterized along different dimensions: workflow processes, user groups, communication styles and interaction modes, system functions, information sources, and technical methodologies:

- Workflow processes include patient administration, accounting and financial reporting, organizational management, quality assurance, and clinical decision support.
- User groups include hospital executives, administrators, clinicians – physicians, nurses, and other clinical support personnel – and patients.
- Communication styles include consulting, critiquing, and monitoring and interaction modes include request-based and automated.
- System functions include information visualization and reporting, cost-benefit analysis, policy planning and analysis, monitoring and control, diagnosis and treatment management, and protocol-based decision support.
- Information sources include data-based, model-based, knowledge/expert-based, and evidence-based.
- Technical methodologies include statistical-based, model-based, decision analytic-based, simulation-based, knowledge- or heuristic-based, other algorithmic-based, and hybrid.

As outlined in Table 1, health decision support can be expressed in terms of a continuum of functionalities, ranging from basic to highly advanced applications.

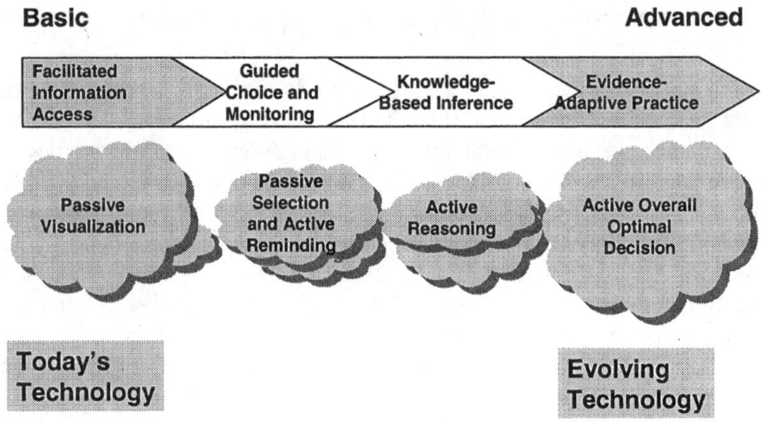

Fig. 1. Continuum of Health Decision Support Functionalities

The most basic form of decision support provides timely access to organizational-, domain-, or individual-specific scientific, clinical, or operational information, statistical summaries, periodic reports, and visualization charts to support decision-making. At the second level, pre-encoded standards of decision guiding templates support selection of preferred alternatives among available options. Predefined rules also generate alerts and reminders by detecting special or abnormal events in biomedical records or business or care settings. Such systems are usually static and do not support real-time updating and context-dependent reasoning. At the next level, diagnostic and remedial advice or critique systems and provide reasoning capabilities to support diagnosis and generate remedial or policy recommendations. At the most advanced level, decision support applications aim to achieve overall improvement in financial and clinical outcomes via analysis of trends, patterns, and data that are then used to guide business and care decisions, identify best practices, and evaluate and refine practice guidelines. This process involves management and assimilation of different types of evidence from possibly multiple sources.

2.2 Technical Foundations

At the heart of an HDSS is a set of inference or analytic methodologies that manipulate the available data, information, and knowledge to produce insights to support effective and efficient decision making. Such insights augments human reasoning by extending the range and expanding the capability of cognitive and decision processes. Health decision making involves both well-structured and precise calculations and ill-structured and uncertain complex problems. The inference or analytic capabilities of HDSSs, therefore, often involve a variety of mathematical and soft-computing techniques to support the major analytical tasks involved.

2.2.1 Major Inference Techniques

There are many general inference methods or analytic "models" applicable in an HDSS; each technique is most suitable for different tasks; each has its own strengths and weaknesses. Hybrid or combination of different techniques may be incorporated in an HDSS. The major inference techniques include:

- Statistical-based techniques, including simple statistics, ratio analysis, on-line analytic processing (OLAP) techniques, regression analysis, Bayesian analysis, time-series analysis.
- Mathematical programming techniques, including model-based techniques such as linear programming, data envelope analysis.
- Decision analytic techniques, including decision tree-based and acyclic directed graph-based modeling frameworks.
- Simulation-based techniques, including Monte-Carlo simulation.
- Heuristic programming or knowledge-based techniques, including rule-based inferences, case-based inferences, frame-based or object-oriented inferences
- Soft-computing techniques, including Bayesian networks, genetic algorithms, neural networks, fuzzy logic, and neuro-fuzzy systems.
- Hybrid techniques, e.g. a decision theoretic artificial intelligence programming language that support time-dependent decision making in multiple perspectives (Leong 1998).

2.2.2 Major Analytical Tasks

The major analytical tasks of an HDSS include summary, discovery, diagnosis, discrimination, optimization, prediction, and scenario analysis, i.e., asking "what-if" questions:

1. Summary – Summarize the salient features or facts from a large set of data or information. Relevant analytical methods include statistical techniques such as sufficient statistics, ratio analysis, regression analysis, etc. Results are presented in terms of short reports comprising graphical or visual aids.

2. Discovery – Perform exploratory analysis on the data to discover interesting trends and unobvious facts. Relevant analytical methods include statistical and data-mining techniques, which in turn deploy neural networks, classification trees, genetic algorithms, Bayesian learning, and other soft-computing techniques. Results are presented in terms of discovered rules and short reports comprising graphical or visual aids.

3. Diagnosis – Suggest and rank hypotheses based on a set of signs and symptoms, and other relevant findings. Relevant analytic techniques include rule-based, case-based, Bayesian reasoning, fuzzy reasoning, neural networks, and other soft-computing techniques. Results are presented in a ranked list hypotheses and possible explanations on how the diagnosis was derived.

4. Discrimination – Choose the best course of action given a fixed set of alternative strategies or action plans. Usually applicable in resource, test, or treatment planning and management. Relevant analytic techniques include rule-based, case-based, simulation, and decision analytic techniques.
5. Optimization – Construct the best course of action with respect to some objective measures, given a set of alternative actions at each decision stage or phase. Usually applicable in resource, test or treatment planning and management. Relevant analytic techniques include rule-based, case-based, linear programming, simulation, and decision analytic techniques.
6. Prediction – Given the history of disease or condition and the current observations, predict the future evolution of the disease state or action effects. Usually applicable in prognostic modeling and analysis. Relevant analytic techniques include statistical modeling, simulation, time series analysis, dynamic decision analysis, and stochastic modeling.
7. Scenario analysis – Also called sensitivity analysis, scenario analysis supports answering various "what-if" questions in operational and clinical processes. Closely tied to prognostic modeling and analysis, the relevant analytic techniques include statistical modeling, simulation, time series analysis, dynamic decision analysis, and stochastic modeling. Some rule-based and case-based techniques could also be applied to support this task.

3. Categorizing Health Decision Support Systems

HDSSs support decision making in both operational and clinical processes in a healthcare organization. The two components are usually separately developed and support different functions. Recent trends, however, indicate the emergence and the importance of systems that integrate both operational and clinical decision support functions in a connected framework, sharing the same data and information sources and analytic paradigms.

3.1 Operational decision support systems

Operational decision support systems, usually built on top of organizational-wide information systems, use corporate data repositories, such as data warehouses of financial and clinical information to perform retrospective evaluation. The system functions support organizational management, patient administration, accounting, utilization review, component cost and clinical performance evaluation, risk management, and policy planning and analysis in a healthcare organization such as a hospital, a clinic, or an integrated delivery network.

Healthcare organizations with quantitative analytic or quality assurance departments usually employ trained statisticians or quantitative analysts to apply standard statistical tools and packages, e.g., SPSS[2], SAS[3], etc., to conduct quantitative analyses to support operational decision making. Recent emphasis on improving quality and reducing cost in healthcare has also driven the application of advanced decision analytic and simulation tools, e.g., DATA[4], Netica[5], ReasonEdge Modeler[6], DecisionPro[7], etc., to conduct cost-effectiveness analysis on various aspects of health organization management.

Healthcare organizations are also increasingly deploying enterprise business intelligence systems, e.g., Business Objects[8], Pro-Clarity[9], Brio[10], etc., to support operational decision making. Such systems include comprehensive functionalities that perform statistical and OLAP of transaction data, and produce executive summaries and reports for health administrators and managers.

3.1.1 Executive Information Systems

An emerging trend in operational decision support systems is to build an executive information system (EIS) on top of the decision support functions. The target users for the EISs are the health policy makers, quality assurance managers, hospital administrators, medical directors, department chiefs, etc.

The EISs usually incorporate a set of easy-to-use graphical user interfaces to access the information and analysis reports produced by the underlying operational decision support systems. The EISs interfaces integrate information from different sources, keep track of internal and external changes, identify and monitor resource utilization, and support risk analysis and risk management. The objectives of such systems are to help healthcare managers to gain high-level perspective on key performance indicators and trends in the organization quickly, so as to achieve the overall strategic vision and mission.

3.2 Clinical decision support systems

Clinical decision support systems use population statistics, expert knowledge, and/or medical databases to support monitoring and control of adverse events, diagnosis and treatment consultation and critiquing, and protocol-based practice guidance and implementation in a healthcare organization. Such systems are usu-

---

[2] http://www.spss.com
[3] http://www.sas.com
[4] http://www.treeage.com
[5] http://www.norsys.com
[6] http://www.reasonedge.com
[7] http://www.vanguardsw.com
[8] http://www.businessobjects.com
[9] http://www.proclarity.com
[10] http://www.brio.com

ally stand-alone applications that provide proactive advice both prospectively and at point of care in specific domains.

Clinical decision support systems incorporate many advanced analytic techniques as mentioned earlier to make inferences and derive useful insights from the information sources. Based on different analytical tasks, different technical inference methodologies, interaction modes and communication styles may be appropriate. Such systems are usually customized and built in-house for different health organization settings, using off –the-shelf software or research system development platforms. Most of the resulting systems are not reusable or portable to another organization.

3.2.1 Monitoring and Control Systems

Monitoring and control systems selectively monitor clinical data continuously, and test such data against predefined criteria to send reminders and alerts upon detection of pre-specified patterns. The objectives are to detect and prevent adverse events, identify alarming laboratory results, raise drug contraindications, and to support critical care monitoring in general. The target users are clinicians, pharmacists, and administrators.

3.2.2 Clinical Diagnosis and Treatment Systems

Clinical diagnosis and treatment systems recommend diagnosis and treatment plans, and critique care management plans. The objectives are to facilitate decision making in routine and complex situations, provide reference and confirmation information, support scenario analyses for better insights, and serve as educational and communication tools. The target users are clinicians, patients, and students.

3.2.3 Protocol-Based Decision Systems

Protocol-based decision systems create, maintain, and implement disease management and best practice guidelines from different information sources. Such systems transform often-ignored guidelines to dynamic programs for real-time patient-specific management and automated recommendations, reminders, alerts, and adjustment of device settings. These systems also support outcomes analysis and outcomes management. The objectives are to promote systematic record keeping, support rational decision making, improve clinician acceptance of best practices, and improve quality and reduce cost of care. The target users are clinicians, patients, and administrators.

3.3 Integrated Health Decision Support Systems

Built on top of the hospital information system, there are few systems that provide both operational and clinical decision support in a healthcare organization. With emphasis on care quality and patient's satisfaction, the healthcare service model is shifting from the traditional provider-centered care to patient-centered care. The ultimate objective is to provide seamless services according to patient needs.

3.3.1 Patient Care Systems

Patient care systems acquire, store, process, and present patient-, organization-, and domain-specific information to support each professional's care of the individual patients. Through appropriate use of patient-specific care requirements, organization-specific responsibilities, policies and procedures and domain-specific practice guidelines, such systems aid the co-ordination of multidisciplinary services for individual patients and the planning and scheduling of the caregivers' work activities (Ozbolt and Bakken 2001). While still in the early stages of development and adoption, patient care systems merge the capabilities and functions of traditional operational and clinical decision systems, and form the basis of next generation HDSSs.

Patient care systems in use today attempt to use single or integrated data repositories to support a variety of individual or team-based operational and clinical functions. The data warehouse and data mart concepts are beginning to take shape in the healthcare setting. To make optimal use of such data, again business intelligence tools, statistical and modeling packages, and knowledge-based system shells are applied to different parts or "views" of the data to derive the necessary insights to support decision making. Some patient-care systems also offer the option to organize care temporally into clinical pathways and practice guidelines and to have variances from the anticipated activities, sequences, or timing reported automatically (Ozbolt and Bakken 2001).

3.4 Examples of Health Decision Support Systems

Many well-known HDSSs have been developed over the years. While not all of these systems have been put into practical use, many of them have been influential and critical for the development and evolution of new decision support and inference paradigms that have impact beyond the healthcare setting. The following discussions summarize a few of these efforts; the lessons learnt from them chart the emerging trends and explicate the success factors of HDSSs in general.

3.4.1 Early Clinical Decision Support Systems

CASNET (Causal-associational network) (Kulikowski and Weiss. 1982) is a model for describing disease processes developed at Rutgers University. CASNET was originally applied in a consultation program for glaucoma diagnosis and therapy (CASNET/Glaucoma). The system was evaluated extensively and applied in other domains. It is a graph-based formalism that attempts to capture the notion of causality in diagnostic and therapeutic processes. All the disorders and the relevant causative factors and possible consequences are represented as nodes in a network. Different scores can be calculated for the nodes, indicating the possibilities of causal pathways that would lead to the particular disorder. The belief in a disorder can be ruled out if every possible pathway to a suspected disorder can be ruled out (Szolovits 1982). Therapy efficacy is modeled in the same manner as disease progression. The semantics and definitions of the causal links in the system, however, are not well-defined and rendered system hard to apply in practice. Nevertheless, this is one of the major efforts that have inspired many graph-based representation frameworks such as the semantic networks (Fahlman 1979), conceptual graphs (Sowa 1984), and Bayesian networks (Pearl 1988) in later years.

MYCIN (Shortliffe 1976) is a consultation system developed at Stanford University for the diagnosis and treatment of infectious diseases. MYCIN popularized the application of production rules as a modular, effective knowledge representation mechanism to support clinical decision making. Each individual rule can be independently created, analyzed, modified, or discarded to incrementally change the behavior of the overall program. Together with an inference engine that operates on the rules in a simple manner, a mechanism to incorporate uncertainty in the inferences, and a set of explanation generation capabilities, MYCIN has led to the development and the wide-spread use of the powerful problem solving paradigm of knowledge-based or expert systems. The MYCIN system itself, however, was never put into routine use.

PIP (Present Illness Program) (Pauker et al. 1976) is a guided history-taking and diagnostic system developed at MIT and the New England Medical Center. The system proposes a cognitive model of clinical reasoning that takes the present illness of a patient with edema. The knowledge is represented in a highly connected set of associations that is used by a set of dynamically generated problem-solving methodologies to check the validity of facts, generate and test hypothesis, and construct a coherent picture of the patient (Szolovits 1982). PIP is one of the first programs that integrates knowledge from computer science, cognitive psychology, and clinical problem solving. It has inspired many second generation knowledge-based systems and paradigms that focus on the integration of such multidisciplinary knowledge, e.g., ABEL (Patil et al. 1982) and other systems that support reasoning in multiple levels of abstraction.

INTERNIST-1 (Miller et al. 1982) is a diagnostic consultation system in internal medicine developed at the University of Pittsburgh. It uses a cognitive-based heuristic to select a set of plausible diseases that would explain the current abnormal findings, and rank them according to how well the diseases could explain the findings. INTERNIST-1 covers a wide scope of the domain; its knowledge base is one of the most comprehensive among all the clinical decision support systems built to-date. The system is extensively evaluated and seems to do well in many cases, including those that involve multiple and complex diagnosis in internal medicine. The INTERNIST-1 consultation system itself, however, was not put into routine use. It's comprehensive knowledge base, however, has inspired the development of education and training systems that would serve as "intelligent textbooks" for healthcare practitioners.

CADIAG (Adlassnig and Kolarz 1985), developed at the University of Vienna, is one of the most famous fuzzy expert systems built to date. Research and development based on the CADIAG framework have gone on for more than two decades and the program is still evolving. CADIAG-1 and CADIAG-2 are medical expert systems with applications in rheumatology, gastroenterology, and hepatology. CADIAG-1 is based on a symbolic logic representation of medical relationships among symptoms, signs, and diseases. Both definite and uncertain relationships are applied to confirm or exclude diagnoses and to establish diagnostic hypotheses. CADIAG-2 employs fuzzy set theory and fuzzy logic to formalize the medical entities and relationships. The CADIAG-2 system was evaluated in various clinical settings and the findings shed light on the power and the limitations of the methodologies used (Kolarz and Adlassnig 1986), (Leitich et al. 2001). The latest MedFrame/CADIAG-4 system (Sageder et al. 1997) still employs fuzzy set theory and fuzzy logic to represent medical knowledge. Findings, diseases, and therapies are organized in hierarchies.

3.4.2 Clinical Diagnostic Systems

Despite the widely recognized strong potential of clinical decision support systems, and the numerous analytic and inference methodologies resulted from their research and development, there are only a limited number of clinical decision support systems in routine use. Most of the deployed systems support diagnostic screening and risk assessment.

The Acute Abdominal Pain (AAP) diagnostic system is based on Bayesian models. This system has been evaluated extensively over a decade in both UK and the European Community. The results consistently indicated an approximately 20% improvement in the diagnostic accuracy of the doctors with almost a 50% reduction in the rates of perforation and negative laparotomy (Adams et al. 1986; McAdam et al. 1990; de Dombal et al. 1997). Despite the remarkable evidence of practical usefulness, however, this system is not widely known outside the European community.

A number of scoring or risk stratification systems that indicate the severity or susceptibility of illnesses have been developed over the years. These are practical systems that use shallow associations between observations and a fixed universe of allowed outcomes to formulate a predictive classification of specific classes of disease, applicable in specific clinical settings (Lisboa 2002). Some common examples include the APACHE and SAPS scores for critical care, and the Parsonnet (Parsonnet et al. 1989) and EuroSCORE (Nashef et al. 1999) scores for cardiac and thoracic surgery. Most of these scoring systems are statistically based. Application of the scoring systems, however, can sometimes be biased or show contra-indicative results in certain settings. As compared to multi-layer perceptron (MLP) implementation, for example, the meaning of a prediction of a particular percentage of deaths is over-estimated by APACHE when it is small and under-estimated when it is large, whereas the MLP over-estimates everywhere except in the top range (Lisboa 2002).

PAPNET is a neural network-based system for diagnosis of cervical cancer from Pap smear slides (Mango 1994; Mango 1997; Mango and Valente 1998). This patented system gained Federal Drug Administration (FDA) approval for secondary screening in 1995. PAPNET identifies a small set of the most atypical single cells and cell clusters for visual inspection among the vast amount of cell images in cervical smears. This system uses a two-tier MLP architecture with different levels of resolution for initial and final classification. Early clinical trials indicated that the system may significantly improve detection accuracy for the most severe cytological abnormalities (Mango and Valente 1998). The system is being considered for primary screening.

Several aspects of the PAPNET system have contributed to its sustained adoption for routine use alongside trained cytologists in clinical laboratories (Lisboa 2002). First, the system addresses an issue where there is an identified need for improvement. Second, the system aims to substantially improve the sensitivity, as compared to trained cytologists, of diagnostic cell identification to reduce the highest-cost errors caused by false negative tests. The resulting poor specificity is managed by referring a small number of images for manual inspection. Third, the system promotes user acceptance by integrating seamlessly into the standard testing protocol (Mango 1997).

3.4.3 Education and Training Systems

While most of the clinical decision support systems remained in the research setting, a number of education and training systems that have evolved from or inspired by the research work have been successfully deployed. Many training systems are still in use today and have shown to be effective in improving the clinical problem solving skills of young clinicians and students (Friedman et al. 1999). Among the most popular education or training systems are QMR, DXplain, and Iliad (Lincoln 1999).

QMR (Quick Medical Reference) (First et al. 1985; Miller et al. 1986) was an outgrowth of the INTERNIST-1 project. The system is commercially available[11] and used by a large community of practitioners and students. The system covers 680 disorders in internal medicine and the data dictionary contains 5000 medical findings. QMR was originally designed to operate in three modes: 1) as an electronic textbook that link patient characteristics to specific diseases, display and stratify disease profiles, review disease relationships, and access literature references retrieved through a link to MEDLINE; 2) as a "spreadsheet" for testing simple diagnostic concepts by combining a few characteristics or diseases and determine the

[11] http://www.firstdatabank.com

implications; and 3) as a expert diagnostic consultant that provides patient-specific diagnostic advice. A case simulation model was added later. QMR has been evaluated extensively and continues to be delivered and deployed through various platforms. The developers of QMR argue that the system's use as an electronic reference is far more important than its use as a consultation program to a clinician (Miller and Masarie 1989). In the current version of QMR, many of the consultation features of INTERNIST-1 have been removed.

DXplain (Barnett et al. 1987) was developed at Harvard Medical School's Laboratory for Computer Science. Its knowledge base contains 5000 diagnostic findings and covers about 2000 diseases in internal medicine. The system uses a modified form of Baye's theorem to process diagnostic findings. DXplain provides diagnostic consultation, case simulation, and a textbook mode for browsing the knowledge base. DXplain has been widely deployed in academic settings for teaching and training purposes.

Iliad (Lincoln 1999) was originally developed as part of the HELP system (Kuperman et al. 1991) at the University of Utah and is commercially available[12]. The Iliad 4.5 knowledge base covers 930 diseases in internal medicine, pediatrics, dermatology, psychiatry, obstetrics and gynecology, peripheral vascular diseases and sleep disorders. Its knowledge base contains 1500 syndromes and treatment protocols and 11,900 findings, including digital images. Iliad's functions include diagnostic consultation, treatment reference, case simulation, and a browsing mode. The diagnostic accuracy and training efficacy of Iliad have been extensively evaluated and shown to be effective in reducing diagnostic errors, especially in uncommon diseases (Friedman et al. 1999; Lincoln 1999).

3.4.4 Successful Integrated HDSSs

The HELP (Health Evaluation through Logical Processes) system (Kuperman et al. 1991) developed at University of Utah has been in use at the LDS Hospital for more than 3 decades. Major software components of the system have also been installed in seven hospitals operated by the Intermountain Health Care. Originally implemented as a hospital information system, HELP has many components that directly support decision making in both operational and clinical settings. The system consists of an integrated clinical database, a frame-based medical decision support system, different programs to support organization-wide administrative functions, and software tools to develop, maintain, and expand these components. The HELP system provides the setting for implementing and testing various decision support tools, including expert systems in blood gas interpretation, intensive care settings, medication monitoring, etc (Haug et al. 1999).

The HELP system includes 4 types of HDSSs (Haug et al. 1999):

The alerting systems monitor and identify specific, well-defined medical conditions; they alert adverse drug events and undesirable or life-threatening laboratory

[12] http://www.amimed.com

results. Such applications use simple decision logics that create large potential impact. In the HELP environment, alerting systems are observed to significantly shorten the length of time in the life-threatening condition and length of stay in the hospital.

The critiquing systems manage the ordering of different procedures and assists in ensuring compliance with criteria for proper use of such procedures. Such systems function by accessing different data sets of the integrated clinical database in the HELP system, e.g., the admitting department, clinical laboratory, surgical scheduling, physicians' and nurses' orders, computerized patient records, etc. Alternate management plans are suggested after analyzing the submitted plans, if necessary.

The suggestion systems implement a set of computerized protocols in specific domains. The systems react to requests for assistance and respond by making concrete suggestions concerning which actions should be taken next, guided by the implemented protocols.

The diagnostic decision support systems assist in identifying different diseases and disorders based on a set of patient characteristics; these include: a) Applications for standardizing various diagnostic categorizations that impact therapeutic decisions using thc HELP's integrated database and analytic methods such as Boolean logic, statistical, rule-based, and probabilistic approaches. Some examples include an effective adverse drug events detection system that has reduced the length of hospital stay and cost; an efficient and highly sensitive hospital-acquired infection diagnostic system; and an antibiotic assistance system that assembles relevant data for physicians to improve efficacy of antibiotic treatment and cut down unnecessary prescriptions. b) Applications for simulating the more extensive and flexible diagnostic behavior of physicians using frame-based Bayesian, Shannon Information Content-based, and other novel analytic approaches. These are usually on-going research work.

The evolution and experience with the HELP system indicate that diagnostic applications that focus on specific, limited decision support goals with recognizable targeted tasks have been more successful. General-purpose diagnostic programs, while addressing many interesting issues, have yet to be accepted into main-stream routine use.

3.4.5 Emerging Integrated HDSSs

Developed at Stanford University, EON is a second generation knowledge-based system that facilitates patient care in accordance with treatment protocols and guidelines (Musen et al. 2001). EON consists of a set of software components that must be embedded within some health information systems containing patient-related data. The major components of EON includes: 1) a set of problem solvers, each addresses specific tasks, such as determining the treatment to be given in a particular setting in accordance with predefined clinical protocol and determining the eligibility of a patient for some specific protocols; and 2) a set of knowledge bases, each encodes a set of clinical protocols that all the problem solvers can examine and operate on.

The EON system components can be mixed and matched to address different protocol-based decision making tasks in different medical domains. For example, Therapy-Helper (T-Helper) consists of EON's therapy-planning component and eligibility-determination component, and the knowledge bases for protocol-based care of AIDS and HIV-related diseases. The same therapy-planning component and eligibility determination component are used with a knowledge base of breast-cancer protocols in another system.

New problem solvers and new knowledge bases can be easily added in the EON architecture. Knowledge acquisition is facilitated by a knowledge-based development environment Protégé (Musen 1998), which processes different models for clinical protocols to construct customized tools for entry and review of specific protocol descriptions. Overall, EON shows how decision-support systems can be embedded within larger health information systems, using emerging standards for network-based communication among software modules that can be separately developed and integrated.

4. Toward A New Decision Management Paradigm

Emerging over 50 years ago, decision support systems have a long history in the health care setting. HDSSs' potential to improve care quality and reduce cost is widely recognized, and tremendous research opportunities, interests, and impact have been generated over the years. Nevertheless, with the exception of a few implementations like the HELP system, most of the systems that are deployed are mainly operational decision support systems, usually consists of specific, well-defined functions built on top of the health information systems.

Lessons learnt from the past decades of HDSS research, development, and implementation have shed light on the desiderata of the next generation systems. Not only must these systems be able to address a large number of decision support tasks, they must also be readily accepted by healthcare organizations and practitioners. These systems should:

- Support decision makers, not replace them
- Be transparent and user-friendly
- Work with multiple information sources
- Possess multiple functions targeting at specific, well-defined tasks
- Contain components that can be modularly created, modified, and discarded
- Possess a range of analytic capabilities to handle different tasks
- Be adaptive to environmental changes, technology advancements, user needs
- Facilitate evidence-based medicine, reduce medical errors, improve clinical outcomes, care quality, and efficiency
- Optimize both clinical and financial outcomes

4.1 Evidence-Adaptive Health Decision Support Systems

Lessons learnt from HDSSs development and deployment over the past decades, the strive to understand and management the voluminous biomedical information, and the global mandate to improve quality and reduce cost in healthcare have led to a recent proposal for evidence-adaptive HDSSs (EA-HDSSs) (Sim et al. 2001). EA-HDSSs, while still very much in the conceptual stage, gives a glimpse of what future HDSSs with possible wide deployment would look like. The EA-HDSSs fit into the current and emerging workflow of healthcare organizations, integrating important information from multiple sources. The multi-faceted functional and analytical capabilities of the systems can be modularly incorporated, updated, modified, and reused. The knowledge bases are derived from and continuously reflect the most up-to-date information from literature-based evidence, practice-based evidence, and patient-based evidence.

Literature-based evidence includes randomized trials, systematic reviews, published practice guidelines, and other public knowledge or databases. While machine-readable literature based evidence currently constitutes only a small fraction of research literature, and some published evidence suffer from problems in the study design and reporting, such literature is still the de facto source of reference for the practice of medicine. Most electronic resources are currently not machine-interpretable, e.g., the Cochrane Library, Best Evidence, Clinical Evidence, the National Guideline Clearing House, etc (Sim et al. 2001). With the Internet and web-based activities becoming a way of life, however, and new methodologies are being developed to automatically or semi-automatically convert specific text-format literature such as guidelines into machine-readable forms (Zhu and Leong 2001), the significance of developing machine-interpretable knowledge bases is increasingly being recognized, and some early efforts include the Trial Bank (Sim et al. 2000), various genomic information databases, etc.

Practice-based evidence includes organization-wide and local databases and data warehouses from registries and repositories, health information systems, electronic medical records, laboratory systems, accounting systems, etc. Such evidence complements and supplements general, literature-based evidence and is required for context-dependent practice guideline development, performance evaluation, policy analysis and planning, and various diagnostic and treatment management recommendations.

Patient-directed evidence includes patients' preferences and concerns through direct interactions and feedback from health-related resources, e.g., websites, surveys, etc. Incorporating such evidence for decision support would increase health care quality through facilitating communication, fostering shared decision making, individualizing care plan, and improving clinical outcomes (Sim et al. 2001).

Successful design and implementation of an EA-HDSS pose many technical and practical challenges; substantial research, development, and policy issues still need to be addressed. In the following sections, we summarize the current understanding and vision, analyze the technical and practical requirements, and propose an architecture for an EA-HDSS based on existing and emerging technology and organizational infrastructure.

4.2 An Evidence-Adaptive Health Decision Support Architecture

Implementing EA-HDSSs requires new network infrastructure to support different decision support applications to capture the operational and clinical processes and events in diverse settings. The knowledge bases involved also need to be interoperable by different information access and analytic applications. In addition to a range of analytical capabilities that must take into consideration multiple perspectives and multiple objectives, mechanisms to incorporate new evidence must be incorporated. The system must be safe-guarded by a set of security protocols and interact with the end users through different delivery platforms. A feasible architecture for an EA-HDSS is shown in Figure 2 below.

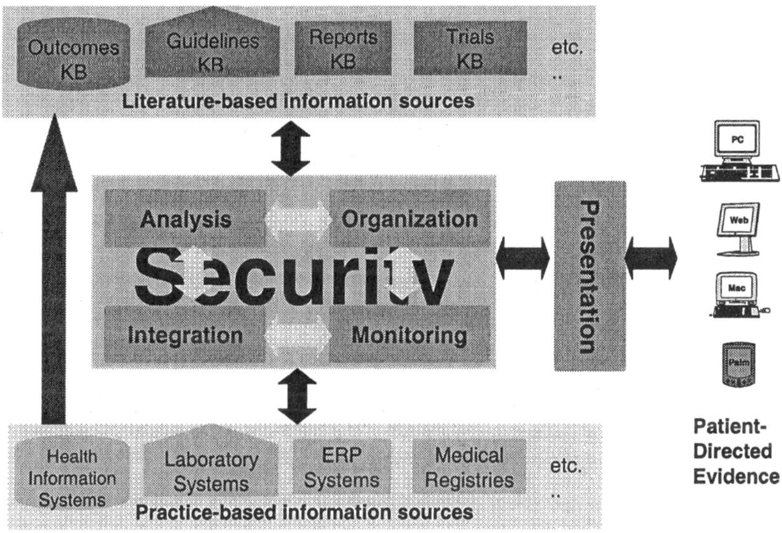

Fig. 2. Evidence-Adaptive Healthcare Decision Support System Architecture

An EA-HDSS consists of a set of core modules and capabilities to provide users with the most efficient and effective access to relevant strategic information to support decision making. This is done by Integrating, Organizing, Analyzing, Presenting, and Monitoring of the relevant information:

Integrating: This module intelligently gathers relevant information from multiple data sources both structured and unstructured, and both on-line and real-time.

Organizing: This module helps to effectively and securely organize information and knowledge for easy reference and auditing, either directly by users or by various analytical tools.

Analyzing: This module applies principal and best-of-breed analytic methodologies to interpret retrospective and real-time information, and generate predictions and optimal recommendations with respect to different scenarios, varying user objectives and possible trade-offs.

Presenting: This module presents specific, tailored response via various delivery platforms to each query and provides users contextual and possibly visual guidance on how their questions can be further refined to further conduct scenario analysis.

Monitoring: This module continually collects and systematically updates analysis results and recommendations based on new user input and data, which significantly reduces the cost and effort in content management.

4.3 Technical Challenges and Potential Solutions

Some of the desired functions of EA-HDSSs can be found in existing enterprise-wide business intelligence systems that include suites of information access and basic analytic functions working on standard-format databases and data warehouses. Such business intelligence systems, however, do not meet the full functional requirements of EA-HDSSs. Customization and enhancement of the component functions are difficult because the business intelligence systems are usually not interoperable and contains only limited analytic power. The operating concept and basic infrastructure of the business intelligence systems, however, form the basis of future development of EA-HDSSs. The specific technical challenges (Sim et al. 2001) and some potential solutions are as follows.

- Infrastructure for communication networks

To properly incorporate the different types of evidence, an effective communication network must be in place for broadband transmission of data and information that may be located in distributed locations across departments, organizations, cities, regions, countries, and continents. Security considerations and protocols need to be incorporated for different levels of authority and authentication for information access and transmission within the system and between the system and its information sources.

- Vocabularies for both operational and clinical decision support

Over the years many conceptual and workflow models of clinical decision support have been developed. In contrast, conceptual and workflow models of operational decision support are lacking; most of the decision support functions involve direct application of statistical or on-line analytic processing methods and simple information visualization functions. Integrated conceptual and workflow models for the full-spectrum of operational and clinical decision making in a healthcare organization must be worked out.

- Designs and interfaces for interoperable registries and repositories

One of the major challenges in porting existing databases and registries within a healthcare organization to another is the interoperability, or the lack there of, of such databases. Many existing health information systems incorporate different data formats without standardized interfaces. New guidelines and incentives must be put in place to create databases, registries, and repositories in standard formats and interoperable interfaces, both within individual healthcare organizations, and at the public level where most of the literature-based evidence is stored.

- Methods for multi-source information syntheses and updating

To integrate and process information from multiple sources, a set of tools and techniques are needed to interface with the various databases, registries, and repositories in a directed and effective manner to access the relevant information. Some exchange formats need to be in place to facilitate the integration. Many database management systems, for example, support the XML format for data specification and exchange, and enterprise application integration (EAI) tools and frameworks are getting wide acceptance to integrate information from different sources. There are, however, no universal adopted strategies for such exchanges and interfaces. Conversion from and integration with many legacy information systems and repositories still remain a major challenge. Moreover, a set of effective knowledge acquisition tools also need to be in place to elicit the relevant information and knowledge from experts and patients. The integrated knowledge from both systems and humans needs to be represented and validated appropriately before put into use for decision support. Both representation and validation remain major challenges in the development of HDSSs.

- Wide range of analytic functionalities

Single problem solving paradigms or algorithms are not sufficient to support the wide range of decision support functions across the operational and clinical processes in a healthcare organization. Such functions may also change or evolve as new evidence is available and the workflow processes change in an organization. A set of analytic functions need to be incorporated into the system; such functions need to be able to support a "plug and play" mode of operation to be packaged into different modular decision support applications within the system.

- Models of multi-level, multi-perspective and multi-objective decision making

Most existing HDSSs support decision making from a single perspective, usually the healthcare practitioners' or the policy makers'. With the incorporation of different types of evidence, ranging from the most general literature-based evidence to the most specific patient-directed evidence, decision making models that incorporate multiple levels of details, multiple decision makers' perspectives, and multiple objectives such as both lowering the cost and improving the care qulity, must be developed. Theories of multiple-objective decision making have been developed over the years in the decision theoretic approaches, recent advances have started to lead to applications in practice (Keeney and Raiffa 1976; Haimes and Steuer 2000).

- Multiple delivery platforms

The system must gather its inputs and present its results through different delivery platforms, including various system interfaces, stand-alone and client-server mode customized screen displays, web-based interactions, wireless transmission, fax, and SMS.

5. Practical Considerations

Besides technological challenges, there are practical considerations that need to be addressed in EA-HDSSs implementation. Some of the current understandings, based on lessons learnt from past experiences with HDSSs, are as follows.

5.1 Management Challenges

The key objective of HDSS development and implementation is to "manage" reality by integrating computer and information technology, organizational workflow, and user modeling.

To gain wide acceptance and make significant impact in a healthcare organization, an EA-HDSS must fit into the organization in terms of workflow, users' level of expertise, information technology infrastructure. There must also be a match between technology and information needs in healthcare operations. The EA-HDSS functions must be compatible with the values and professional norms, institutional settings, communication patterns, organizational culture, and cognitive processes. The main reasons for reluctance to adopt many past HDSSs are due to a poor fit between the HDSS developers' goals and healthcare practitioners' cultural values (Kaplan 2001a).

Critical factors for the effectiveness and success of EA-HDSSs, therefore, are that such systems must help in: 1) enhancing mental models or cognitive capacities of the decision makers; 2) saving in time, i.e., to increase the rate at which executives react to changes in the organizational environment; and 3) directing attention to focus on the most relevant information in decision making. Some recommendations for system planning and management include (Tan et al. 1998):

1. Adopting a general top-down planning model to achieve a strategic fit between organizational vision, philosophy, and culture with information technology vision, philosophy, and culture.
2. Incorporating within the system the elements of 1) fitting information technology vision within the organizational vision; 2) developing possible future scenarios; 3) articulating the organizations' current information technology reality; and 4) anticipating planning for future system implementation.

3. Developing a technology learning culture within the organization so that all stakeholders can expect, understand, anticipate, and welcome rapid changes.
4. Implementing a detailed action plan to achieve the desired end results.

5.2 Implementation

Implementing a successful EA-HDSS needs the collaboration of all necessary participants – Organizational champions, end users, facilitators, developers, and technical support personnel. Clear objectives and missions should be conveyed to all stakeholders. Measures should be instituted to ensure deployment and acceptance. Applicability of different implementation approaches, such as "turnkey" and "plug-and-play" for different parts of the system must be evaluated before adoption. Implementation methodologies and system components must be customized to fit the local contexts and conditions within the organization. The "human" touch must not be ignored. When there is conflict between technology and organization culture or value, the "human" factor should be given the higher priority in the final decision.

5.3 Evaluation

One of the major obstacles in gaining wide acceptance of HDSSs is the difficulty in conducting proper assessment and evaluation of the efficacy, effectiveness, and efficiency of such systems. To date, especially for the clinical decision support systems, most evaluation studies were experimental or randomized controlled clinical trial designs (RCT) that measure system performance and changes in clinical performance affecting patient care. These measures adopt a rationalist, objectivist, epistemiologic perspective to measure the quantitative performance of economic and technical factors without contextual considerations. There were very few field tests. Almost none of the studies used naturalistic designs in routine clinical settings. These studies did not shed light on why the HDSSs and CDSSs under study may or may not be effective. Methodological pluralism is lacking, and many useful measures, e.g., ethnographical field studies, record and playback techniques, socio-ethical analyses, etc., could potentially contribute significantly (Kaplan 2001b).

In general, an EA-HDSS evaluation needs to clearly define, at each stage of the development and deployment processes, the evaluation criteria and, if possible, the objective reference or "gold standard." Major issues that need to be addressed include whether large validated databases or qualitative performance references are available, sufficient amount test data and evaluation data can be accessed and properly separated, different software components can be systematically and modularly evaluated, integrated multi-center studies can be independently analyzed, and there is sufficient iterative feedback to reduce inter-observer variability.

5.4 Promotion

Successful development and deployment of EA-HDSSs depend not only on the organization and the development team, but also on public policies and regulatory authorities that oversee population-based healthcare to which the organization contributes. Government, industry, and professional organizations should develop appropriate financial and reimbursement policies for successful EA-HDSS implementations with respect to measurable quality goals, e.g., number of patients served satisfactorily for a given community or X amount of dollars saved over a 3 year period. Such entities should also provide coordination and leadership to promote the relevant incentive and coordination policies, and make sure that the promotions are effectively communicated through all levels of the healthcare organizations (Sim et al. 2001).

5.5 Legal and Ethical Issues

To ensure smooth implementation and maintenance of EA-HDSSs, regulations and monitoring policies must be instituted both at the governmental and the organizational levels. Currently, the Food and Drug Administration (FDA) of the United States do not regulate decision support software whose results will be interpreted and used by a trained practitioner before putting into practice; such software is also not subject to government regulations in other countries. To make sure that HDSSs are properly used and maintained, extensive training programs must be conducted for the users. Responsibilities of the regulatory agencies, organizations, users, and vendors in ensuring the professional integrity of the technologies, knowledge bases, and the operational and clinical process models deployed must be clearly defined and distinguished. Legal remedies must be sought when the systems result in harm or appear to cause harm, or when the desired results are not achieved because the users fail to use or improperly use the systems.

6. Conclusions

Judging from the information technology spending patterns of healthcare organizations and research activities in academic and commercial institutions, opportunities and promises of HDSSs abound. With the increasing pressure that healthcare providers and practitioners are facing in the vast amount of information needed to practice medicine, and the global mandate to improve quality of care and practice cost-effective medicine - consider carefully clinical utility and reliability of tests, procedures, and treatments which are expensive or risky (Musen et al. 2001), computerized decision support in healthcare is a necessity, not a myth!

The lessons learnt in HDSSs development and implementation in the past few decades have indicated that EA-HDSSs, which can incrementally adapt new evidence into context-dependent decision support, show significant promise that would lead to wide acceptance and deployment. While in the early stage of conceptualization, we have shown how a feasible EA-HDSS architecture could evolve from the basis of a set of existing technologies in BI and EAI. New capabilities can be incrementally added, as the underlying supporting technology and infrastructure become available, to enhance the scope and effectiveness of the EA-HDSS. Instead of being constrained by a single, specific analytic framework, a wide range of soft computing other advanced analytic techniques for addressing a spectrum of decision support functionalities will be central to the success of an EA-HDSS.

Acknowledgment

This work is supported by a research grant no. R-252-000-111-112/303 from the Ministry of Education and the Agency for Science, Technology, and Research (A*Star) in Singapore.

References

Adams, I., M. Chan, et al. (1986). Computer aided diagnosis of acute abdominal pain: a multicentre study. *British Medical Journal* 293(6550): 800-4.

Adlassnig, K. P. and G. Kolarz (1985). Present state of the medical expert system CADIAG-2. *Medical Information* 24: 13-20.

Barnett, G. O., J. J. Cimino, et al. (1987). DXplain: An evolving diagnostic decision-support system. *Journal of the American Medical Association* 258: 67-74.

Beliakov, G. and J. Warren (2001). Fuzzy logic for decision support in chronic care. *Artifcial Intelligence in Medicine* 21(2001): 209-213.

Bell, R. E. and L. A. Zadeh (1970). Decision making in a fuzzy environment. *Management Science* 17(1970): 151-169.

Berner, E. S., Ed. (1999). *Clinical Decision Support Systems: Theory and Practice.*, Springer-Verlag.

Cote, R. A., D. J. Rothwell, et al., Eds. (1993). *The Systematized Nomenclature of medicine: SNOMED International.* Northeld, IL, College of American Pathologists.

de Dombal, F., S. E. Clamp, et al. (1997). Measuring surgical performance in acute abdominal pain: some reflections from international studies. *European Journal of Surgery* 163(5): 323-329.

Fahlman, S. E. (1979). *NETL: A System for Representing and Using Real-World Knowledge.* Cambridge, MA, The MIT Press.

First, M. B., L. J. Soer, et al. (1985). QUICK (quick index to Caduceus knowledge): Using the Internist-1/Caduceus knowledge base as an electronic textbook of medicine. *Computers and Biomedical Research* 18: 137-165.

Friedman, C. P., A. S. Elstein, et al. (1999). Enhancement of clinicians' diagnostic reasoning by computer-based consultation: a multisite study of 2 systems. *Journal of the American Medical Association* 282(19): 1851-6.

Haimes, Y. Y. and R. E. Steuer, Eds. (2000). *Research and practice in multiple criteria decision making : proceedings of the XIVth International Conference on Multiple Criteria Decision Making (MCDM), 1998*, Springer.

Haug, P. J., R. M. Gardner, et al. (1999). Hospital-based decision support. *Clinical Decision Support Systems: Theory and Practice*. E. S. Berner, Springer-Verlag: 77-104.

Hripcsak, G., P. Ludemann, et al. (1994). Rationale for the Arden syntax. *Computers and Biomedical Research* 27: 291-324.

Hudson, D. L. and M. E. Cohen (1994). Fuzzy logic in medical expert system. *IEEE Engineering in Medicine and Biology Magazine*: 693-698.

Humphreys, B. L., Ed. (1990). *UMLS Knowledge Sources - First Experimental Edition Documentation*. Bethesda, MA, National Library of Medicine.

Kaplan, B. (2001a). Evaluating informatics applications - clinical decision support systems literature review. *International Journal of Medical Informatics* 64: 15-37.

Kaplan, B. (2001b). Evaluating informatics applications - some alternative approaches: Theory, social interactionism, and call for methodological pluralism. *International Journal of Medical Informatics* 64: 39-56.

Keeney, R. L. and H. Raiffa (1976). *Decisions with Multiple Objectives: Preferences and Value Tradeoffs*. New York, John Wiley and Sons.

Kohn, L. T., J. M. Corrigan, et al. (2000). To err is human; building a safer health system, National Academy Press.

Kolarz, G. and K. P. Adlassnig (1986). Problems in establishing the medical expert systems CADIAG-1 and CADIAG-2 in rheumatology. *Journal of Medical Systems* 10(4): 395-405.

Kulikowski, C. A. and S. M. Weiss. (1982). Representation of expert knowledge for consultation: The casnet and expert projects. *Artificial Intelligence in Medicine*. P. Szolovits. Boulder, Colorado, Westview Press,. 51.

Kuperman, G. J., R. M. Gardner, et al. (1991). *HELP: A Dynamic Hospital Information System.*, Springer-Verlag.

Ledley, R. S. and L. B. Lusted (1959). Reasoning foundations of medical diagnosis. *Science* 130: 9-21.

Leitich, H., H. Kiener, et al. (2001). A prospective evaluation of the medical consultation system CADIAG-II/RHEUMA in a rheumatological outpatient clinic. *Methods of Information in Medicine* 40(3): 213-20.

Leong, T. Y. (1998). Multiple perspective dynamic decision making. *Artifcial Intelligence* 105(1-2): 209-261.

Lichtig., L. K. (1986). *Hospital Information Systems and Case Mix Management.*, John Wiley and Sons.

Lincoln, M. J. (1999). Medical education applications. *Clinical Decision Support Systems: Theory and Practice.* E. S. Berner, Springer-Verlag: 105-138.

Lisboa, P. J. G. (2002). A review of evidence of health benefit from articial neural networks in medical intervention. *Neural Networks* 15(2002): 11-39.

Lusted, L. B. (1960). Computer programming of diagnostic tests. *IRE Transactions on Medical Electronics* ME-7: 255-258.

Mango, L. J. (1994). Computer-assisted cervical cancer screening using networks. *Cancer Letters* 77: 155-162.

Mango, L. J. (1997). Clnical validation of interactive cytologic rescreening: automating the search, not the interpretation. *Acta Cytologica* 41(1): 93-97.

Mango, L. J. and P. T. Valente (1998). Neural-network-assisted analysis and microscopic rescreening in persumed negative cervical cytologic smears: A comparison. *Acta Cytologica* 42(1): 227-232.

McAdam, W., B. Brock, et al. (1990). Twelve years' experience of computer-aided diagnosis in a district general hospital. *Ann R Coll Surg Engl* 72(1): 140-6.

Miller, R., F. E. Masarie, et al. (1986). Quick medical reference (QMR) for diagnostic assistance. M.D. Computing. 3: 34-48.

Miller, R. A. (1994). Medical diagnostic decision support systems - past, present, and future: A threaded bibliography and commentary. *Journal of the American Medical Informatics Association* 1: 8-27.

Miller, R. A. (1996.). Evaluating evaluations of medical diagnostic systems. *Journal of the American Medical Informatics Association* 3: 429-431.

Miller, R. A. and F. E. Masarie (1989). Use of the Quick Medical Reference (QMR) program as a tool for medical education. *Methods of Information in Medicine* 28: 340-345.

Miller, R. A., H. E. J. Pople, et al. (1982). Internist-1, an experimental computer-based diagnostic consultant for general internal medicine. *New England Journal of Medicine* 307: 468-476.

Musen, M. A. (1998). Domain ontologie in software engineering: Use of PROTEGE with the EON architecture. *Methods of Information in Medicine* 37(4-5): 540-550.

Musen, M. A., Y. Shahar, et al. (2001). Clinical decision support systems. *Medical Informatics: Computer Applications in Health Care and Biomedicine*. E. H. Shortliffe, L. E. Perreault, G. Widerhold and L. M. Fagan, Springer: 573-609.

Nashef, S. A. M., F. Roques, et al. (1999). European system for cardiac operative risk evaluation (EuroSCORE). *European Journal of Cardio-thoracic Surgery* 16(1999): 9-13.

Ohno-Machado, L., J. H. Gennari, et al. (1998). The guideline interchange format: a model for representing guidelines. . *Journal of the American Medical Informatics Association* 5(4): 357-72.

Ozbolt, J. G. and S. Bakken (2001). Patient care systems. *Medical Informatics: Computer Applications in Health Care and Biomedicine*. E. H. Shortliffe, L. E. Perreault, G. Widerhold and L. M. Fagan, Springer: 421-442.

Parsonnet, V., D. Dean, et al. (1989). A method of uniform stratification of risk for evaluating the results of surgery in acquired adult heart disease. *Circulation* 79(6 Pt 2): I3-12.

Patil, R. S., P. Szolovits, et al. (1982). Modeling knowledge of the patient in acid-base and electrolyte disorders. *Artificial Intelligence in Medicine*. P. Szolovits. Boulder, Colorado, Westview Press,. 51: 187-222.

Pauker, S. G., G. A. Gorry, et al. (1976). Towards the simulation of clinical cognition: Taking a present illness by computer. *American Journal of Medicine* 60: 981-996.

Pearl, J. (1988). *Probabilistic Reasoning in Intelligent Systems: Networks of Plausible Inference.* San Mateo, CA, Morgan Kaufmann.

Peleg, M., A. A. Boxwala, et al. (2001). Sharable representation of clinical guidelines in GLIF: relationship to the Arden Syntax. *Journal of Biomedical Informatics* 34(3): 170-81.

Sageder, B., K. Boegl, et al. (1997). The knowledge model of MedFrame/CADIAG-IV. *Stud Health Technol Inform* 43 Pt B: 629-33.

Shortliffe, E. H. (1976). *MYCIN: Computer-based Medical Consultations.* New York, American Elsevier.

Sim, I., P. Gorman, et al. (2001). Clinical decision support systems for the practice of evidence-based medcine. *Journal of the American Medical Informatics Association* 8(6): 527-534.

Sim, I., D. Owens, et al. (2000). Electronic trial banks: a complementary method for reporting randomized trials. *Med Decis Making.* 20(4): 440-50.

Sowa, J. F. (1984). *Conceptual Structures: Information Processing in Mind and Machine.*, Addison-Wesley.

Szolovits, P. (1982). Artificial Intelligence in Medicine. *Artificial Intelligence in Medicine*. P. Szolovits. Boulder, Colorado, Westview Press,. 51: 1-19.

Tan, J. J. H., Ed. (1998a). *Health Decision Support Systems.*, Aspen.

Tan, J. J. H. (1998b). An introduciton to health decision support systems: Definition, evolution, and framework. *Health Decision Support Systems.* J. J. H. Tan, Aspen: 3-32.

Tan, J. J. H., D. Uyeno, et al. (1998). The art of macro-micro view: Strategic planning for health decision and executive support systems. *Health Decision Support Systems.* J. J. H. Tan, Aspen: 265-285.

Teodorescu, H. N., A. Kandel, et al., Eds. (1999a). *Fuzzy and Neuro-Fuzzy Systems in Medicine*, CRC Press.

Teodorescu, H. N., A. Kandel, et al. (1999b). Fuzzy Logic and Neuro-Fuzzy Systems in Medicine and Bio-Medical Engineering: A Historical Perspective. *Fuzzy and Neuro-Fuzzy Systems in Medicine.* H. N. Teodorescu, A. Kandel and L. C. Jain, CRC Press: 3-16.

WHO (1992). International classiffcation of diseases index. Tenth revision. Volume 1: Tabular list. Geneva, The World Health Organization,.

Zadeh, L. A. (1969). Biological applications of thc theory of fuzzy sets and systems. *Biocybernetics of the Central Nervous System.* L. D. Proctor. Boston, MA, Little Brown & Co.: 199-212.

Zhu, A. L. and T. Y. Leong (2001). Automating dynamic decision model construction to support clinical practice guideline development. *Computer-Based Support for Clinical Guidelines and Protocols.* B. Heller, M. Loffler, M. Musen and M. Stefanelli, IOS Press: 133-148.

A View of Public and Private Sectors for Taiwan's BOT Project Financing Using Fuzzy Multi-Criteria Methods

Jao-Hong Cheng

Department and Graduate School of Management Information System
National Yunlin University of Science and Technology
Yunlin, Taiwan, ROC
E-mail: jhcheng@mis4k.mis.yuntech.edu.tw

Abstract. This chapter discusses financial criteria of build-operate-transfer (BOT) projects from the public and the private sectors. The private sector includes sponsor companies, contractors, lenders and investors. Depending on the size and complexity of the BOT transportation projects in Taiwan, the financing phases of projects are divided into six segments for discussion. They are "lifetime phase", "development phase", "contract specified phase", "construction phase", "operation phase" and "transfer phase". To choose theoretical financial criteria of the BOT projects in uncertain situations, this chapter suggests a two-stage selection procedure:

(1) Selecting criteria: we employ the "fuzzy Delphi method via fuzzy integration" to expert advice to select twenty-two important criteria.
(2) Estimating the weight of the segments and criteria: we use the "fuzzy analytic hierarchy process" to manage the uncertainty in the decision environments and to calculate the relative importance of segments and criteria. The results show that the public and private sectors weigh segments and criteria differently which is helpful when trying to understand their distinct concerns.

Keywords: BOT project, financial risk, fuzzy delphi method, fuzzy integration, fuzzy analytic hierarchy process

1. Introduction

The build-operate-transfer (BOT) model of project development is being increasingly used by governments across a number of sectors in their drive to privatize major public projects. It gives governments the best of both worlds—the benefit of more infrastructure projects being built, without the burden of additional public borrowing [20]. The acronym "BOT" was first coined in the early 1980s by Turkey's late Prime Minister Targut Ozal. Adoption of the concept dates back to the mid-1950s in Hong Kong when a privatized cross-harbor vehicular tunnel was first proposed. Since the end of the nineteenth century and throughout most of the twentieth century most of the financing sources of large public "infrastructure" projects in the industrialized world have been from the private sectors. A UK-based journal has constantly monitored privatization matters and valued some US$54 billion worth of concessions in 1994 compared with US$23 billion in 1990 [21].

Over the last decade, it became clear that governments globally had major fiscal shortcomings in funding public works. Total privatization could provide a good solution for BOT projects in certain countries. The concept lends itself to whole or complete sectors of a country's infrastructure such as power or water supply and distribution, ports, railways, airports or motorways.

Table 1. Different phases of the BOT projects in the literature

Phase / Researchers	Develop-ment	Contract specified	Construction	Operation	Transfer	Lifetime
Tiong (1990)			√	√		
Beidleman, et al. (1990)	√		√	√		
Woodward, et al. (1992)						√
Yaworsky (1994)	√	√	√	√	√	
Wong(1997)	√	√	√	√	√	
Chen(1997)	√	√	√	√	√	√

The BOT project financing analysis traditionally embraces qualitative and quantitative techniques under conditions of certainty. Qualitative and quantitative analyses are the means of identifying the main risk sources or factors. A list of them is prepared as an assessment of the consequences on cost, risk and finance. But group decisions on BOT projects inherently reflect the subjectivity, fuzziness and imprecision of human thought and judgment in the face of uncertainty in the decision environments. Extant approaches such as mathematical programming and simulation models cannot solve these problems satisfactorily. In order to resolve the difficulty of choosing financial criteria theoretically in the face of uncertainty in the decision environments, this chapter utilizes fuzzy multi criteria methods.

An idea of how such a list can be compiled may be drawn from the classification schemes suggested by [2, 4, 19, 22, 23, 24], given in Table 1. Depending on the size and complexity of the project, it is desirable to divide each project into assessment phases. In this chapter, the financial criteria of the BOT transportation projects are divided into six phases for discussion. They are "lifetime phase", "development phase", "contract specified phase", "construction phase", "operation phase" and "transfer phase".

2. Literature Review

2.1 Financial factors of BOT projects

Numerous financial factors may affect the success of BOT projects. For example, the level of proposed tolls, financial return to the government, equity contribution by the sponsors and financial strengths of the sponsors [21]. In addition, the government and the private sectors can take into account various risk factors in BOT project financing analysis. Examples of such factors include credit risk, currency risk, funding risk and cost overrun risk [2, 19, 23]. Based on the previous literature and the suggestions from lead-experts, a questionnaire was designed containing fifty-nine questions to capture the concept of public and private sectors. The schematic layout of the relationships of the principal parties in a BOT project is shown in Figure 1.

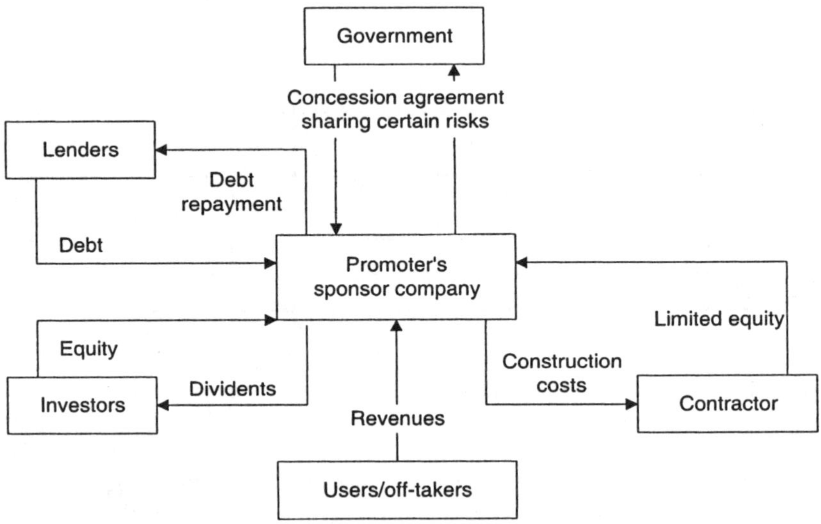

Figure 1. The schematic layout of the relationship of the principal parties of a BOT project (source: Walker and Smith, 1995)

The promoter's sponsor company that is a government department calling to make submissions can be included in a BOT project. A contractor approaches the government with the concept or, alternatively, is a financial group, often instigated by a merchant bank. The private company formed to accept the franchise is named the 'sponsor' and their primary mission is to manage the BOT project

2.2 Methodology

2.2.1 Fuzzy Delphi method via fuzzy integration

Fuzzy set theory has been proven a powerful modeling tool for coping with fuzziness [25]. It has received increasing attention from researchers in different fields. The traditional Delphi method is an effective method of enabling the selection of important criteria by converging a possibility value through the feedback mechanism of the questionnaires. However, the more surveys we conduct, the more costly they become. A fuzzy Delphi method that takes into account fuzziness in selecting important criteria has been proposed. This method applies a three-point estimation to find a given convergence criterion. Furthermore, the fuzzy Delphi method via fuzzy integration employs the professionalism of the experts [9]. For practical reasons, this chapter replaces the concept of maximum possibility and minimum possibility value [9] by maximum (optimistic) and minimum (pessimistic) value in the possibility range.

Depending on the size and complexity of Taiwan's BOT transportation projects, we divided the projects into six phases. Considering the professional nature of financial criteria required by experts, the fuzzy Delphi method via fuzzy integration is used to select criteria considered important by the experts.

A step-by-step description of the algorithms for obtaining values via fuzzy integration is as follows:

Step 1. Construct membership functions

For each criterion obtained from the questionnaire, we construct a maximum (optimistic) membership function $h_t(x)$ and a minimum (pessimistic) membership function $h_u(x)$. $h_i(x)$ is the synthetic function of $h_t(x)$ and $h_u(x)$, where $h_i(x)=\min\ (h_t(x), h_u(x))$, i stands for the ith criterion, as shown in Figure 2.

Step 2. $h_i(x)$ is rearranged in decreasing order

$$h_i(x)=\min\ (h_t(x), h_u(x)) \tag{1}$$

Step 3. $H_i = g_i + H_{i-1} + \lambda g_i \bullet H_{i-1},\ H_1 = g_1$ (2)

is to be computed where:

g_i : degree of professionalism of each expert , i =1,2,3,...,n , i stands for the ith criterion.

H_i : accumulating commitment degree of g_i.

Step 4. Calculate fuzzy integration value u_i due to.

$$u_i = \int h_i \bullet H_i = \bigvee_{i=1}^{n} [h_i \wedge H_i] \tag{3}$$

Repeating Step 1 to step 3, calculate fuzzy integration values of each criterion. The membership function x is the maximum importance value when choosing the maximum fuzzy integration value u_i .

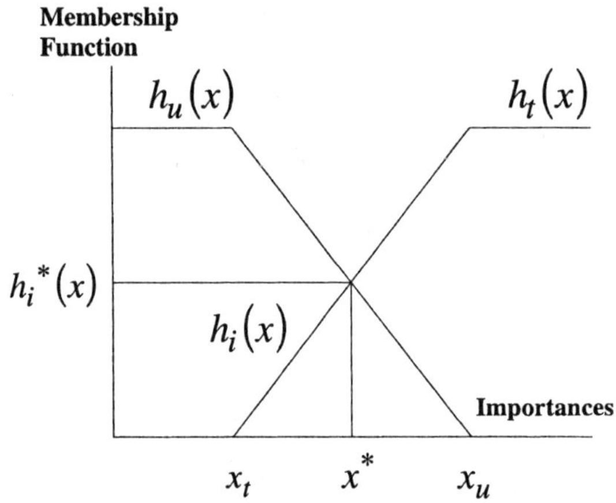

Figure 2. Membership function of the ith criterion

2.2.2 Fuzzy Analytic Hierarchy Process (AHP)

A method of weighting BOT's important financial criteria is developed from the fuzzy analytic hierarchy process suggested by [3, 16]. A step-by-step description of the fuzzy analytic hierarchy process algorithms is as follows:

Step 1. Construct the hierarchy structure

The hierarchy structure is constructed on three levels. The first level (objective) is the best BOT transportation project financing. The second level (segment) consists of different phases. The third level (criterion) consists of the important criteria of the BOT transportation project financing.

Step 2. Develop the pairwise comparison matrix

In any segment of the second level, the experts evaluate the relative importance value of the pairwise criterion. Develop the pairwise comparison matrix by calculating the relative importance values.

Step 3. Calculate the triangular fuzzy number

Owing to subjectivity, fuzziness and the imprecision of the experts' thoughts, it is suitable to ascertain the comprehensive opinion of experts by fuzzy number in pairwise comparison matrix. Robbins suggests combining the experts' opinions by the triangular fuzzy number, being appropriately describing experts' the fuzzily relative important value of pairwise criteria. Calculate the triangular fuzzy number, by

$$\tilde{\alpha}_{ij} = \left(\alpha_{ij}, \beta_{ij}, \delta_{ij}\right)_{L-R} \tag{4}$$

where:

$\tilde{\alpha}_{ij}$: triangular fuzzy number

α_{ij} : minimum value of the jth criterion of the ith segment

β_{ij} : geometric average of the jth criterion of the ith segment

δ_{ij} : maximum value of the jth criterion of the ith segment

L-R: fuzzy range of triangular fuzzy number

Step 4. Construct the fuzzy positive reciprocal matrix

Using the result of step 3, we construct the fuzzy positive reciprocal matrix A

$$\begin{aligned} A &= \left(\overline{\alpha}_{ij}\right) \\ \tilde{\alpha}_{ij} &= \left(\alpha_{ij}, \beta_{ij}, \delta_{ij}\right) \end{aligned} \tag{5}$$

Step 5. Calculate the fuzzy weights of the fuzzy positive reciprocal matrix

Gets critical value from experts and combines the experts' opinions by calculating geometric averages [17]. Because needlessly inputted experts' fuzzy value, it easily considered consistency and normalization of group decision. Calculate the fuzzy weights by equation 6.

$$Z_i = (\tilde{\alpha}_{i1} \otimes ...... \otimes \tilde{\alpha}_{in})^{1/n}, \forall i \tag{6}$$
$$\overline{W_i} = Z_i \otimes (Z_1 \otimes ...... \otimes Z_n)^{-1}$$

where: Z_i: geometric average of the triangular fuzzy number

Step 6. Defuzzification

To translate the aggregate fuzzy set obtained by the inference process into an executable action, a defuzzification method is required. The defuzzification of a fuzzy set results in a single scalar value. Although there exists no optimal method, we can choose an appropriate one based on an axiomatic and/or empirical justification and the computational effort required [25]. The center of gravity method or centroid method is used because (1) it has been predominant in the literature [10], and (2) it is computationally fast. Calculate, when the weight is a triangular fuzzy number

$$W_i = \frac{W_{\alpha i} + W_{\beta i} + W_{\delta i}}{3} \tag{7}$$

where:

$W_{\alpha i}$: the lower bound of the fuzzy weight

$W_{\beta i}$: the fuzzy weight which sums to 1

$W_{\delta i}$: the upper bound of the fuzzy weight

Step 7. Normalization

To compare easily the importance of the various segments and the criteria, this chapter normalizes the weight after defuzzification step by

$$NW_i = \frac{W_i}{\sum_{i=1}^{i=n} W_i} \tag{8}$$

Step 8. Hierarchy connection

After completing step 1 through step 7, we calculate the weight of the ith segment and let the sum of the weights of the entire segments equal 1. We also calculate the weights of the pth criterion and let the sum of the weights of all the criteria equal 1. Calculate the weight of the pth criterion under the second level, by hierarchy connection using

$$NW_k = NW_i \times NW_{ip} \tag{9}$$

3.Experimental Results

A questionnaire was sent out to thirty-five experts in Taiwan. Twenty-five effective responses were received from six professors, six government officials, seven members of the sponsor companies and the contractors, and six members of the lenders and the investors groups. It is suggested by [16] that the number of experts for a group decision context is between five and seven persons. The qualifications of the experts are shown in Table 2.

Table 2. Qualification of the experts

Category	Qualification
The academic	a. Professor. b. Experienced in BOT transportation projects.
Government	a. Director or Vice-director. b. Experienced in BOT transportation projects.
Sponsor companies and contractors	a. Manager. b. Experienced in BOT transportation projects.
Lenders and investors	a. Manager. b. Experienced in BOT transportation projects.

3.1 Important financial criteria for BOT transportation projects

The intervals of possible values obtained from the experts are used to construct their importance values. The importance values are greater than 90 for selecting important criteria that are obtained by using the fuzzy integration method algorithm.

Using inflation as an example, we explain the fuzzy integration method algorithms as follows. The triangular membership function is used to describe the expert's opinion. In this chapter, the lower bound is a linguistic estimation of "least important", and the upper bound is a linguistic estimation of "very important".

Step 1. Establish a membership function for each subject with respect to the importance value

Each identified membership function is shown as $h_t\ (x)$ *and* $h_u\ (x)$, respectively. Thus their synthetic function is $h_i\,(x)$=min $(h_t\,(x),\ h_u\,(x))$. This chapter has constructed triangular membership functions from twenty-five responses to the inflation items, as shown in Figure 3.

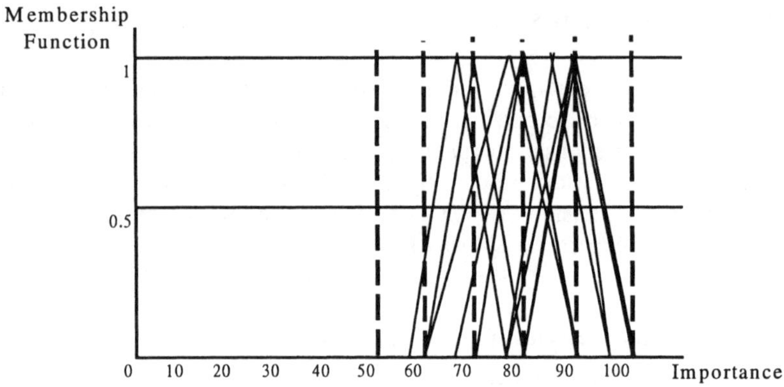

Figure 3. Triangular membership function obtained from the twenty-five experts

The following membership function values are extracted from the triangular membership functions when we apply these triangular membership functions to fuzzy integration, as shown in Table 3.

Table 3. Membership function values extracted from triangular mmbership functions

Importance	Membership function value
50	0
60	0.5,0
70	1,0.64,0.5,0.43,0
80	1,0.8,0.7,0.51,0.3,0
90	1,0.8,0.69,0.50,0
100	0

Step 2. Calculate the maximum importance value of each criterion

Calculate a fuzzy integration value of each criterion and choose the maximum importance value. The algorithm is as follows:

a. Arrange the membership function value $h_i(x)$ in decreasing order.
b. Change Professional degree "H" of experts to [0 1] and order H increasingly.
c. Calculate the fuzzy integration value u_i, by equation 3.

The membership function at the intersection of the two curves is 0.67. It is the solution, as shown in Figure 4.

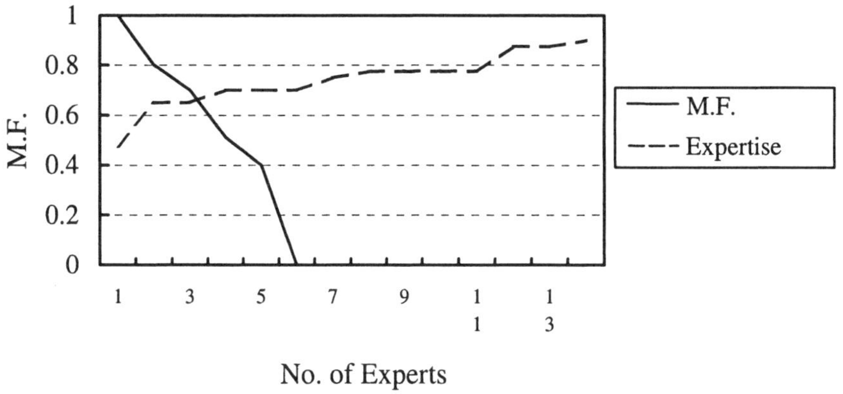

Figure 4. The two curves intersection point when membership value is 80

Repeating a. to c., calculate fuzzy integration values of each criterion. The membership function x is the maximum importance value when choosing the maximum fuzzy integration value u_i. After step 1 to 3, the importance value of inflation scores at "90", as shown in Figure 5. Table 4 shows the results of each criterion on the important value 90.

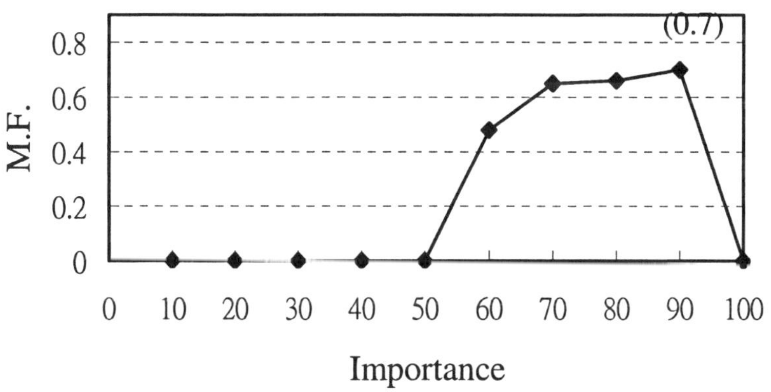

Figure 5. Fuzzy integration values of inflation

Table 4. Important financial criteria of the BOT transportation projects

Objective	Segments	Important financial criteria
The best BOT transportation project financing	Lifetime phase	Inflation
		Foreign exchange rate fluctuations
		Financial risk
		IRR
		ROI
		Sponsor company self compensation
		Social /economics efficiency
	Development phase	Level of proposed toll
		Willingness of loan bank
		Sponsor company's professional finance ability
		Source of loan
		Long-term loan
		Currency risk
		Financial availability analysis
	Contract specified phase	Regulation and audit of financial ratio
		Credit risk
		Funding risk
		Fund of share
		Tax benefit and deduction
	Construction phase	Completion guarantees
	Operation phase	Contractor financial distress in the beginning of operation
	Transfer phase	Cost overrun risk

3.2 Weights of segments

The financial criteria of the BOT transportation projects could be divided into six segments for discussion. They are "lifetime phase", "development phase", "contract specified phase", "construction phase", "operation phase" and "transfer phase". The relative importance values of the pair segments and criteria are obtained by interviewing the experts. The fuzzy analytic hierarchy process, as indicated in Table 5, acquires the results.

According to Table 5, the academic, sponsor companies and contractors most emphasize the "development phase". Both the lenders and the investors place the emphasis on "lifetime phase". The government stresses the "transfer phase". In aggregation, all related parties of the BOT transportation projects emphasize "development phase".

3.3 Weight of important financial criteria

The lifetime phase of the BOT transportation projects can be divided into seven criteria for discussion. They are "inflation", "foreign exchange rate fluctuations", "financial risk", "IRR", "ROI", "sponsor company self compensation", and "social /economics efficiency". The fuzzy analytic hierarchy process determines the relative importance values, shown in Table 6.

Table 5. The relative importance value of the BOT transportation project's segments

Category / Phase	The academic	Government	Sponsor companies and contractors	Lenders and investors	All
Lifetime	0.129	0.114	0.119	0.432	0.163
Development	0.252	0.091	0.240	0.227	0.195
Contract specified	0.135	0.178	0.163	0.060	0.152
Construction	0.208	0.127	0.196	0.110	0.170
Operation	0.201	0.159	0.163	0.120	0.187
Transfer	0.075	0.331	0.119	0.051	0.133
Aggregation	1	1	1	1	1

Table 6. The relative importance value of lifetime phase's criteria

Category / Criteria	The academic	Government	Sponsor companies and contractors	Lenders and investors	All
Inflation	0.201	0.380	0.123	0.323	0.177
Foreign exchange rate fluctuations	0.104	0.143	0.152	0.128	0.149
Financial risk	0.167	0.063	0.122	0.035	0.144
IRR	0.106	0.096	0.182	0.197	0.148
ROI	0.125	0.068	0.157	0.067	0.119
Sponsor company self compensation	0.180	0.078	0.124	0.152	0.145
Social /economics efficiency	0.117	0.172	0.140	0.098	0.118
Aggregation	1	1	1	1	1

According to Table 6, the academic, government, lenders and investors emphasize "inflation". The sponsor companies and contractors emphasize "IRR". In aggregation, all the related parties of the BOT transportation projects stress "inflation".

The development phase of the BOT transportation projects can be divided into seven criteria for discussion. They are "level of proposed toll", "willingness of loan bank", "sponsor company's professional finance ability", "source of loan", "long-term loan", " currency risk", and "financial availability analysis". The fuzzy analytic hierarchy process determines the relative importance values, shown in Table 7.

Table 7. The relative importance value of development phase's criteria

Category / Criteria	The academic	Government	Sponsor companies and contractors	Lenders and investors	All
Level of proposed toll	0.118	0.389	0.252	0.247	0.217
Willingness of loan bank	0.125	0.093	0.127	0.082	0.125
Sponsor company's professional finance ability	0.241	0.157	0.093	0.062	0.142
Source of loan	0.216	0.091	0.102	0.164	0.147
Long-term loan	0.069	0.101	0.147	0.191	0.115
Currency risk	0.120	0.087	0.085	0.116	0.121
Financial availability analysis	0.111	0.082	0.194	0.138	0.133
Aggregation	1	1	1	1	1

According to Table 7, the academic emphasizes the "sponsor company's professional finance ability". The government, sponsor companies and contractors, lenders and investors emphasize the "level of proposed toll". In aggregation, all the related parties of the BOT transportation projects stress the "level of proposed toll".

The contract specified phase of the BOT transportation projects can be divided into five criteria for discussion. They are "regulation and audit of financial ratio", "credit risk", "funding risk", "fund of share" and "tax benefit and deduction". The fuzzy analytic hierarchy process determines the relative importance values, shown in Table 8.

Table 8. The relative importance value of contract specified phase's criteria

Category / Criteria	The academic	Govern-ment	Sponsor companies and contractors	Lenders and investors	All
Regulation and audit of financial ratio	0.241	0.202	0.244	0.313	0.255
Credit risk	0.291	0.250	0.205	0.114	0.201
Funding risk	0.272	0.127	0.179	0.280	0.195
Fund of share	0.111	0.093	0.130	0.207	0.184
Tax benefit and deduction	0.086	0.328	0.242	0.086	0.165
Aggregation	1	1	1	1	1

According to Table 8, the academic emphasizes "credit risk". The government emphasizes "tax benefit and deduction". The sponsor companies and contractors, lenders and investors emphasize "regulation and audit of financial ratio". In aggregation, all the related parties of the BOT transportation projects stress "regulation and audit of financial ratio".

The construction phase of the BOT transportation projects has only one criteria, "completion guarantees", for discussion. The relative importance value, shown in Table 5, is equal to the weight of the construction segment.

The operation phase of the BOT transportation projects has only one criteria, "contractor financial distress in the beginning of the operation" for discussion. The relative importance value, shown in Table 5, is equal to the weight of the operation segment.

The transfer phase of the BOT transportation projects has one criteria, "cost overrun risk", for discussion. The relative importance value, shown in Table 5, is the same as the weight of the transfer segment.

4. Conclusion

The interest groups of the BOT transportation projects include academic, sponsor companies and contractors, and lenders and investors. Group decisions on the BOT transportation projects inherently reflect the subjectivity, fuzziness and imprecision of human thought and judgment in the face of uncertainty in the decision environments. Extant approaches such as mathematical programming and simulation models cannot deal with these problems satisfactorily.

This chapter uses the fuzzy multi-criteria methods to resolve the difficulty of choosing financial criteria theoretically in the face of uncertainty in the decision environments. Significant contributions achieved by this chapter include the following:

- Depending on the size and complexity of Taiwan's BOT transportation projects, they are divided into six phases for discussion, including "lifetime phase", "development phase", "contract specified phase", "construction phase", "operation phase" and "transfer phase".
- This chapter employs the professionalism of the experts and has then used the fuzzy Delphi method via fuzzy integration to select twenty-two important criteria from the experts.
- It solves the difficulty of choosing criteria theoretically in a decision environment replete with uncertainty and then calculates the relative importance of segments and criteria by using the fuzzy analytic hierarchy process.
- The results demonstrate that the government and the private sectors weigh segments and criteria differently. This is helpful when trying to understand the distinct concerns of the public and private sectors. As the analysis structure of this chapter becomes better understood, it should become increasingly easy to pre-determine factors that create success.
- This chapter calculates the most important criteria from all the members of BOT transportation projects. The results are useful for BOT transportation project financing.

References

1. Al-Bahar, J.F. and Crandall, K.C. (1990), "Systematic Risk Management Approach for Construction Projects", Journal of Construction Engineering and Management, 116(3), pp. 533-546.
2. Beidleman, C.R. *et al.* (1990), "On Allocating Risk: the Essence of Project Finance", Sloan Management Review, pp. 47-55.
3. Buckly, J. J. (1985), "Fuzzy Hierarchical Analysis", Fuzzy Sets and Systems, 17, pp. 233-247.
4. Chen, Y.H. (1997), Financial Operation and Case Studies of Private Participation in Public Construction (BOT model), Dissertation of the Graduate Institute of Management Science, National Chiao Tung University.
5. Haley, G. (1992), "Private Finance for Transportation and Infrastructure Projects", International Journal of Project Management, 10(2), pp. 63-68.
6. Hayes, R. W., Perry, J.G., Thompson, P.A., and Willmer, G. (1986), Risk Management in Engineering Construction. London: Thomas Telford.
7. Hayfield, F. (1986), "Project Successes and Failure", Project Management in Progress: Tool and Strategies for the 90s, Edited by M.C. Grool, C. Visser, W.J. Vriethoff, and G. Wijnen, (North-Holland: Elsevier Science Publishers B. V.), pp. 31-36.

8. Hensley, M.L. and White, E.P. (1993), "Integrating Build-Operate-Own and Build-Operate-Transfer Techniques with in the national Privatization Strategy", Columbia Journal of World Business, Spring, pp.70-82.
9. Ishikawa, A., Amagasa, T., Tamizawa, G., Totsuta, R. and Mieno, H. (1993), "The Max-Min Delphi Method and Fuzzy Delphi Method via Fuzzy Integration", Fuzzy Sets and Systems, 55, pp.241-253.
10. Klir, G.J. and Yuan, B. (1995), "Fuzzy Sets and Fuzzy Logic-Theory and Applications", Prentice Hall.
11. Mustafa, M.A. and Al-Bahar, J.F. (1991), "Project Risk Assessment Using the Analytic Hierarchy Process", IEEE Transactions on Engineering Management, 38(1), pp. 46-52.
12. McCarthy. S.C., and Tiong, R.L.K. (1991), "Financial and Contractual Aspects of BOT Projects", International Journal of Project Management, .9(4), pp. 222-227.
13. Petry , G.H. (1993), and James Sprow, "The Theory and Practice of Finance in the 1990s", The Quarterly Review of the Economics and Finance, 33(4), pp. 359-381.
14. Perry, J.G. and Hayes, R.W. (1985), "Risk and its Management in Construction Projects", Proceedings of Civil Engineering, Part 1, 7. (June), pp. 499-521.
15. Pindyck, R.S. (1993), "Investments of uncertain cost", pp. 53-76.
16. Robbins, S.P. (1994), Management, Prentice-Hall.
17. Saaty, T.L. (1980),The Analytic Hierarchy Process, McGraw-Hill.
18. Saaty, T.L. and Vargas, L.G. (1982), Decision Making for Leaders, McGraw-Hill.
19. Tiong, R.L.K. (1990), "BOT Projects: Risks and Securities," Construction Management and Economics, 8, pp. 315-328.
20. Tiong, R.L.K., Yeo, K.-T., and McCarthy, S.C. (1992),"Critical Success Factors in Winning BOT Contracts", Journal of Construction Engineering and Management, ASCE, 118 (2), pp. 217-228.
21. Walker, C. and Smith, A.J. (1995), Privatized Infrastructure: The Build Operate Transfer Approach, Thomas Telford.
22. Wong, P.C. (1997), A Study on Concession Agreement and Risk Management of the BOT Model-Private Sectors, Dissertation of the Graduate Institute of Business Administration, National Taiwan University.
23. Woodward, D. *et al.* (1992), Risk Management of BOT Projects, Project, pp. 10-11.
24. Yaworsky, R.A. (1994), "Risk Planning for Large and BOT Projects: A Holistic Framework", University of British Columbia, Canada, pp. 66-74.
25. Zimmermann, H.J. (1996), Fuzzy Set Theory and Its Applications-Third edition, Boston: Kluwer Academic Publishers.

Relational Structures for the Analysis of Decision Information in an Electronic Market

Bartel van der Walle

Information Systems Department
New Jersey Institute of Technology
College of Computing Sciences
Newark, NJ 07102
USA.
E-mail: bartel@njit.edu

Abstract. Electronic markets are web-based information systems that provide communication and decision support for online business activities of buyers and sellers. This paper focuses on a relatively small but active electronic job market for students (the buyers) and companies (the sellers) who had registered for a forthcoming university career fair. In advance to the career fair, registered students and companies were invited to log on to the job market, provide personal contact information, and create personalized software agents. The purpose of a software agent is twofold: to supply information to the other agents and, simultaneously, to demand information from these same agents. The information supplied or requested by an agent consists of actual values on pre-defined decision criteria. Every software agent is equipped with a multi-criteria decision model to process the information it has obtained from the other agents and produce a rank-ordered list of students or job offers. The purpose of this paper is to present a technique to analyze and synthesize the decision information available in the market, based on fuzzy relational compositions first proposed by Bandler and Kohout. We derive new relational structures among the decision information from the participants' use and exchange of the information in the market. We motivate the strategic importance of these compositions to achieve a better understanding of the marketplace, and develop a practical case to illustrate the analytical power of the relational compositions and the rich semantics they embed.

Keywords: electronic markets; decision support systems; multi-criteria decision analysis; fuzzy relational compositions.

1. Introduction

Electronic commerce, or ecommerce, covers the processes of buying, selling, or exchanging products, services, or information over computer networks, including the Internet (Turban et al. 1999). Electronic marketplaces, or emarkets, are ecommerce sites on the Internet that allow communities of buyers and suppliers to meet and trade with each other. They present ideal structures for commercial exchange, with the potential of achieving new levels of market efficiency by tightening and automating the relationships between buyer and seller (Ariba 2000).

Horizontal emarkets integrate across industries, providing generic products and functional services like office supplies, maintenance, and surplus asset liquidations. While eprocurement received most media coverage and stock markets' attention, the benefits of ecommerce and emarkets extend beyond the purchasing departments of organizations into several other functional areas, which include sales and marketing, inventory management, human resources and IT management (Weller 2000). Vertical markets focus on the whole value chain of a single industry, integrating over the entirety or parts of the chain from raw materials to the finished product. Setting up vertical markets requires a deep knowledge of the business area, and typically such markets will only thrive if a critical mass of sellers and buyers actively participates. Achieving liquidity indeed is one of the major challenges vertical markets currently face.

Markets, whether they are electronic or not, 'clear' by matching supply and demand. The process of matching buyers' demands with sellers' products has three main components: determining product offerings, search and price discovery (Bakos 1998). It is well known that in order to obtain and process this information, buyers incur search costs. A market with high search costs may prevent buyers from finding the best priced products. An electronic market however can provide electronic search tools to its participants, so that market information can be easily retrieved at a marginal search cost and buyers can swiftly compare prices of products offered by competing sellers. Therefore, electronic markets may lead to increasing price competition among sellers, potentially to the point of a price war in which most sellers ultimately have to abandon the market (Bakos 1997). One common strategy among sellers to counter the reduction in buyer search costs is to provide product information in addition to the price, such as the seller's reputation, a product quality label or product delivery times. As an effect of this strategy, buyers are now confronted with information overload making it difficult to assess and compare the value of sellers' products.

The increasing complexity of product offerings we witness in many electronic markets or ecommerce sites raises many important issues (Beam et al. 1999). For instance, how should complex information be visualized and interfaces designed to guide buyers in their search for the best products (Nielsen 2000). Or, closer to the topic of this volume, what decision support tools should a market provide to its participants so that they can define what really is the 'best' product for them in the first place.

In the following section, Section 2, we describe an electronic job market whose participants are students looking for a position and companies who are interested in recruiting them. The job market provides decision and negotiation support in the form of software agents who continuously scan the market for opportunities and assist in the negotiated exchange of additional information. We outline how this electronic market operates on a particular RFQ (Request for Quotes)/RFP (Request for Proposals) mechanism through which students and companies specify their quotes (resumes, resp. job offers) in terms of specific criteria, and how this information is used in the decision and negotiation support tools provided by the market.

We discuss the exchange of criteria-based information among buyers and sellers in Section 3, and point out the crucial importance of matching both parties' criteria. We illustrate why conventional matching may lead to limited information only, and we introduce the concept of non-symmetrical matching first proposed by Bandler and Kohout.

Section 4 provides the theoretical backbone of this paper. We present Bandler and Kohout's seminal development of symmetrical and non-symmetrical relational compositions for classical (or crisp) and fuzzy relations. A summary of the mathematics involved is given, leading to four important relational compositions. We present examples of such compositions in the case of our electronic job market, and indicate the surprisingly rich semantics of these compositions.

Section 5 provides a synthesis of the decision criteria as they are adopted by students in their resume descriptions. We develop a detailed computational account of the relational compositions, focusing on the use of criteria and the consequences of their selection in finding suitable job offers. We conclude in Section 6 with a summary of our findings and an indication of ongoing and future research on the behavioral analysis and synthesis of electronic markets.

2. Multi-Criteria Decision Support for an Electronic Job Market

For their annual career fair organized in November 2001, the New Jersey Institute of Technology (NJIT) decided to set up an electronic job market for the students and companies participating at the fair. Companies were invited to post job offers, while NJIT students could submit their resumes. The goal of the electronic job market was to allow both parties to explore their mutual offerings – jobs and resumes – in advance of the career fair. Students could identify those companies whose booth they would benefit from visiting, and companies could identify the students they would benefit from interviewing. In total, close to 350 students posted their resumes, while 33 companies posted 51 jobs. At the day of the career fair, more than 1,000 students attended and more than 80 companies participated. As these numbers indicate, a significant percentage of students and companies actively participated in the electronic job market.

Upon registering for the electronic job market, every market participant can provide basic contact information that can be seen by the other participants. They can also create software agents to aid them in their personal decision efforts. We distinguish two types of software agents: resume agents and job offer agents. Clearly, a resume agent is created by a student and represents his or her resume, while a job offer agent is created by a company and represents a particular job offer. While companies can create multiple job offer agents, a student can only specify one resume agent.

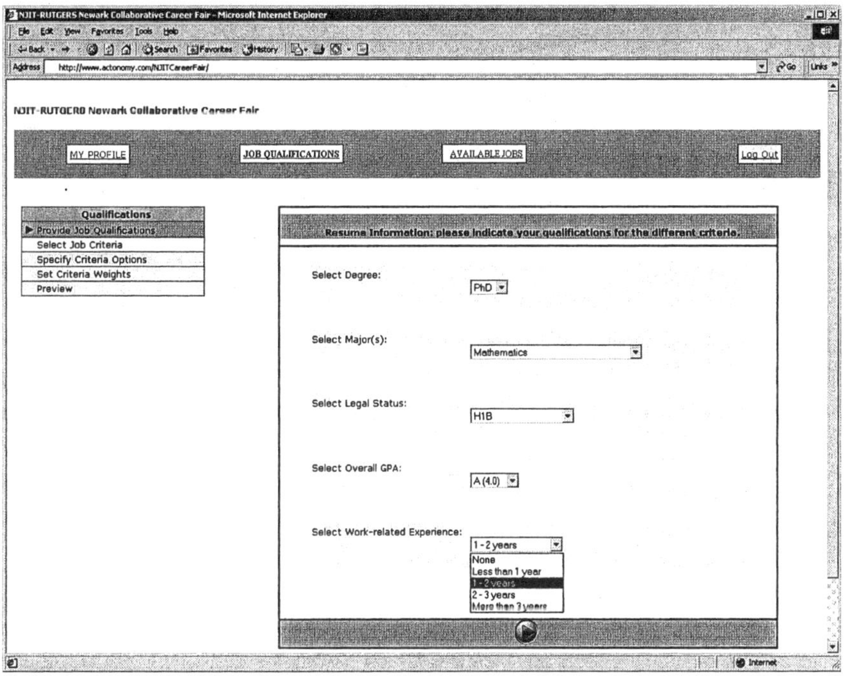

Figure 1. A student provides resume information while creating a resume agent

The purpose of a software agent is twofold: to supply information to the agents 'of the other kind' and, simultaneously, to demand information from these same agents. The information supplied or requested by an agent consists of actual values on pre-defined criteria. A resume agent supplies a student's values on resume criteria and demands a company's values on job offer criteria, and vice versa for a job offer agent. The market's resume and job offer criteria are given in Table 1.

For example, a student can decide not to provide information on the resume criteria *GPA* or *Degree*, but indicate his or her *Major* and *Work-related Experience* (Figure 1). The student could also choose to ask companies for information on the job offer criteria *Salary* and *Position Type*. Similarly, a company could provide *Salary* information, and ask for information on students' *GPA* or *Major* (Figure 2).

Table 1. The criteria available to student and job offer agents

Resume Criteria	Job Offer Criteria
Degree	Salary
Major	Industry Sector
Legal Status	Position Type
GPA (Grade Point Average)	
Work-related Experience	

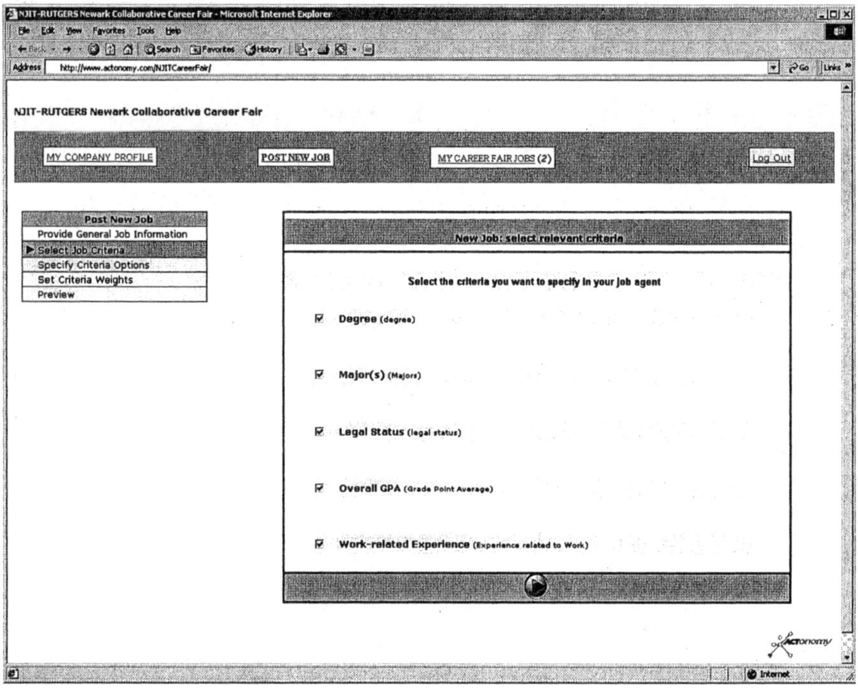

Figure 2. A company indicates on what criteria it needs information from students while creating a job offer agent

Every software agent is equipped with a multi-criteria decision model to process the information it has obtained from the other agents and produce a rank-ordered list of students or job offers.

Once created, the software agents roam the market to gather information on the criteria their creators have specified from all agents of the opposite kind. As such, the electronic job market can be viewed as an agent-based marketplace governed by a multi-criteria RFQ/RFP negotiation protocol (Lomuscio et al. 2000; Ströbel 2000; Van de Walle et al. 2001).

The software agents can be updated by their creators at any time to reflect changing preferences, with every modification immediately leading to a revised ranking. Often, we observed that students modified their preferences and used their software agents as a flexible tool to carry out a sensitivity analysis of the job offers in the market. This reflects how agents may influence human choice behavior in market-based activities (Maes et al. 1998).

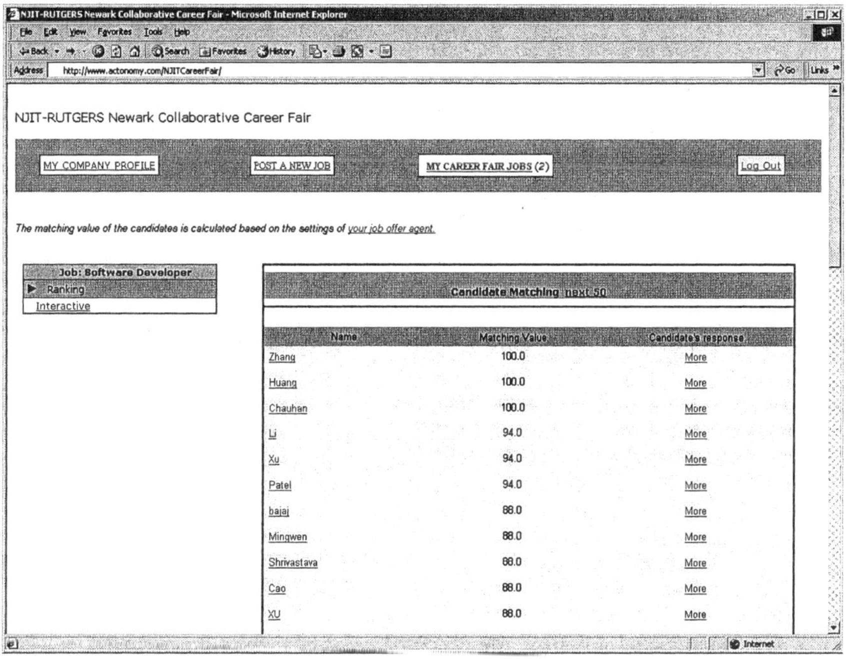

Figure 3. A company's view of the ranking of students produced by its job offer agent

As we can see in Figure 3, the list of rank-ordered companies or students generated by the agents contains hyperlinks to the specific job offer made or to the student contact information. In addition to the rank-ordered list view, the users can also launch an interactive Java applet through which the students or companies can change the weights of their resume and job offer criteria and instantly observe the changes in the ranking of job offers or students. This interactive decision support tool proved to be widely used by both students and companies to explore job offers or resumes.

3. Matching Resume and Job Offer Agents: An Information Retrieval Perspective

As markets need to clear by matching supply and demand, student resumes must be matched to job offers and vice versa. An electronic market such as NJIT's job market can in fact be viewed as an information system that uses the forces of supply and demand to coordinate business processes (such as deciding to interview a student, or deciding to visit a company booth at the career fair) among its participants (Malone 1987).

As indicated in Section 2, a software agent calculates the ranking of the opposing party's offers or resumes using a multi-criteria decision model. This information gives the participant an indication of how well his or her resume (or job offer) matches with the job offers (or resumes) in the market. Clearly, this matching indication is highly dependent on the overlap and number of criteria selected by the by both parties and the weight assigned to each of these criteria. For instance, if a company only requests information on the criterion *GPA* to rank order the students' resumes, only students with a high *GPA* will be at the top of the company's ranking. Or, conversely, even if a student has a low *GPA*, it does not matter much to companies who assign a low weight to the *GPA* criterion, and consequently the student may still end up high in these companies' ranking.

Therefore, providing additional information to the market participants about the criteria that are being used in their opponents' agents may be critically important to learn how well their (resume or job offer) agent matches with the other (job offer or resume) agents in the market. The problem we are confronted with here is in essence an information retrieval problem, as it is the goal of an information retrieval system to inform the user on the existence (or non-existence) of documents relating to his or her request.

To better appreciate the newly found nature of our problem, let us first present some basic information retrieval principles and express them in terms of set theoretical concepts (Bandler and Kohout 1984). Basically, information retrieval deals with a set of documents D, and the information contents of any document d belonging to D is characterized in a standard way, usually by means of some index terms (or keywords) selected from a term set T. A user may then formulate a query q to which the answer is a subset D_q of D, satisfying the information need expressed by this question. Mapping these concepts to the area of our concern, the matching of resumes and job offers, we introduce the following notation. We deal with a set of agents D, partitioned into a set of "Job offer agents" D_J and a set of "Resume agents" D_R, with obviously $D_J \cap D_R = \varnothing$. The term set T consists of criteria, and every agent in D can be completely described by T. In our case, either a resume agent in the set D_R or a job offer agent in the set D_J can be viewed as a query as well: when a candidate or a company creates an agent, the agent reports back a list of job offers or a list of student resumes satisfying to a varying degree the agent's information needs expressed in terms of the keywords in the set T.

In order to determine the subset of the set of all documents relevant to a user's query, we need to construct a retrieval matching function, i.e., a mapping from the set of queries to the corresponding documents. There are many examples of matching functions in the IR literature (e.g., Van Rijsbergen 1979), among which the following functions are most commonly used.

Let P be the subset of resume criteria specified in a resume agent, and Q the subset of resume criteria used by a job offer agent. Assume that the resume agent 'queries' the job offer agents, then some of the most popular matching functions M are:

the simple matching coefficient $|P \cap Q|$; Dice's coefficient $\frac{2|P \cap Q|}{|P|+|Q|}$; Jaccard's coefficient $\frac{|P \cap Q|}{|P \cup Q|}$; or the Overlap coefficient $\frac{|P \cap Q|}{\min(|P|,|Q|)}$. The latter can all be considered as normalized versions of the simple matching coefficient.

All of these matching functions are representing a symmetric relation. However, we may be interested in the containment of terms (criteria) that are not symmetric: while a resume agent may contain all of the criteria T_C of a job offer agent, the job offer agent need not necessarily contain all the criteria of the resume agent as given by T_C. This indicates that the commonly used symmetric matching functions are rather special relations, which restricts the retrieval in a rather undesirable fashion (Bandler and Kohout 1984), as we will illustrate in Section 5.

We introduce symmetric and non-symmetric relational compositions in the following section.

4. Fuzzy Relational Compositions

4.1 The composition of crisp relations

Assume for instance that we are dealing with the relation $\boldsymbol{R}$ from a set X to another set Y, and the relation $\boldsymbol{S}$ from the set Y to a third set Z. The circular (or round) composition of $\boldsymbol{R}$ and $\boldsymbol{S}$, denoted as $\boldsymbol{R} \circ \boldsymbol{S}$, is defined as

$$\boldsymbol{R} \circ \boldsymbol{S} = \{ (x,z) \mid \text{there exists an element } y \text{ in } Y \text{ such that } x\boldsymbol{R}y \text{ and } y\boldsymbol{R}z \}.$$

Recall that the $\boldsymbol{R}$-afterset of an element x of X, denoted $x\boldsymbol{R}$, consists of all elements of Y that are $\boldsymbol{R}$-related to x, and the $\boldsymbol{R}$-foreset of an element y of Y, denoted $\boldsymbol{R}y$, consists of all elements of X that are $\boldsymbol{R}$-related to y. We can then easily rewrite the composition condition in terms of fore- and aftersets as

$$R \circ S = \{ (x,z) \mid xR \cap Sz \neq \emptyset \},$$

or, in words, the intersection of the R-afterset of x and the S-foreset of z is not empty. The non-emptiness of the intersection of xR and Sz obviously is not a very strong condition. This makes it on the one hand easy to satisfy, but on the other hand it is not very informative since we don't know how many elements belong to the intersection or how "strong" the composition really is. Bandler and Kohout have introduced other types of compositions which each impose stronger conditions on the intersection $xR \cap Sz$, and hence are more difficult to satisfy, but when doing so, are much more informative. The first such composition is the so-called triangular sub-composition of R and S, denoted as $R \triangleleft S$, and defined as:

$$R \triangleleft S = \{ (x,z) \mid xR \cap Sz \neq \emptyset \text{ and } xR \text{ is a subset of } Sz \}.$$

In addition to the non-emptiness condition of the circular composition, the triangular sub-composition (or shortly sub-composition hereafter) requires that the R-afterset of x is a subset of the S-foreset of z. In the same spirit, the (triangular) super-composition of R and S, denoted as $R \triangleright S$, is defined as:

$$R \triangleright S = \{ (x,z) \mid xR \cap Sz \neq \emptyset \text{ and } Sz \text{ is a subset of } xR \}.$$

The super-composition switches the roles of xR and Sz, now requiring that the S-foreset of z is a subset of the R-afterset of x. Finally, the square composition of R and S, denoted as $R \square S$, is defined as:

$$R \square S = \{ (x,z) \mid xR \cap Sz \neq \emptyset \text{ and } xR = Sz \}.$$

Clearly, the square composition is the intersection of the super-composition and the sub-composition, i.e.,

$$R \square S = (R \triangleleft S) \cap (R \triangleright S),$$

meaning that the square composition is the most restrictive composition.

4.2. The composition of fuzzy relations

A fuzzy set is a very natural extension of a classical set, in that it extends binary set membership, i.e., “no membership” (0) or “membership” (1), to membership degrees taking values in the interval [0,1]. Classical or crisp sets are therefore special cases of fuzzy sets. Similarly, a fuzzy relation extends a classical or crisp relation, allowing for objects being related to a partial degree in the interval [0,1]. Our motivation to use fuzzy relations for our analysis is that, rather than only considering ordered pairs that are related or not, we are really interested in degrees of “relatedness”.

Let $\boldsymbol{R}$ be a fuzzy relation from X to Y, i.e., a mapping from $X \times Y$ into [0,1]. For every ordered pair $(x,y) \in X \times Y$ the quantity $\boldsymbol{R}(x,y)$, taking values in [0,1], is to be interpreted as the strength of the existing (fuzzy) $\boldsymbol{R}$-relation between x and y. As such, the $\boldsymbol{R}$-afterset of x associates to every $y \in Y$ the degree in which x is $\boldsymbol{R}$-related to y; while the $\boldsymbol{R}$-foreset of y associates to every $x \in X$ the degree in which x is $\boldsymbol{R}$-related to y;

Technically speaking, there are different ways to define the composition of two fuzzy relations. The most popular extension of the classical circular composition to the fuzzy case is the so-called max-min composition defined as follows. Let X, Y and Z be ordinary non-empty and finite sets, $\boldsymbol{R}$ a fuzzy relation from X to Y and $\boldsymbol{S}$ a fuzzy relation from Y to Z. The max-min composition of $\boldsymbol{R}$ and $\boldsymbol{S}$ is the fuzzy relation $\boldsymbol{R} \circ \boldsymbol{S}$ from X to Z defined by

$$x(\boldsymbol{R} \circ \boldsymbol{S})z = \max_{y \in Y} \min(\boldsymbol{R}(x,y), \boldsymbol{S}(y,z)),$$

for all $(x,z) \in X \times Z$. This means that the strength of the relation between x and z is defined as the strongest of all possible "connections" between x and z, where the strength of a connection is equal to the strength of its weakest connecting part.

The extension of the sub-, super- and square composition requires the introduction of additional mathematical operators. Indeed, recall for instance the definition of the sub-composition of the relation $\boldsymbol{R}$ from X to Y and the relation $\boldsymbol{S}$ from Y to Z:

$$\boldsymbol{R} \triangleleft \boldsymbol{S} = \{\, (x,z) \mid x\boldsymbol{R} \cap \boldsymbol{S}z \neq \emptyset \text{ and } x\boldsymbol{R} \text{ is a subset of } \boldsymbol{S}z \}.$$

Consider now the case where the relations $\boldsymbol{R}$ and $\boldsymbol{S}$ are fuzzy. Then the $\boldsymbol{R}$-afterset of x, $x\boldsymbol{R}$, and the $\boldsymbol{S}$-foreset of z, $\boldsymbol{S}z$, obviously are fuzzy sets in Y. The common definition of inclusion of the fuzzy set $x\boldsymbol{R}$ in Y in the fuzzy set $\boldsymbol{S}z$ in Y is given by

$$x\boldsymbol{R} \subseteq \boldsymbol{S}z \Leftrightarrow (\forall\, y \in Y)(\, x\boldsymbol{R}(y) \leq \boldsymbol{S}z(y)).$$

However, this definition represents an unconscious step backward to the realm of dichotomy as Bandler and Kohout already noted. Far more in accord with the spirit of fuzzy set theory, would be the assignment of a degree of subsethood of $x\boldsymbol{R}$ in $\boldsymbol{S}z$. One approach to achieve that objective is to define the binary fuzzy inclusion relation *Inc* for any two fuzzy sets A and B in X as follows:

$$Inc(A,B) = \frac{1}{n} \sum_{x \in X} \min(1 - A(x) + B(x), 1),$$

with n the cardinality of X (De Baets and Van de Walle, 1994). These fuzzy inclusion relations have been applied in the area of decision analysis and multi-criteria negotiations (Van de Walle 2002; Van de Walle and Faratin 2001; Van de Walle et al. 2001). Using this definition of fuzzy inclusion, we easily obtain the following definitions for the remaining compositions:

$$x_i(R \triangleleft S)z_j = \frac{1}{n}\sum_{y\in Y}\min(1 - x_iR(y) + Sz_j(y),1);$$
$$x_i(R \triangleright S)z_j = \frac{1}{n}\sum_{y\in Y}\min(1 + x_iR(y) - Sz_j(y),1);$$

with n the cardinality of Y, while for the square product it still holds that

$$x_i(\boldsymbol{R} \,\square\, \boldsymbol{S})\, z_j = x_i\, (\boldsymbol{R} \triangleleft \boldsymbol{S})\, z_j \cap x_i\, (\boldsymbol{R} \triangleright \boldsymbol{S})\, z_j,$$

with the intersection usually modeled by the minimum operator, i.e.,

$$x_i(\boldsymbol{R} \,\square\, \boldsymbol{S})\, z_j = \min(x_i\, (\boldsymbol{R} \triangleleft \boldsymbol{S})\, z_j,\, x_i\, (\boldsymbol{R} \triangleright \boldsymbol{S})\, z_j), \text{ or hence}$$

$$x_i(\boldsymbol{R} \,\square\, \boldsymbol{S})\, z_j = \frac{1}{n}\sum_{y\in Y}\min(\min(1 - x_iR(y) + Sz_j(y),1), \min(1 + x_iR(y) - Sz_j(y),1)).$$

The sub-composition can be interpreted as giving the mean degree to which the fuzzy afterset $x_i\boldsymbol{R}$ is contained in the fuzzy foreset $\boldsymbol{S}z_j$, and similar for the super-composition.

4.3. Possible relational compositions and their semantics

Following the example of Bandler and Kohout, we clarify the above constructions in this section by showing how the relational compositions allow to retrieve important information for participants in the electronic job market. Recall that in our market, we are dealing with the following key entities: students, companies, resume agents, job offer agents, and criteria.

Students create a resume agent to post their resume and find suitable job offers, and similarly companies create job offer agents in order to announce job openings and find suitable candidates. Both resume and job offer agents are fully characterized by criteria selected from a pre-defined list.

Formally, we have the set of students $\boldsymbol{A}$, the set of companies $\boldsymbol{O}$, the set of resume agents $\boldsymbol{E}$, the set of job offer agents $\boldsymbol{J}$, and finally the set of criteria $\boldsymbol{C}$. We will now construct various relations between these sets, and indicate how the relational compositions reveal important information regarding the sets from which they were constructed.

4.3.1. Example 1: the use of criteria in resume agents

Let $\boldsymbol{R}$ be a relation from the set of criteria $\boldsymbol{C}$ to the set of resume agents $\boldsymbol{E}$, and let $a\boldsymbol{R}b$ carry the meaning that criterion a is selected or used by resume agent b. Then

$\boldsymbol{S} = \boldsymbol{R}^{t}$, with $\boldsymbol{R}^{t}$ the transpose of $\boldsymbol{R}$, is a relation from the set of resume agents $\boldsymbol{E}$ to the set of criteria $\boldsymbol{C}$, with $b\boldsymbol{S}c$ meaning that resume agent b has selected criterion c.

The $\boldsymbol{R}$-afterset of a criterion a, $a\boldsymbol{R}$, is the subset of $\boldsymbol{E}$ consisting of those resume agents who use criterion a. Similarly, $\boldsymbol{S}c$ or the $\boldsymbol{S}$-foreset of criterion c is the subset of $\boldsymbol{E}$ consisting of those resume agents who use criterion c. One easily verifies that the relational compositions $\boldsymbol{R} \circ \boldsymbol{S}$, $\boldsymbol{R} \lhd \boldsymbol{S}$, $\boldsymbol{R} \rhd \boldsymbol{S}$, $\boldsymbol{R} \square \boldsymbol{S}$, are all binary relations in the set of criteria $\boldsymbol{C}$. The various compositions will allow us to derive information on how different criteria are used in the students' resume agents.

In this case, the relational compositions carry the following meaning:

- $a(\boldsymbol{R} \circ \boldsymbol{S})c$: there exists at least one resume agent using both criteria a and c;

- $a(\boldsymbol{R} \lhd \boldsymbol{S})c$: criterion c is used in every resume agent in which criterion a is used, while the opposite is not necessarily true;

- $a(\boldsymbol{R} \rhd \boldsymbol{S})c$: criterion a is used in every resume agent in which criterion c is used, while the opposite is not necessarily true;

- $a(\boldsymbol{R} \square \boldsymbol{S})c$: every resume in which criterion a is used, also uses criterion c, and vice versa.

4.3.2. Example 2: the simultaneous use of criteria in job offer and resume agents

Let $\boldsymbol{R}$ be a relation from the set of resume agents $\boldsymbol{E}$ to the set of criteria $\boldsymbol{C}$, and assume that $a\boldsymbol{R}b$ means that resume agent a uses criterion b. Let $\boldsymbol{S}$ be a relation from the set of criteria $\boldsymbol{C}$ to the set of job offer agents $\boldsymbol{J}$: $b\boldsymbol{S}c$ iff criterion b is used in job offer agent b. The relational compositions $\boldsymbol{R} \circ \boldsymbol{S}$, $\boldsymbol{R} \lhd \boldsymbol{S}$, $\boldsymbol{R} \rhd \boldsymbol{S}$, $\boldsymbol{R} \square \boldsymbol{S}$, now are all relations from the set of resume agents $\boldsymbol{E}$ to the set of job offers $\boldsymbol{J}$. The various compositions will allow us to derive information on how the criteria used in a resume agent are also used in a job offer agent.

Here, the relational compositions carry the following meaning:

- $a(\boldsymbol{R} \circ \boldsymbol{S})c$: there exists at least one criterion that is used in resume agent a and used in job offer agent c;

- $a(\boldsymbol{R} \lhd \boldsymbol{S})c$: whatever criteria resume agent a uses, they are also used in job offer agent c, while the opposite is not necessarily true;

- $a(\boldsymbol{R} \rhd \boldsymbol{S})c$: whatever criteria are used in job offer agent c, they are also used in resume agent a, while the opposite is not necessarily true;

- $a(\boldsymbol{R} \square S)c$: represents a symmetrical matching of using criteria: whatever criteria are used in resume agent a, they are also used in job offer agent c, and vice versa.

4.3.3. Example 3: the interest of candidates in job offers

Let $\boldsymbol{R}$ be a relation from the set of students $\boldsymbol{A}$ to the set of job offers $\boldsymbol{J}$, and assume that $a\boldsymbol{R}b$ means that student a is interested in job offer b. Then, $\boldsymbol{S} = \boldsymbol{R}^{t}$ is a relation from $\boldsymbol{J}$ to $\boldsymbol{A}$: $b\boldsymbol{S}c$ iff job offer b is of interest to student b. The relational compositions $\boldsymbol{R} \circ \boldsymbol{S}$, $\boldsymbol{R} \lhd \boldsymbol{S}$, $\boldsymbol{R} \rhd \boldsymbol{S}$, $\boldsymbol{R} \square \boldsymbol{S}$, now are all binary relations in the set of students $\boldsymbol{A}$. The various compositions will allow us to derive information on how different students are interested in the market's job offers.

Here, the relational compositions carry the following meaning:

- $a(\boldsymbol{R} \circ S)c$: there exists at least one job offer in which both students a and c are interested;

- $a(\boldsymbol{R} \lhd S)c$: whatever job offers student a is interested in, student c is also interested in, while the opposite is not necessarily true;

- $a(\boldsymbol{R} \rhd S)c$: whatever job offers student c is interested in, student a is also interested in, while the opposite is not necessarily true;

- $a(\boldsymbol{R} \square S)c$: represents a symmetrical matching of job offer interests: whatever job offers student a is interested in, student c is also interested in, and vice versa.

The attentive reader will have noted the crisp nature of the relational compositions in the examples above. Indeed, the use of fuzzy relations $\boldsymbol{R}$ or $\boldsymbol{S}$ modifies the semantic descriptions to refer to degrees of inclusion. For instance in Example 2, the round composition $a(\boldsymbol{R} \circ S)c$ then should express *the degree* to which there exists at least one criterion used in resume agent a and job offer agent c, etc. The following section will further clarify the workings of fuzzy relational compositions.

5. Case Study: The Adoption of Criteria Among Students

In this case study, we focus on a set of five students $\boldsymbol{A}$, each having created a resume agent in the job market. Every resume agent contains a number of criteria on which the candidate provides information about his or her resume. Recall that the criteria are taken from the resume criteria list given in Table 1, and that the actual selection may differ from one student to another. Assume, for the sake of computational simplicity, that the set of resume criteria $\boldsymbol{C}$ contains four criteria only. In addition, we have seven job offer agents in the job market, each of which contains requests for information on a number of criteria from the same set $\boldsymbol{C}$. As in the previous section, we denote as $\boldsymbol{E}$ the set of resume agents, and $\boldsymbol{J}$ the set of criteria agents.

Assume that we have a relation $\boldsymbol{R}$ from the set of resume agents $\boldsymbol{E}$ to the set of criteria $\boldsymbol{C}$ and a relation $\boldsymbol{S}$ from the set of job offer agents $\boldsymbol{J}$ to the set of criteria $\boldsymbol{C}$:

$$R = \begin{pmatrix} 1 & 1 & 0 & 1 \\ 0 & 1 & 0 & 1 \\ 1 & 1 & 1 & 1 \\ 1 & 1 & 0 & 0 \\ 1 & 1 & 0 & 1 \end{pmatrix}, \text{ and } S = \begin{pmatrix} 0 & 1 & 1 & 1 \\ 1 & 1 & 0 & 0 \\ 0 & 1 & 0 & 0 \\ 1 & 1 & 1 & 0 \\ 0 & 1 & 0 & 1 \\ 0 & 1 & 1 & 1 \\ 1 & 1 & 1 & 1 \end{pmatrix}.$$

5.1 Analysis of the students' criteria

We here focus on relational compositions of the transpose $\boldsymbol{R}^t$ of $\boldsymbol{R}$ and the relation $\boldsymbol{R}$ itself, as in Example 1 above. The relational compositions all are binary relations in $\boldsymbol{C}$.

- *The sub- and supercomposition*

The matrix representation of the subcomposition of $\boldsymbol{R}^t$ and $\boldsymbol{R}$, which we will denote by $\boldsymbol{D}$, is given by:

$$\boldsymbol{D} = \begin{pmatrix} 1 & 1 & 0.4 & 0.8 \\ 0.8 & 1 & 0.2 & 0.8 \\ 1 & 1 & 1 & 1 \\ 0.8 & 1 & 0.4 & 1 \end{pmatrix}.$$

In order to further analyze the fuzzy relation $\boldsymbol{D}$, we need to ensure that $\boldsymbol{D}$ is transitive. Since this is not the case here, we need to take the transitive closure of $\boldsymbol{D}$ first, leading to the following transitive relation $\boldsymbol{Q}$, the transitive closure of $\boldsymbol{D}$:

$$
\boldsymbol{Q} = \begin{pmatrix} 1 & 1 & 0.4 & 0.8 \\ 0.8 & 1 & 0.4 & 0.8 \\ 1 & 1 & 1 & 1 \\ 0.8 & 1 & 0.4 & 1 \end{pmatrix}.
$$

Note that imposing transitivity upon $\boldsymbol{D}$ by taking the transitive closure has increased the membership degree of $\boldsymbol{D}(2,3)=0.2$ to $\boldsymbol{Q}(2,3)=0.4$, or in other words, the degree to which the presence of criterion 2 implies the presence of criterion 3 has been 'artificially' increased. This can certainly be considered as a drawback of the transitive closure technique, and alternative techniques have recently been proposed (De Baets and De Meyer 2002). Since every alpha-cut of $\boldsymbol{Q}$ is a crisp quasi-order relation, we can analyze $\boldsymbol{Q}$ at its relevant alpha-cuts and provide a graphical Hasse diagram representation as shown in Figure 4. For a more detailed account on the processing of quasi-order relations we refer to earlier work (De Baets and Van de Walle 1994; Van de Walle et al. 1995).

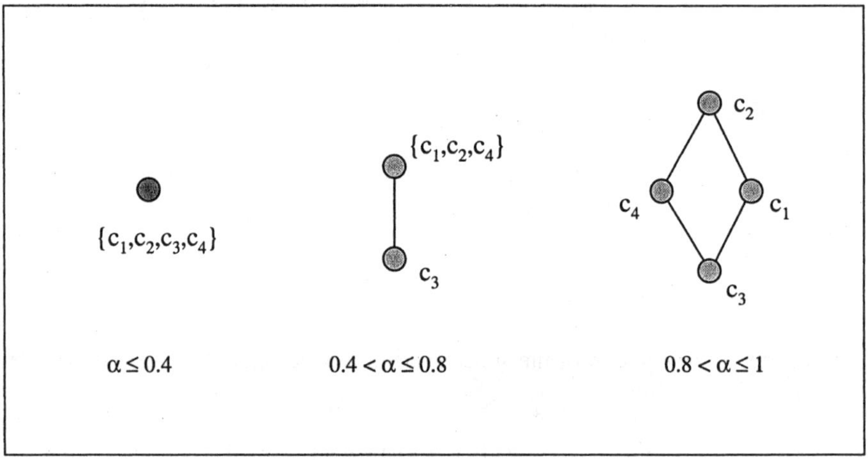

Figure 4. Hasse diagrams for the subcomposition of $\boldsymbol{R}'$ and $\boldsymbol{R}$

At the highest cut level, i.e. alpha strictly greater than 0.8, we observe the following. First, when criterion 3 is used in a resume agent, so are all other criteria. The use of criteria 1 or 4 implies the use of criterion 2, but we cannot say anything definite about the co-use of criteria 1 and 4: they are incomparable. Reducing the cut level 'softens' the differences between the usage patterns of the criteria: the use of criterion 3 implies all other criteria to be used at this level. Clearly, this is a less precise view on the information presented by $\boldsymbol{Q}$: the small differences in occurrence between these criteria are now considered indistinguishable.

The analysis of the super-composition is very similar to this analysis, and we leave the computational details to the willing reader.

- *The square composition*

The square composition expresses the degrees to which the criteria match exactly among the resume agents. The matrix representation of the square composition of $\boldsymbol{R}^t$ and $\boldsymbol{R}$ which we again denote by $\boldsymbol{D}$ is:

$$\boldsymbol{D} = \begin{pmatrix} 1 & 0.8 & 0.4 & 0.6 \\ 0.8 & 1 & 0.2 & 0.8 \\ 0.4 & 0.2 & 1 & 0.4 \\ 0.8 & 0.8 & 0.4 & 1 \end{pmatrix},$$

and its transitive closure relation $\boldsymbol{Q}$ is given by:

$$\boldsymbol{Q} = \begin{pmatrix} 1 & 0.8 & 0.4 & 0.8 \\ 0.8 & 1 & 0.4 & 0.8 \\ 0.4 & 0.4 & 1 & 0.4 \\ 0.8 & 0.8 & 0.4 & 1 \end{pmatrix}.$$

Note that $\boldsymbol{D}$ and $\boldsymbol{Q}$ are symmetric matrices.

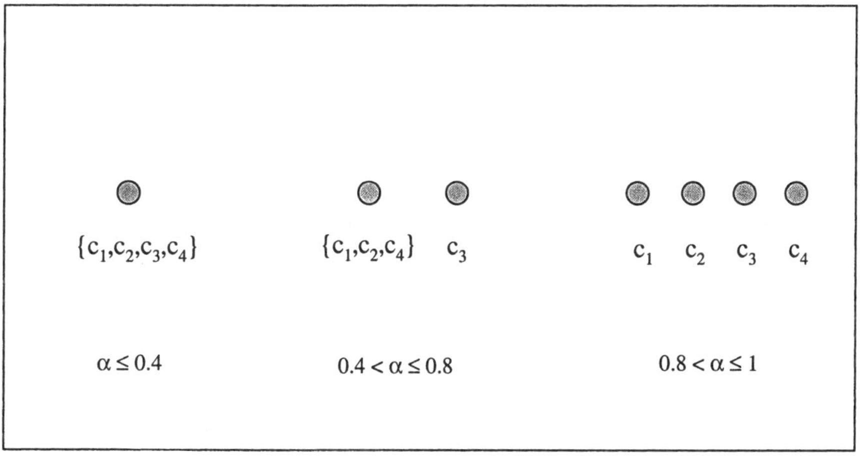

Figure 5. Hasse diagrams for the square composition of $\boldsymbol{R}^t$ and $\boldsymbol{R}$

The Hasse diagrams of the alpha-cuts of $\boldsymbol{Q}$ are shown in Figure 5. At the highest cut levels, for alpha strictly greater than 0.8, all criteria are incomparable: there are no exact matches on how the criteria are used in the resume agents. In other words, the use of one criterion may lead to the use of another criterion among all resume agents, but the converse then does not hold for these agents. Lowering the cut level below 0.8 leads to a grouping of criteria 1, 2 and 4, with criterion 3 incomparable. Finally, at cut levels below 0.4, all criteria are grouped and all distinctions disappear. We here clearly observe the demanding nature of the square composition, and the effects of its symmetry: we only have indifferences or incomparabilities.

5.2. Analysis of the match of criteria in resume and job offer agents

We here focus the composition of the relation $\boldsymbol{R}$ and the relation $\boldsymbol{S}^t$.

- *The sub- and the supercomposition*

The matrix representation of the subcomposition of $\boldsymbol{R}$ and $\boldsymbol{S}^t$, which we as usual denote by $\boldsymbol{D}$ is given by:

$$\boldsymbol{D} = \begin{pmatrix} 0.75 & 0.75 & 0.75 & 0.75 & 0.75 & 0.75 & 1.0 \\ 1.0 & 0.75 & 0.75 & 0.75 & 1.0 & 1.0 & 1.0 \\ 0.75 & 0.5 & 0.25 & 0.75 & 0.5 & 0.75 & 1.0 \\ 0.75 & 1.0 & 0.75 & 1.0 & 0.75 & 0.75 & 1.0 \\ 0.75 & 0.75 & 0.5 & 0.75 & 0.75 & 0.75 & 1.0 \end{pmatrix}.$$

$\boldsymbol{D}$ is a fuzzy relation from the set of resume agents $\boldsymbol{E}$ to the set of job offer agents $\boldsymbol{O}$. We observe that $\boldsymbol{D}(i, 7)=1$, for $i \in \{1, \ldots, 7\}$. This implies that whatever criteria are mentioned in any of the resume agents, they are all used in job offer agent 7. Conversely, the single criterion used in job offer agent 3 is mentioned in every resume agent. It is interesting to note that resume agent 3, which mentioned all of the criteria, leads to the most variation in its match with job offer agents, ranging from 0.25 with job offer agent 3, 0.5 with job offer agents 2 and 5, to a high of 0.75 with job offer agents 1, 4 and 6.

Generally speaking, the membership degrees reflect the degree to which the criteria specified in a resume agent are also used in a job offer agent. The lower the membership degree is, the more criteria are specified in a resume agent that are not asked for by the job offer agent.

We again invite the willing reader to analyze the supercomposition of $\boldsymbol{R}$ and $\boldsymbol{S}^t$.

- *The square composition*

Clearly, the square composition of $\boldsymbol{R}$ and $\boldsymbol{S}^{t}$ will produce lower membership degrees for the co-usage of criteria in resume and job offer agents. For example, denoting the composed relation as $\boldsymbol{D}$, we have that $\boldsymbol{D}(1,1)=0.25$, implying that only to a low degree the criteria mentioned in the first resume agent are used in the first job offer agent, and vice versa. We invite the reader to verify the other values of the composition.

5. 3 Analysis of the subcomposition $R^{t} \lhd R$ with R a fuzzy relation

We now turn to a more complex scenario, in which a student or a company can assign an importance to a criterion, as an indication of the relevance of that criterion to his or her decision. Assuming that these importances are taking values in the interval [0,1], we can define a fuzzy relation $\boldsymbol{R}$ from the set of resume agents $\boldsymbol{E}$ to the set of criteria $\boldsymbol{C}$, with $\boldsymbol{R}$'s membership degrees indicating the degree to which a particular criterion is important to a student.

For the remainder of this section, we will focus exclusively on $\boldsymbol{R}$, a fuzzy relation from the set of resume agents $\boldsymbol{E}$ to the set of criteria $\boldsymbol{C}$. Assume that $\boldsymbol{R}$ has the following matrix representation:

$$R = \begin{pmatrix} 0.6 & 0.9 & 0.2 & 0.8 \\ 0.3 & 1.0 & 0.4 & 0.8 \\ 0.7 & 0.9 & 1.0 & 0.8 \\ 0.6 & 0.5 & 0.4 & 0.4 \\ 1.0 & 0.3 & 0.1 & 0.9 \end{pmatrix}.$$

The matrix representation of the subcomposition of $\boldsymbol{R}^{t}$ and $\boldsymbol{R}$ which we denote by $\boldsymbol{D}$ is given by:

$$\boldsymbol{D} = \begin{pmatrix} 1 & 0.84 & 0.7 & 0.94 \\ 0.76 & 1 & 0.68 & 0.9 \\ 0.92 & 0.98 & 1 & 0.96 \\ 0.84 & 0.88 & 0.64 & 1 \end{pmatrix},$$

and its transitive closure relation $\boldsymbol{Q}$ is given by:

$$\boldsymbol{Q} = \begin{pmatrix} 1 & 0.88 & 0.7 & 0.94 \\ 0.84 & 1 & 0.7 & 0.9 \\ 0.92 & 0.98 & 1 & 0.96 \\ 0.84 & 0.88 & 0.7 & 1 \end{pmatrix}.$$

The Hasse diagrams of the alpha-cuts of Q are presented in Figure 6. This figure illustrates how the relationships between the criteria evolve from complete indifference for alpha levels smaller than or equal to 0.7 to rather complex configurations for increasing levels. Initially, we observe the existence of linear relationships among the criteria. For values of alpha in the interval]0.7,0.84], we have that whenever criterion 3 is considered important or relevant in a resume agent, the other criteria are considered even more important or relevant. Increasing alpha to levels in the interval]0.84,0.88] learns that in this interval it also holds that whenever criterion 1 is considered important, criteria 2 and 4 are considered as even more important or relevant. The first incomparability appears for values of alpha in the interval]0.88,0.9]: criterion 1 and 2 become incomparable, implying that we cannot any longer draw conclusions about the importance implications among both criteria.

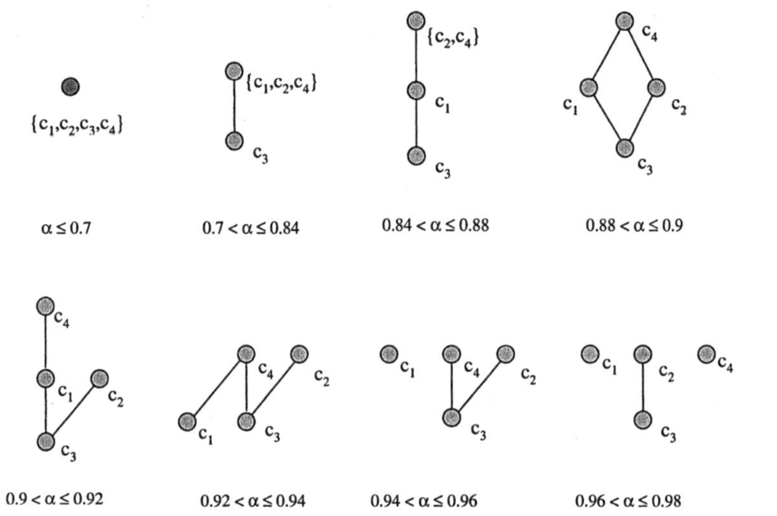

Figure 6. Hasse diagrams for the subcomposition of $\boldsymbol{R}^{t}$ and $\boldsymbol{R}$

Increasing the alpha levels introduces additional incomparabilities: among criteria 2 and 4 in]0.9,0.92], criteria 1 and 3 in]0.92,0.94], criteria 1 and 4 in]0.94,0.96], criteria 3 and 4 in]0.96,0.98] and finally we are left with criteria that are all incomparable. This evolution of incomparabilities can be interpreted as a gradual 'break-away' of one criterion after the other: first, we have that criterion 2 'breaks away from' (i.e., becomes incomparable with) criteria 1 and 4; then we have that criterion 1 breaks away from criteria 3 and 4, and finally we have that criterion 3 breaks away from criteria 2 and 4, with the relationship among criteria 3 and 2 being the strongest among these until the alpha level becomes too high at the very end when at the highest cut levels, i.e. for alpha strictly larger than 0.98, all criteria become incomparable.

6. Conclusions and Future Research

We have presented an electronic job market as an advanced information system offering decision support for its participating students and companies. The market participants create software agents through which they post their resume or job offer described in terms of multiple decision criteria. Every agent contains criteria-based information as well as a multi-criteria decision model. The objective of the criteria-based information is dual: to provided information to other agents, and to request information from these same agents. The objective of the multi-criteria decision model is to calculate a ranking of the other agents on the basis of the information obtained from them. This ranking information provides a first clue to the participants on how well their resume or job offer 'fits' within the market. We have argued that this information is strategically important, provided that we obtain a better understanding of how the decision criteria are adopted among the participants in the job market. To achieve this deeper understanding, we have proposed the use of fuzzy relational compositions, inspired by the work of Bandler and Kohout. We have shown in a few short examples and a longer case study how these relational compositions provide a semantically rich interpretation of the adoption of decision criteria among students and companies. Future work will focus on the application of the technique of fuzzy relational compositions to provide an analysis and synthesis of the entire job market, building upon recent work analyzing the cognitive maps of decision makers (Kwon et al. 2002). At the same time, we hope to develop visual interfaces in which the market participants can navigate through more complex Hasse diagrams to better understand the market in which they operate.

Acknowledgments

The author acknowledges support for part of this research from the NJIT Faculty SBR Grant 2001. He also gratefully thanks ACTONOMY NV (Belgium) for the use of their HRDirector software product, and Veerle Van der Sluys for her assistance in setting up the job market. The encouragement and support from the Career Services Department at NJIT, and especially from Marcelle Jackson, Gregg Mass and Rosalind Newton, was essential to the job market's success.

References

1. Ariba (2001) B2B marketplaces in the new economy. Ariba Inc., Mountain View, California.

2. Bakos Y (1997) Reducing buyer search costs: implications for electronic marketplaces. Manage. Sci. 43 (12).
3. Bakos Y (1998) The emerging role of electronic marketplaces on the internet. Commun. ACM 41, 8:35–42
4. Bandler W, Kohout L (1980) Fuzzy relational products as a tool for analysis and synthesis of the behavior of complex natural and artificial systems. In: Wang, SK, Wang PP (eds) Fuzzy Sets: Theory and Application to Policy Analysis and Information Systems. Plenum Press New York and London, pp 341–367
5. Bandler W, Kohout L (1984) Automatic documentary information retrieval by means of fuzzy relational products. TIMS/Studies in the Management Sciences 20:383 – 404
6. Beam C, Segev A, Bichler M, Krishnan R (1999) On Negotiations and Deal Making in Electronic Markets. Information Systems Frontier 1, 3:241–258
7. De Baets B, De Meyer H (2002) Transitive approximation of fuzzy relations by alternating closures and openings. Soft Computing (to appear)
8. De Baets B, Van de Walle B (1994) Dependencies among alternatives and criteria in MCDM problems. In: Proceedings of CIFT'94, Fourth International Workshop on Current Issues in Fuzzy Technologies, pp 80–85
9. Kwon H, Im I, Van de Walle B (2002) Are you thinking what I am thinking? – A comparison of decision makers' cognitive maps by means of a new similarity measure. In: Proceedings of 35th Hawaii International Conference on System Sciences HICSS 2002, pp
10. Lomuscio AR, Wooldridge M, Jennings NR (2000) A classification scheme for negotiation in electronic commerce. In: Dignum F, Sierra C (ed) Agent-Mediated Electronic Commerce: A European Perspective. Springer-Verlag Berlin Heidelberg New York, pp 19–33
11. Maes P, Guttman RH, Moukas AG (1999) Agents that buy and sell: Transforming commerce as we know It. Commun. ACM 42 (3):81–91
12. Malone T, Yates J, Benjamin R (1987) Electronic markets and electronic hierarchies. Commun. ACM 30:484–497
13. Nielsen J (2000) Designing Web usability: the practice of simplicity. New Riders Publishing, Indianapolis.
14. Ströbel M (2000) Effects of Electronic Markets on Negotiation Processes. In: Proceedings of the ECIS 2000 Conference, pp 445–452
15. Turban E, Lee J, Kind D, Chung HM (1999) Electronic Commerce. A Managerial Perspective. Prentice Hall, Upper Saddle River, New Jersey.
16. Van de Walle B (2002) A relational analysis of the decision makers' preferences. International Journal of Intelligent Systems, to appear.
17. Van de Walle B, Faratin P (2001) Fuzzy preferences for multi-criteria negotiation, AAAI Technical Report FS-01-03, pp 116–119
18. Van de Walle B, De Baets B, Kerre EE (1995) Fuzzy multi-criteria analysis of cutting techniques in a nuclear reactor dismantling project. Fuzzy Sets and Systems 74:115–126

19. Van de Walle B, Heitsch S, Faratin P (2001) Coping with one-to-many multi-criteria negotiations in an electronic marketplace. In: Proceedings of the eNegotiations Workshop at DEXIA'01, pp 747–751
20. Van Rijsbergen CJ (1979) Information Retrieval, 2nd edn. Butterworths, London.
21. Weller T (2000) B2B eCommerce – the rise of emarketplaces. Legg Mason Wood Walker Equity Research Report.

A Fuzzy Evaluation Model: A Case for Intermodal Terminals in Europe

Rita A. Ribeiro*, Ana M. D. Moreira** and Eddy Declercq***

*Management Department, Universidade Lusíada
192 Rua da Junqueira
1349-001 Lisboa, Portugal

E-mail: rar@uninova.pt

**Departamento Informática
Faculdade Ciências Tecnologia, Universidade Nova Lisboa
2825 Monte da Caparica, Portugal

E-mail: amm@di.fct.unl.pt

***Strateco
Floris JespersBuilding
1, Plantinkaai Box 8
2000 Antwerp, Belgium

E-mail: e.declerq@strateco.org

Abstract. Freight transport services have evolved substantially in the last decade and encountered a continuously increasing demand for speed, reliability, reduced transportation costs and value added services. To minimize the potential negative effects of the expected increase of commercial transport, the development of friendly transport concepts such as intermodal terminal selection is required.

The goal of our work is to develop a decision support system (DSS) for selecting the intermodal terminal that best fits required conditions and criteria. The DSS includes a simulation model developed using fuzzy logic techniques to handle imprecise concepts. The end-result is intended for decision-makers, i.e., managers that need to choose intermodal destination terminals to be used for the transportation of their goods.

Keywords: intermodal terminal, decision support system, fuzzy multiple-attribute decision making

1. Introduction

The objective of this work is to develop a decision support system to find the best terminals in a given country (and optionally in a given city), which satisfy certain restrictions and criteria. In order to accomplish this, the following assumptions are made:

- The distance between terminals are not considered;
- the terminal of origin is not important, it is only important the destination terminal and respective conditions;
- intermediate terminals are not handled, since here we only consider the set of destination terminals;
- the model is developed both for passengers and cargo transport;

It should be noted that the information regarding the type of terminal required by the user also determines its intermodal combination. The types of modes the decision maker can select are, for example, road haulage, railway transport, inland waterway, maritime transport and airway transport. In this work we divided the terminals into passengers terminals and cargo terminals and they include as different modes: road, rail, sea, inland waterway and air. However, since we only deal with destination terminals we do not handle intermodal combinations.

Our system is a decision support system developed for non-computer experts, using computer-assisted data input, processing wizards and other applications that facilitate its use. Our system lowers the threshold to use computer modeling in the strategic decision making process. It contains a multiplicity of data from different fields of interest. The information provides insights into the economic, social and infrastructure image of a country and is necessary at a level of strategic decisions.

The developed DSS contains the three classical modules of a DSS [1]: database, simulation and interface. Its main features are:

1. It ranks the alternative destination terminals according to criteria values (cost, flexibility and reliability) and respective importances;
2. It is a flexible system for handling terminal features, such as modifications and new insertions;
3. It contains a sophisticated, but simple, algorithm for solving the rating and ranking of terminals, using a fuzzy multiple attribute model;
4. The fuzzy multiple attribute model includes fuzzy attributes (also denoted criteria) and fuzzy importances/weights and uses a weighted average aggregation procedure;
5. The construction of the fuzzy attribute membership functions is transparent to the user since the system extracts automatically the values from the database, to build them. This feature implies that with the increase of cases (terminals) in the database the attributes will better represent them;

6. It is interactive and user-friendly in all dialogue with the user. For example, it contains two levels for criteria importances to allow the user to detail criteria importances only when wanted;
7. It incorporates a simulation facility, where the user can compare what are the results of changes in his/her criteria weights in the ranking of terminals.

The paper is structured in three sections. Section 1 is this introduction. Section 2 is the main body of this paper and it describes the three modules that compose our decision support system: the database module, the simulation module and the interface module. Section 3 presents the conclusions.

2. The Decision Support System

As mentioned, our DSS includes three modules:

- a conceptual data model and a data dictionary defining the structure of a relational database;
- the simulation module, which contains two main processes, (a) a pre-selection of terminals, according to the user specific needs and (b) a fuzzy multiattribute analysis to rank those terminals, according to the user preferences (weights);
- the interface to access the database and to establish the dialog with the user. The idea of wizards will be used to define the steps required to access the database.

The conceptual model of the database was obtained using structured analysis techniques [2] and implemented using ACCESS. The simulation module uses fuzzy multiple attribute (criteria) concepts [3] [4]. The interface wizard will drive the user through the process of entering data necessary for defining the transportation framework requirements, the conditions, the criteria and the weights. The Wizard's last screen will give the user the possibility of choosing how to present the results: a combination of charts, maps and reports.

2.1 Database module

As mentioned before, the conceptual data model was obtained by using structured analysis techniques [2]. The main entities used in our model are represented in the entity relationship model depicted in Figure 1. Each entity is described in the terms of its attributes, both identifiers (preceded by @) and descriptors, as represented in the following data dictionary:

Terminal	= @coordinate + terminal_type + terminal_modes + terminal_name + address + services
Cargo_T	= @coordinate + DGHC + storage_capacity + cargo_types + specials + fridge + cranes + cargo_price
Passenger_T	= @coordinate + capacity + price + infrastructures
Connection	= @coordinate_origin + @coordinate_destiny + connection _mode
Country	= @country_code + country_name + GNP + labour_cost_hour + tax_level + extra + index_infrastructure + union + social_status + stability
City	= @city_code + city_name

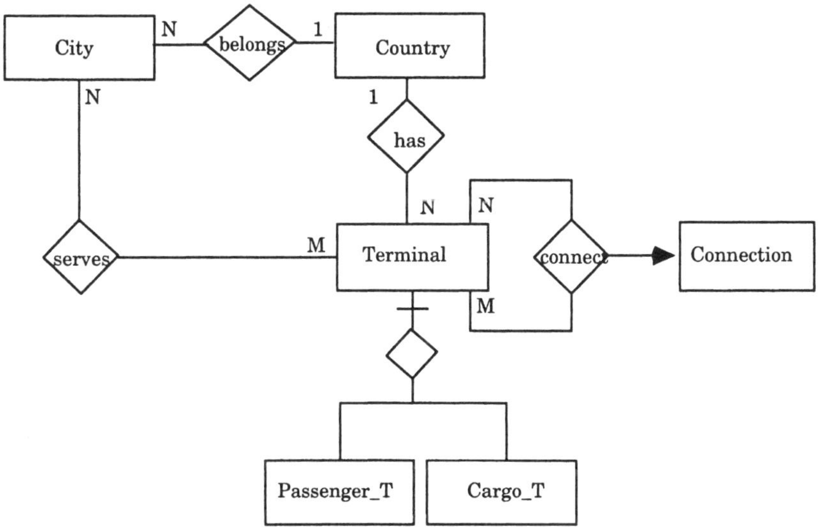

Figure 1: Entity relationship model for the intermodal terminals

The skeleton database tables we can derive from that entity relationship model are shown in Figure 2.

The database was filled in using the information (real values) available from the AND Mapping B.V. database, a company from Netherlands.

The database Wizard will allow the user to pick any relational database connected to the application and retrieve and modify data where and when appropriate. Access to data stored in the relational database is implemented through the ODBC (Open DataBase Connectivity) mechanism. This will allow to use other databases, from different vendors in a transparent mode, which makes the application more flexible and open. Moreover, with this strategy the application will be ready for a real client/server architecture with few adjustments.

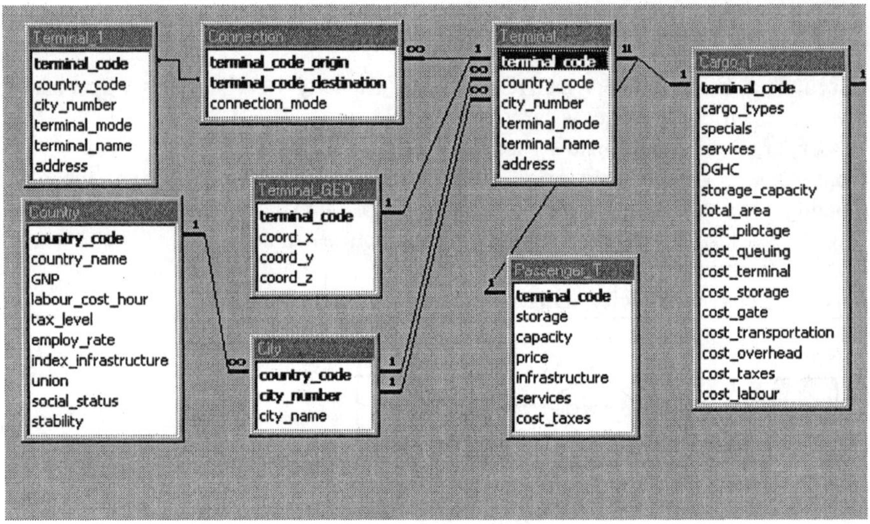

Figure 2: Tables implemented in the relational database

2.2 Simulation module

The simulation model is the module of the decision support system, which handles the selection and classification of terminals to recommend to the decision maker. This model includes two main tasks: (a) a pre-selection of terminals, according to user specific requisites and (b) a multicriteria analysis to rank those terminals, according to the user preferences (weights). The general description is shown in Figure 3.

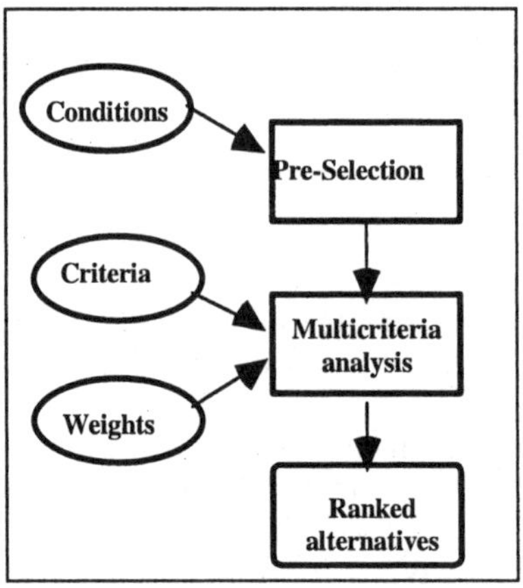

Figure 3: Main components of the simulation module

2.2.1 *Pre-selection*

The information needed to perform the pre-selection is taken both from the users and the databases. The users provide the conditions they need for the transportation of their goods and the system matches them with information on existing destination terminals, which fulfill those needs. These condition are the restrictions of our DSS.

For example, we can choose conditions for the capacity (e.g. minimum value for lifting capacity of cranes) and also for prices (e.g. maximum value to spend in pilotage costs). Then, the pre-selection process matches these conditions with the data from the existing terminals and selects the set that fulfills the required condtions. The pre-selection process will be further detailed in the Interface Module section.

From this pre-selection process a set of possible alternatives is passed to the multicriteria analysis. In the multicriteria analysis, users select the criteria they want to be used for terminal selection and provide the weights for those criteria. With this information a multicriteria analysis, using fuzzy logic, is performed to rank the terminals.

2.2.2 *Multicriteria analysis*

We will start the development of the multicriteria model by defining the main features the multicriteria analysis should include. Secondly we will analyze the type of attributes defined and how to represent them in the system. Thirdly we will define the most appropriate decision making algorithm to handle those attributes. Finally we will define the weighting procedure to allow the user to specify which are his/her preferences (relative importance) regarding the criteria and the attributes.

The main features of the multicriteria analysis model will be: (1) it accepts and manipulates both types of criteria, crisp and fuzzy; (2) it is interactive and user-friendly in the sense that all dialogue with the user is guided by scroll-bar menus and dialogue-boxes, from the 'construction' of specifications to the manipulation and selection of query-types; (3) it is a flexible system in terms of the handling of criteria and weights, such as new insertions and modifications of their respective values. (4) it incorporates simulation facilities, providing an efficient way for the user to change preferences-weights, enabling the tuning of the system and the verification of the variables sensitivity.

In order to achieve an optimal ranking of terminals, fuzzy set theory [5] will be used because this problem involves many imprecise criteria such as stability, prices and infrastructure capabilities. Fuzzy set theory main objective is to solve problems in which descriptions of activities and observations are imprecise, vague and uncertain. It is a theory of graded concepts where everything is a matter of degree. For example the criteria stability can be expressed as "good", "average" and "bad". Formally, if **X** is a collection of components denoted generically by **x** then a fuzzy set **A** in **X** is a set of ordered pairs **A** = {**x**,μ_A(**x**) | **x** , **X**}, where μ_A(**x**) is called the membership value or grade of membership of **x** in **A**. Another important aspect of decision making in a fuzzy environment is the reasoning process that takes place when a decision is made. When in the presence of imprecise concepts, this reasoning is usually denoted as approximate reasoning [6].

The fuzzy multicriteria analysis model used in the simulation module is based in the MAF-DSS (Multiple-Attribute Fuzzy Decision Support System) developed by [7]. The main entities to be considered in the multicriteria analysis are alternatives, criteria, attributes and weights. Alternatives are the set of possible terminals determined by the pre-selection procedure. Criteria are the features used to classify the terminals, such as reliability, flexibility and cost. Attributes are the criteria elements; they can be either crisp or fuzzy, as for example costs of pilotage, country social status and the stability of criteria reliability. Weights express the relative importance, attached by the user, for each criterion and their respective attributes. By default the weights are considered equivalent.

The multicriteria analysis (triggered by the run option on the menu *Simulation*) includes the following elements: (1) three evaluation criteria (cost, flexibility, reliability) with their respective attributes; (2) weights for the three evaluation criteria and also for the respective attributes; (3) Fuzzy attributes; (4) an aggregation algorithm for rating the terminals; (5) a raking process for the selection of terminals (from best terminal to worst, of the set of possible ones). A summary of each element is described in the following subsections.

Three evaluation criteria and respective attributes

There are three criteria in the DSS system and each criterion includes a set of subcriteria, here called attributes. The criteria and respective attributes are the following:

- *Costs* = overhead + pilotage cost + queuing time + terminal cost + storage + gate cost + taxes + labour cost + GNP + transportation
- *Flexibility* = storage + cranes + fridge + connection + TEN accessibility + infrastructure
- *Reliability* = unions + social status + stability

Importance/weights

The weights for the three criteria (cost, reliability and flexibility) and their respective attributes are defined as discrete fuzzy sets with the following labels:

[*irrelevant, low; important, very important, extremely important*]

The DSS presents a scale-bar to the user with the above labels, where each tick corresponds to the above membership labels. The user introduces his/her weights for the three upper level criteria (Costs, Flexibility, Reliability) and, when desired, for each of the detailed attributes (there are three windows for the detailed attributes). By default the value is *irrelevant*.

Fuzzy attributes

There are two types of attributes: crisp and fuzzy. Within the fuzzy attributes we can have discrete or continuous. The construction of the continuous attributes is as follows: (1) the *min* and the *max* values of each attribute is extracted from the database; (2) the type of function for that criteria (increasing or decreasing) is obtained from a database table. The *min* and *max* values correspond to the lowest value and highest value, contained in the database, for all terminals specific attributes. The type of function depends on the criteria, as for example: the attributes of cost criterion are always considered as decreasing (the lower the value the better). Note that if new terminals are included in the database, the new information will automatically change the *min* and the *max* from each attribute, if the values are higher or lower. Furthermore, new fuzzy sets are automatically re-calculated for each attribute when there new terminal inclusions.

The corresponding fuzzy sets are built with open-end triangular functions and the membership value $\mu(x)$ for each attribute, is calculated with the respective database values for that attribute, x, as follows:

- Increasing linear function:

$$\mu(x) = \begin{cases} 0 & x \leq \min \\ \dfrac{x - \min}{\max - \min} & \min \leq x \leq \max \\ 1 & x \geq \max \end{cases}$$

- Decreasing linear function:

$$\mu(x) = \begin{cases} 0 & x \geq \max \\ \dfrac{\max - x}{\max - \min} & \min \leq x \leq \max \\ 1 & x \leq \min \end{cases}$$

It should be noted that this construction is transparent to the users beacuse the fuzzy attributes are built-in. All that is required from the users is the introduction of weights for the criteria and/or attributes.

The membership values $\mu(x)$ obtained for the attributes, for each terminal, is later used to calculate the final evaluation aggregation procedure for rating the terminals (using the respective weights for criteria evaluation).

Aggregation procedure

The aggregation algorithm implemented in the simulation module is based in a fuzzy weighted average and uses a philosophy of "role model" or "ideal" terminal. This phase of the evaluation module corresponds to the *Rating* phase of the Fuzzy Multiple Attribute (FMA) approach [3], [4].

Formally, the aggregation formulas are:

$$Rating(C_j) = \sum_i W_i * Attr_i$$

$$FuzzyRating(Term_k) = \sum_j W_j * C_j$$

where W_i is the importance of an attribute, $Attr_i$ is the membership value of the attribute, C_j corresponds to criteria (Cost, Flexibility or Reliability), W_j is the importance of criterion Cj and $Term_k$ is the terminal being evaluated. Furthermore, the weights are normalized and sum up 1.

The operators used for the fuzzy aggregation are the algebraic product and algebraic sum, which belong to the set of *t-norms* and *t-conorms* [8]. These type of operators satisfy the axioms of boundary, associatively, monotone non-decreasing and commutative (similarity). Many other operators from the class of t-norms and even compensatory operators could have been used, but for reasons of simplicity we only used these ones.

Another interesting aspect of our DSS is the notion of distance from an "ideal" terminal. This measure shows how far away from the ideal any choice really is. The "ideal" is calculated using the sum of real values contained in the database for the attributes of a terminal and considers that all the attribute weights are 1 (*extremely-important*). Then we use the real values of attributes multiply them by the weights, given by the decison maker, to obtain the weighted average of the terminal evaluation. Finally, we divide the two values to obtain the relation from a weighted terminal and its "ideal". This is an original measure, proposed in this project, which accounts for the importances given by the user to each attribute and criteria, to obtain a final evaluation that shows how far away from the ideal each choice is. Hence, the "ideal" measure, per terminal, is obtained as follows:

$$Evaluation(Term_k) = \sum_j W_j * ValueAttribute_j / Ideal$$

This measure could also have been done using the membership values previously obtained. However, we choose to use the real values to highlight the relation between the sum of real values and the weighted sum.

Hierarchization–ranking

The final ordering of the terminals, which corresponds to the *Ranking* phase of the FMA approach [3], [4], is the descending order of the values obtained in the *Rating(Term)*. That is, the best terminal is the one with the highest rating value, the second best is the one with the second highest value and so forth.

The implemented system shows, in a summary report, both the *Evaluation* value and the *Rating(Term)* to help the decision making process. With the results obtained, the user has a rating and ranking of all possible terminals (the ones selected with the pre-selection) and if he is not satisfied he can always change some of the weights to perform a "what-if" analysis (simulation capability).

2.3 Interface module

We will describe the interface dialog using an example with a cargo terminal, as depicted in Figure 4.

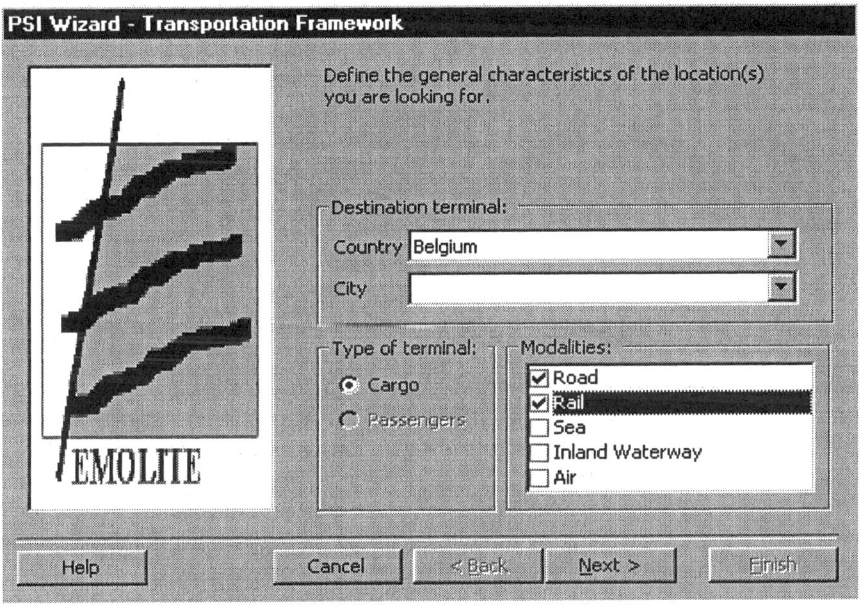

Figure 4: Transportation framework

The application (through a wizard) pops up the window shown in Figure 4, which allows specifying the destination country of our goods. Optionally, we can also choose a destination city in the selected country. Still in this window we should also choose the type of terminal (passenger or cargo) and a set of alternative modes of transportation (railway, road, train, maritime, inland waterway, airway). As mentioned in the introduction, the modes are related with the intermodal terminals.

If we choose to continue, by clicking *Next*, the window depicted in Figure 5 appears:

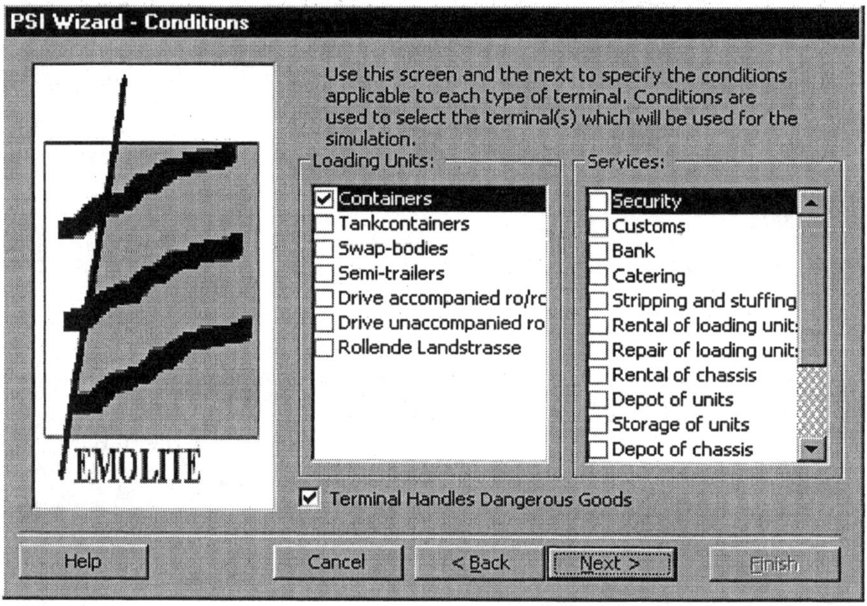

Figure 5: Terminal conditions

Figure 5 allows the user to specify the conditions applicable to each type of terminal. In other words, the user is able to narrow the set of terminals, evaluated for the intended purpose, through some conditional parameters. This step corresponds to the Pre-selection process (section 2.2.1).

The user may also select one or more types of loading units as well as services for the requested terminals. If the user selects more than one option, the possible terminals extracted from the database will have all the capabilities selected.

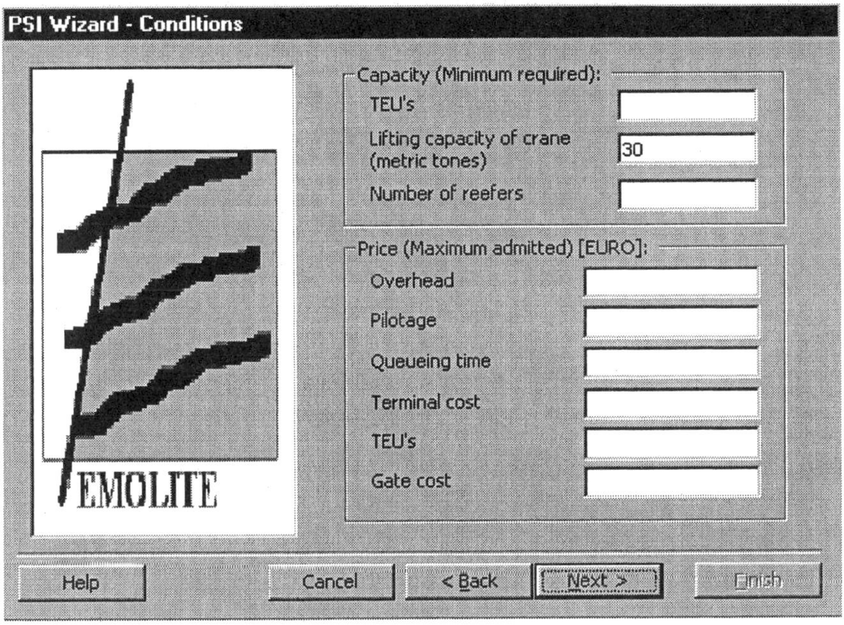

Figure 6: Terminal conditions (continuation)

After clicking *Next* in window 5, the window displayed in Figure 6 will appear. This window allows the user to specify additional conditions applicable to each type of terminal. This step still corresponds to the "Pre-Selection" procedure. The user may enter minimum values for terminals' capacities as well as maximum admitted costs for various items. If the user enters more than one value, the set of possible terminals extracted from the database will obey to all the restrictions specified, for instance if the user specifies a maximum price of 2000 Euros for the pilotage costs , only the terminals that satisfy that constraint will be extracted.

Now, clicking *Next* allows us to address the multicriteria analysis of our model (see Figure 7):

Figure 7 corresponds to the main interface of the multicriteria analysis (see section 2.2.2) for the specification of criteria weights that will apply to the set of possible terminals extracted from the pre-selection process.

The user may choose which criteria will be considered in the analysis by checking, or not, the desired criteria. For each criterion checked the user must define the weight (relative importance) that he assigns to that criteria, using the scroll bar, as shown in the respective window. For example, in Figure 7 the criterion flexibility has assigned and importance of *very-important*. The weights assigned at this (high) level will affect the criteria as a whole as explained in the Simulation Module section.

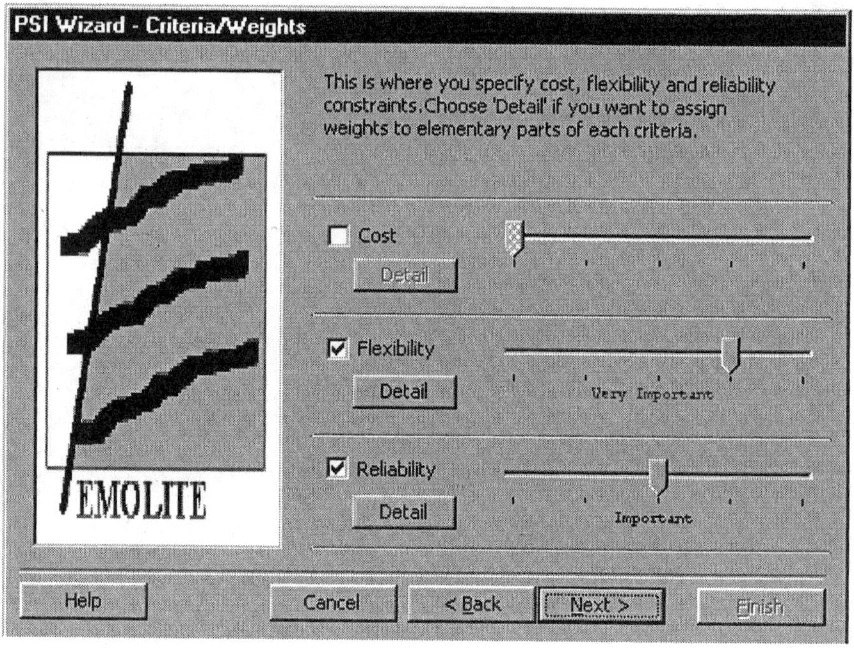

Figure 7: Multicriteria analysis

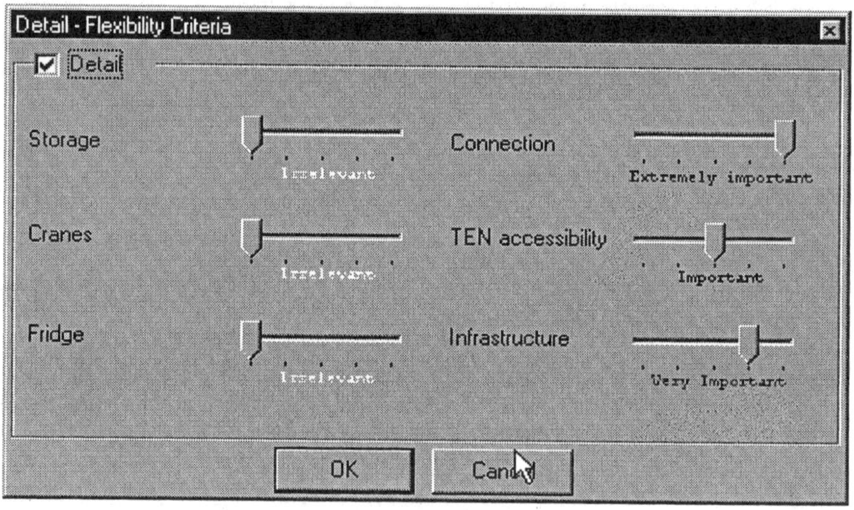

Figure 8: Definition of detailed attributes for flexibility criteria

Summary Report

Number of locations found: 33

Rank	Terminal	Evaluation	Fuzzy	Final
1	Antwerpen-Noord Natie-Delwaidedok	29,920	0,124	1,000
2	Antwerpen-ACT-6th Harbour Dock	29,339	0,122	0,981
3	Antwerpen-ACT-7th Harbour Dock	28,762	0,119	0,961
4	Zeebrugge-Ocean Containerterminal Hessena	28,680	0,119	0,959
5	Antwerpen-Noord Natie-Noordzee Terminal	28,674	0,119	0,958
6	Antwerpen-Seaport-Hansadok	28,301	0,117	0,946
7	Antwerpen-Hessenatie-Scheldeterminal	28,076	0,116	0,938
8	Renory-Container Terminal	27,939	0,116	0,934
9	Antwerpen-Hessenatie-Delwaidedok	27,881	0,116	0,932
10	Antwerpen-Seaport-Delwaidedok	27,881	0,116	0,932
11	Antwerpen-Allied Stevedores-Terminal Delwai	27,881	0,116	0,932
12	Athus	27,834	0,115	0,930
13	Antwerpen-Interferry	27,782	0,115	0,929
14	Antwerpen-Hessenatie-Churchilldok	27,589	0,114	0,922
15	Antwerpen-Seaport-Vrasenedok	27,545	0,114	0,921
16	Bressoux (Liège)	27,543	0,114	0,921
17	Antwerpen-Seaport-Churchilldok	27,377	0,114	0,915
18	Antwerpen-Westerlund	27,340	0,113	0,914
19	Mouscron-DPMLI	27,282	0,113	0,912
20	Zeebrugge-Flanders Container Terminal	27,274	0,113	0,912
21	Zeebrugge-Sea-Ro Terminal	27,251	0,113	0,911
22	Antwerpen-Mexico Natie-Berths 8-11	27,251	0,113	0,911
23	Antwerpen-Cirkeldyk	27,198	0,113	0,909
24	Gent-Farman	27,198	0,113	0,909

Figure 9: Summary report

The next step of the complete selection process is to assign weights to the attributes of each of the checked criteria. There are three detailed windows, which correspond to the criteria cost, flexibility and stability. The process to give the weights for the attributes is identical to the one of giving weights to criteria. Note that when a criterion is checked it is mandatory to assign at least one weight to its elementary criteria (i.e. the attributes). For reasons of space we only show one example (see Figure 8), which depicts the detailed costs for flexibility.

When the assignment of weights for the detailed criteria is finished a final window is presented to the user to allow choosing the preferred presentation style: Full Report, Summary Report, Graph, Map. This final dialog window is not shown here for reasons of space but the resulting windows are shown in Figures 9 and 10.

Hence, after running the simulation the results are depicted in the summary report as shown in Figure 9. These results show an example of the ranking of the possible set of terminals, after being evaluated with the fuzzy multiple attribute simulation model. They depict the global evaluation (*Evaluation*), the respective aggregated membership value (*Fuzzy*) and the normalized fuzzy membership value (*Final*). The results obtained refer to terminals of the selected country, Belgium. It should be pointed that the global evaluation value is the relation between the ideal value and the value obtained for the terminal, as described in section 2.2.2. It is also possible to double-click on every terminal presented in Figure 9. If this happens, the system will open a window showing detailed information about the terminal we just selected in the summary report.

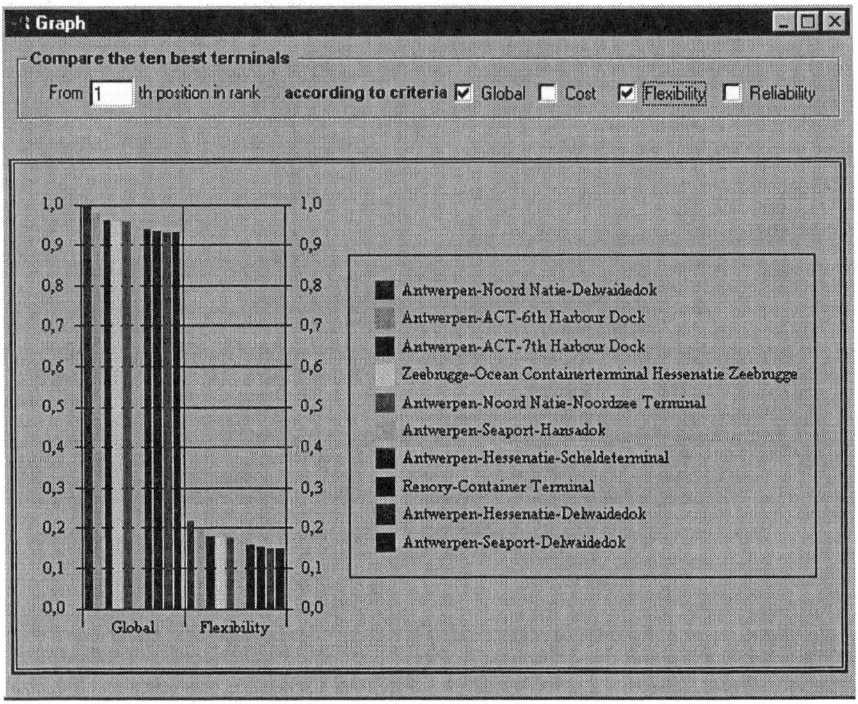

Figure 10: Graphical representation of 10-top terminals ranked

The final widow of our prototype is window 10 that includes a comparative graph of the terminals. Observing the graph in Figure 10 we can see the global evaluation and the flexibility criteria evaluation for the first ten terminals ranked. The presentation of the final results in graphical format improves the readability of the user, as well as to have a comparison between terminals. This window is updated every time a simulation runs. For readability reasons, only ten terminals are presented at one time.

Another important aspect of our prototype is the ability to save the results of any simulation. This feature allows the decision maker to perform an analysis over various scenarios (simulations) and select the one that best suits his/her purpose. In addition, the results obtained from any simulation can also be used future decision analysis.

3. Conclusions

The prototype described in this work is the final product of a research project financed by the EU. We believe this prototype is an interesting tool to be used for decision makers that need to use intermodel terminals, to transport their products. Our work joined efforts from several different experts (e.g. transports and logistics, computing science) to produce an evaluation model to identify the optimal intermodal destination terminals in Europe. The objective was to provide a ranked list of possible destination terminals to a decision maker that wishes to send his goods to a European country.

The decision support system built has been developed following a rapid prototyping approach. We started building the user interfaces during the requirements analysis phase. This helped us understanding the requirements and having a tool to communicate within our team.

The development techniques used come from classical structured techniques, such as structured analysis, to artificial intelligence concepts, such as fuzzy set theory.

The main advantages of our system are: it is an interactive and user-friendly system; it is a flexible system for handling intermodal terminal features (conditions, criteria and weights); it incorporates a simulation facility, where the users can compare what are the results of changes in their criteria weights in the ranking of terminals; it includes various types of result forms to support the decision maker's decision process.

An interesting future work to improve this prototype would be the inclusion of intermediate terminals in the multicriteria model; this would greatly improve the scope and versatility of the decision support tool.

Acknowledgements

This work has been realized under the support of the European Commission, project EMOLITE, contract nr. ST-96-SC.1018.

References

[1] Turban, E. (1996). *Decision Support and Expert Systems*, Macmillan Publishing Company.
[2] Yourdon, E. (1989). *Modern Structured Analysis*, Prentice-Hall.
[3] Ribeiro R. A. (1996). Fuzzy Multiple Attribute Decision Making: A Review and New Preference Elicitation Techniques. *Fuzzy Sets and Systems* Vol. 78, pp. 155-181.
[4] Zimmermann, H.-J. (1987). *Fuzzy Sets, Decision Making, and Expert Systems*. Kluwer Academic Publishers.

[5] Zadeh, L. A. (1965). Fuzzy Sets, *Information and Control*. Vol. 8, pp. 338-353.
[6] Zadeh, L. A. (1975). Fuzzy Logic and Approximate Reasoning. *Syntheses*, Vol. 30, pp. 407-428.
[7] Ribeiro R. A. (1993). Application of Support Logic Theory to Fuzzy Multiple Attribute Decision Problems. PhD Thesis, University of Bristol, UK.
[8] Klir, G. J. and T. A. Folger (1988). *Fuzzy Sets, Uncertainty, and Information*, Prentice-Hall International.

Application of Kemeny's Median for Group Decision Support

Hanna Bury and Dariusz Wagner

Systems Research Institute
Polish Academy of Sciences
ul. Newelska 6, 01-447 Warsaw, Poland
E-mails: bury@ibspan.waw.pl; D.Wagner@ibspan.waw.pl

Abstract. Some general problems concerned with the choice of a method of group judgement for expert opinions given in the form of preference orders are discussed. A special attention is paid to Kemeny's median and its modification proposed by B.G. Litvak, because of their important properties. Basic notions related to those methods are given. Heuristic algorithms for computing a median for both cases under consideration are proposed. Numerical examples illustrating the application of these algorithms are presented. A real-life example is also described.

1. Introduction

In 1959 John Kemeny in his book "Mathematics without numbers" [9] formulated the rule for determining a preference order that is the "closest" to the wishes of the voters. In his approach the problem considered is reduced "to one, which is analogous to those of classical statistics" [10] with the use of the concept of a distance between any two preference orders. This rule is also called the median procedure and Slater's rule in the literature on tournaments [20]. J. Kemeny concentrated his analysis on the problem of distance minimization, paying less attention to the fact that his rule is a decision function possessing the important properties. Kemeny's approach has been widely accepted, however real applications have been limited due to computing difficulties.

The growing interest paid to Kemeny's rule considered as a method of determining group judgement results from the fact that this rule enjoys remarkable properties and does not exhibit pathological behaviour typical for other methods, for example it has consistency in group judgement when candidates are dropped [4], [14], [19].

The aim of the paper is to discuss basic definitions concerning Kemeny's median as well as its modification proposed by Litvak and present some algorithms making it possible to determine the median.

The first section of the paper is concerned with characteristic features of methods, which can be used for group judgement purposes. In the second section the classical definition of median proposed by Kemeny is discussed in detail. All basic notions related to this definition are presented and examined. Properties of the median important for computational purposes are analysed. A heuristic algorithm for computing Kemeny's median is presented and its application is exemplified with a numerical example. The third section is concerned with a modification of Kemeny's median concept proposed by Litvak [11]. All basic notions associated with this modification are presented and discussed. A heuristic algorithm for computing Litvak's median is proposed and its use is exemplified with a numerical example. In the fourth section an example of application of Litvak's median to determine the ranking of university teachers on the basis of students' judgements expressed in anonymous questionnaires is presented.

2. Methods of Group Judgement

In general methods used to determine group judgement can be divided into two categories. The first category comprises so called positional methods derived from the method of Borda[1)] [3]. Borda called his method as "election by order of merit". Using these methods group judgement is determined taking into account positions taken by alternatives in judgements of particular experts.

The second one includes methods based on pairwise comparisons. They are derived from the method of Condorcet [5]. The method of Condorcet is intuitively based on the assumption that the best (according to a judgement of given group of experts) alternative is the alternative that is ranked higher than any other alternative by a majority of experts. Such an alternative is called the Condorcet winner [12], [13]. A difficulty with the Condorcet winner is that in general it need not exist.

There has been a lot of attempts to unify both methods. One of the earliest is due to Black [2] who suggested choosing the Condorcet winner if it exists and if such an alternative does not exist to use the Borda method.

---

1) It should be noted that Borda and Condorcet works concerned designing voting rules for an assembly. It is worth noting that both were members of l'Académie Royale de Sciences.

With all troubles associated with Condorcet' principle it has been generally assumed that this method should be regarded as the standard one. However, recent works of Saari [17], [18], suggest that rather then serving as a standard the Condorcet winner must be held in suspect.

There is a lot of methods of both categories. They have advantages as well as disadvantages, with some of them so called voting paradoxes are associated [13], [15]. In order to overcome these difficulties two approaches can be applied. The first one consists in devising – again and again – new methods which do not possess particular disadvantages or exhibit desired features. The second one is based on determining properties that group judgement should be characterized by and analysing existing methods from such a point of view. This approach is presented e.g. in papers of Nurmi [16], the book of Kelly [8] or Aizerman and his group [1]. The aroused interest in Kemeny's rule as a method of determining group judgements stems from such an approach.

Young and Levenglick [21] proved that Kemeny's rule is the unique preference function that is neutral, consistent and Condorcet. [1)] However, to be fair it has to be admitted that Kemeny's rule is not that elusive "ideal social choice mechanism" [19]. Such conclusion, of course, should be expected. First of all Kemeny's median is not unique. Moreover, according to Saari "it achieves its consistency by weakening the crucial assumption about the individual rationality of the voters" [19]. Saari also proved [19] that Kemeny's rule violate the natural continuity assumption when small changes in expert opinion translate into small changes in expert judgement.

The problem considered in the paper is the following. There is a set of n alternatives $\mathcal{O} = \{O_1, \ldots, O_n\}$ to be ranked by the group of K experts from the point of view of a given criterion or a set of criteria. It is assumed that their judgements have the form of preference orders and that ties can occur. One has to determine a group judgement being the aggregation of experts' opinions. For aggregation purposes two types of median are used: that proposed by Kemeny [9] and that introduced by Litvak [11].

[1)] A group judgement having the form of preference order is called neutral if it is symmetric in its treatment of alternatives. A preference function is consistent if two subgroups of experts meeting separately arrive – with the use of this function – at the same group judgement, then in the case when they work together the reached group judgement should be the same. A preference function is Condorcet if from the fact that the number of experts in whose judgement $O_i \succ O_j$ is greater than the number of those in whose judgement $O_j \succ O_i$ it follows that in the preference order being group judgement obtained with the use of this preference function the alternative O_j cannot precede O_i [21]. The notation $O_i \succ O_j$ means that from the point of view of comparison criterion (the set of criteria) the i-th alternative O_i is better than the j-th one. The notation $O_i \approx O_j$ means that both alternatives are considered to be equivalent.

3. Kemeny's Median

3.1 Basic notions

Given a preference order of n alternatives presented by the k-th expert (k=1,...,K)

$$P^k = (O_{i_1}, O_{i_2}, \ldots, O_{i_n}) \ .$$

For such a preference order the following matrix of pairwise comparison can be constructed

$$\mathbf{A}^k = \begin{bmatrix} a^k_{11}, & \ldots, & a^k_{1n} \\ \vdots & \ddots & \vdots \\ a^k_{n1}, & \ldots, & a^k_{nn} \end{bmatrix} \text{ where } a^k_{ij} = \begin{cases} 1 & \text{for } O_i \succ O_j \\ 0 & \text{for } O_i \approx O_j \\ -1 & \text{for } O_i \prec O_j \end{cases} \tag{1}$$

Definition 1 [11]. Assume that two preference orders P^{k_1} and P^{k_2} are given. The distance between these two preference orders can be expressed as follows

$$d(P^{k_1}, P^{k_2}) = \frac{1}{2}\sum_{i=1}^{n}\sum_{j=1}^{n}\left|a^{k_1}_{ij} - a^{k_2}_{ij}\right| \tag{2}$$

It can be proved [11] that the distance defined in such a way satisfies all the axioms determining the measure of "closeness" in a unique way. These axioms are as follows:

Axiom 1. $d(P^{k_1}, P^{k_2}) \geq 0, \quad d(P^{k_1}, P^{k_2}) = 0 \iff P^{k_1} = P^{k_2}$.

Axiom 2. $d(P^{k_1}, P^{k_2}) = d(P^{k_2}, P^{k_1})$.

Axiom 3. $d(P^{k_1}, P^{k_2}) \leq d(P^{k_1}, P^{k_3}) + d(P^{k_3}, P^{k_2})$.

Let $P^{k_1'}$ and $P^{k_2'}$ are preference orders obtained from P^{k_1} and P^{k_2} by a change of element enumeration. Then:

Axiom 4. $d(P^{k_1'}, P^{k_2'}) = d(P^{k_1}, P^{k_2})$.

Assume that in P^{k_1} and P^{k_2} only the order of alternatives placed in positions from t+1 to t+s (t+s $\leq$ n) is different. Let $T(P^{k_1})$ and $T(P^{k_2})$ denote the preference orders obtained from P^{k_1} and P^{k_2} by deleting alternatives placed in positions from 1 to t and from t+s+1 to n or from 1 to t when t+s=n.

Axiom 5. $d[T(P^{k_1}),T(P^{k_2})]=d(P^{k_1},P^{k_2})$.

Axiom 6. The minimum distance between preference orders is equal 1.

It is said that the preference order P^{k_3} is between preference orders P^{k_1} and P^{k_2} if elements of corresponding pairwise comparison matrices $\mathbf{A}^{k_1}=[a_{ij}^{k_1}]$, $\mathbf{A}^{k_2}=[a_{ij}^{k_2}]$ and $\mathbf{A}^{k_3}=[a_{ij}^{k_3}]$ satisfy the following inequality

$$\min\{a_{ij}^{k_1},a_{ij}^{k_2}\}\le a_{ij}^{k_3}\le\max\{a_{ij}^{k_1},a_{ij}^{k_2}\} \tag{3}$$

for any pair of indices (i, j).

This fact is denoted as $[P^{k_1},P^{k_2},P^{k_3}]$.

Axiom 7. If $[P^{k_1},P^{k_2},P^{k_3}]$, then $d(P^{k_1},P^{k_2})=d(P^{k_1},P^{k_3})+d(P^{k_3},P^{k_2})$.

This axiom is a special case of Axiom 3.

Given a set of preference orders $\{P^{(k)}\}$ = $\{P^1,\ldots, P^K\}$. The distance of some preference order P from this set is defined as follows

$$d\left(P,P^{(k)}\right)=\frac{1}{2}\sum_{i=1}^{n}\sum_{j=1}^{n}\sum_{k=1}^{K}d_{ij}\left(P,P^k\right)=\frac{1}{2}\sum_{i=1}^{n}\sum_{j=1}^{n}\sum_{k=1}^{K}d_{ij}^k=\frac{1}{2}\sum_{i=1}^{n}\sum_{j=1}^{n}\sum_{k=1}^{K}\left|a_{ij}^k-a_{ij}^P\right| \tag{4}$$

It follows from definition of coefficients a_{ij}^k (1) that

$$\left|a_{ij}^k-a_{ij}^P\right|+\left|a_{ji}^k-a_{ji}^P\right|=2\left|a_{ij}^k-a_{ij}^P\right|=2\left|a_{ji}^k-a_{ji}^P\right| \tag{5}$$

Hence the expression (4) can be rewritten as follows [6]

$$\begin{aligned}d\left(P,P^{(k)}\right)&=\frac{1}{2}\left(\sum_{\substack{i\\(i,j)\in I_P^{(1)}}}\sum_{j}\sum_{k=1}^{K}d_{ij}^k+\sum_{\substack{i\\(i,j)\in I_P^{(3)}}}\sum_{j}\sum_{k=1}^{K}d_{ij}^k\right)+\frac{1}{2}\left(\sum_{\substack{i\\(i,j)\in I_P^{(2)}}}\sum_{j}\sum_{k=1}^{K}d_{ij}^k+\sum_{\substack{i\\(j,i)\in I_P^{(2)}}}\sum_{j}\sum_{k=1}^{K}d_{ji}^k\right)\\&=\sum_{\substack{i\\(i,j)\in I_P^{(1)}}}\sum_{j}\sum_{k=1}^{K}d_{ij}^k+\sum_{\substack{i\\(i,j)\in I_P^{(2)}}}\sum_{j}\sum_{k=1}^{K}d_{ij}^k\end{aligned} \tag{6}$$

where:

$I_P^{(1)}$ -the set of indices (i,j) for which $O_i\succ O_j$ in the preference order P or - in other words - the set of indices for which $a_{ij}^P=1$,

$I_P^{(2)}$ -the set of indices (i,j) for which $O_i \approx O_j$ in the preference order P or - in other words - the set of indices for which $a_{ij}^P = 0$,

$I_P^{(3)}$ -the set of indices (i,j) for which $O_i \prec O_j$ in the preference order P or - in other words - the set of indices for which $a_{ij}^P = -1$.

It is evident that $(i,j) \in I_P^{(1)} \cup I_P^{(2)} \cup I_P^{(3)}$ is the whole set of indices occurring in the sum (6).

If $(i,j) \in I_P^{(1)}$, then $(j,i) \in I_P^{(3)}$ and if $(i,j) \in I_P^{(3)}$, then $(j,i) \in I_P^{(1)}$.
If $(i,j) \in I_P^{(2)}$, then $(j,i) \in I_P^{(2)}$.

Assume that in the preference order P $O_i \succ O_j$, i.e. $a_{ij}^P = 1$. In order to determine the distance of this preference order from a given set $\{P^{(k)}\}$, one can make use of coefficients defined as follows [11]

$$r_{ij} \underset{a_{ij}^P=1}{=} \sum_{k=1}^{K} d_{ij}\left(P, P^k\right) \underset{a_{ij}^P=1}{=} \sum_{k=1}^{K} \left|a_{ij}^k - a_{ij}^P\right| = \sum_{k=1}^{K} \left|a_{ij}^k - 1\right| \quad i, j = 1,\ldots n. \tag{7}$$

where r_{ij} are called the loss coefficients and the matrix $R=[r_{ij}]$ is called the loss matrix. It is assumed that $r_{ij}=0$ for all i=j. It should be noted that elements of the matrix R depend upon the form of preference orders P^k (k=1,...,K) only.

Making use of the coefficients r_{ij} (i, j = 1,...,n) the formulae (6) can be rewritten in the following form [6]:

$$d = \sum_{\substack{i \\ (i,j)\in I_P^{(1)}}} \sum_{j} r_{ij} + \sum_{\substack{i \\ (i,j)\in I_P^{(2)}}} \sum_{j} \sum_{k=1}^{K} \left|a_{ij}^k\right| \tag{8}$$

Definition 2 [11]. A preference order P^M such that

$$M\left(P^1, \ldots, P^K\right) = \arg\min_P d\left(P, P^{(k)}\right) \tag{9}$$

is called the *median* of a set $\left\{P^1, \ldots P^K\right\}$.

In other words it is such a preference order that in sense of the distance (8) is the "closest" one to all the preference orders of the set $\{P^{(k)}\}$.

To simplify considerations it is assumed that the set $I_P^{(2)}$ is empty. In other words it is assumed that equivalent alternatives do not occur in the median. Hence, one has to consider only the first term of the expression (8), i.e.

$$d = \sum_{i}\sum_{j} r_{ij} \quad (i,j) \in I_P^{(1)} \tag{10}$$

3.2. Properties of Kemeny's median

In his book Litvak [11] presented some theorems that may facilitate computing of Kemeny's median. First of them concerns the lower bound of the distance (10).

Theorem 1 [11]. The lower bound H of the distance (10) is equal to

$$H = \sum_{i=1}^{n}\sum_{j=1}^{n} \min\left(r_{ij}, r_{ji}\right) \tag{11}$$

This result directly follows from the relation (10).

It is easy to prove [6] that $r_{ji} = 2K\text{-}r_{ij}$, i, j = 1,…,n. Hence, making use of the expressions (7) and (10) it can be shown that the lower bound H is equal to the sum of loss coefficients less or equal K located under or above the matrix diagonal. If $r_{ij}=r_{ji}$, then in the sum H a given element occurs only once.

Also the theorem given in the paper [6] may be of some use. It shows how does the distance change in a case when all the elements in the preference order are reversed.

Theorem 2 [6]. Given a preference order $P=\left(O_{i_1},\ldots,O_{i_n}\right)$ for which the distance from the set of preference orders $\left\{P^{(k)}\right\}$ is equal d and the reversed preference order $\overline{P}=\left\{O_{i_n},\ldots,O_{i_1}\right\}$ for which the distance from this set of preference orders is equal $\overline{d}$. The following relation holds

$$\overline{d} = \frac{(n-1)n}{2} \cdot 2K - d \tag{12}$$

Both theorems make it possible to determine the range of values that the distance of a preference order P from the set of preference orders $\left\{P^{(k)}\right\}$ can assume.

The second group of theorems concerns the relationship between Kemeny's median and the group judgement determined with use of the method of Condorcet [5], [13], [14].

Given the set of experts' preference orders $\{P^{(k)}\} = \{P^1,\ldots,P^K\}$. For this set one can construct the so called outranking matrix S [14]:

$$S = \begin{array}{c|c|c|c|c}
 & O_1 & O_2 & \cdots & O_n \\ \hline
O_1 & - & s_{12} & \cdots & s_{1n} \\ \hline
O_2 & s_{21} & - & \cdots & s_{2n} \\ \hline
\vdots & \vdots & \vdots & \cdots & \vdots \\ \hline
O_n & s_{n1} & s_{n2} & \cdots & s_{nn}
\end{array} \tag{13}$$

where:

s_{ij} – is the number of experts in whose opinion $O_i \succ O_j$

s_{ji} – is the number of experts in whose judgements $O_j \succ O_i$ or $O_j \approx O_i$.

It is clear that $s_{ij}+s_{ji}=K$.

The *Condorcet winner* is an alternative such that $s_{ij} > K/2$ for each $j \neq i$. To avoid ambiguity it is assumed that the number of experts is odd [13].

Theorem 3 [11]. If the loss matrix R corresponding to a given set $\{P^{(k)}\}$ is consistent, i.e. $r_{ij} \le r_{ji}$, $r_{jl} \le r_{lj}$ $\Leftrightarrow$ $r_{il} \le r_{li}$, then a preference order $(O_{i_1},\ldots,O_{i_n})$ is Kemeny's median if and only if

$$r_{i_\nu i_{\nu+1}} \le r_{i_{\nu+1} i_\nu},\ \nu \in [1,\ldots, n-1]. \tag{14}$$

Definition 3 [11]. A set $\{P^{(k)}\}$ of preference orders has the Condorcet property if and only if there exists the Condorcet winner for any subset of this set.

Theorem 4 [11]. If the loss matrix corresponding to a given set $\{P^{(k)}\}$ is consistent, then this set has the Condorcet property. The converse of this theorem also holds true.

Theorem 5 [11]. If a given set of preference orders $\{P^{(k)}\}$ has the Condorcet property, then Kemeny's median of this set is formed by the Condorcet winners determined for subsequent subsets of $\{P^{(k)}\}$ and the distance (10) is equal to its lower bound.

In the paper [7] it is shown that the problem of determining Kemeny's median can be solved as an integer programming problem. The distance of a preference order P from the set of preference orders $\{P^{(k)}\}$ can be written as

$$d = \sum_{i=1}^{n}\sum_{j=1}^{n} y_{ij} r_{ij} \tag{15}$$

where:

$$y_{ij} = \begin{cases} 1 & \text{if } (i, j) \in I_P^{(1)} \\ 0 & \text{if } (i, j) \notin I_P^{(1)} \end{cases} \tag{16}$$

Hence, the problem of determining the Kemeny median for a given set $\{P^{(k)}\}$ can be formulated as follows

$$\min_{y_{ij}} \ d$$

subject to

$$y_{ij} + y_{ji} = 1, \quad \text{for } i \neq j \qquad \sum_{i=1}^{n}\sum_{j=1}^{n} y_{ij} = \frac{n(n-1)}{2}, \quad i, j = 1,\dots,n, \quad i \neq j \tag{17}$$

Moreover, the y_{ij} variables should be chosen in such a way that alternatives O_i (i=1,…,n) form a preference order. This can be written as

$$\text{there exist: } \begin{cases} i_1: & y_{i_1 j} = 1, \quad j = 1,\dots,n; j \neq i_1 \\ i_2 \neq i_1: & y_{i_2 j} = 1, \quad j = 1,\dots,n, \ j \neq i_1, i_2 \\ i_a \neq i_1, i_2, \dots i_{a-1}: & y_{i_a j} = 1, \quad j = 1,\dots,n, \ j \neq i_1, i_2, \dots i_{a-1}, i_a \\ & a = 2,\dots,n-1 \end{cases} \tag{18}$$

3.3. A heuristic algorithm for computing Kemeny's median

Generally, it is possible to compute Kemeny's median by solving the integer programming problem (17 - 18) using available software like CPLEX, LINDO packages or Solver library in MS Excel. The disadvantage of such approach is that it is rather complicated to apply this software in specialized interactive group decision support systems. So it is necessary to work out some heuristic algorithms making it possible to compute the median. That proposed is as follows. It consists of the following steps:

Step 1° For given expert judgements P^k (k=1,...K) the pairwise comparison matrices A^k are determined and the matrix R of r_{ij} coefficients (7) is computed.

Step 2° For each of two pairs (i,j) and (j,i) the coefficients $r_{ij}^* = \min(r_{ij}, r_{ji})$ are determined. The pair of elements (i,j) for which $r_{ij} = r_{ij}^*$ is denoted as $(i,j)^*$.

Step 3° A matrix $T=[t_{ij}]$ of elements $t_{ij} = \begin{cases} 1 & \text{for } r_{ij} = r_{ij}^* \\ 0 & \text{otherwise} \end{cases}$

and the vector $t_i=[t_1,...t_n]$ of elements $t_i = \sum_{j=1}^{n} t_{ij}$ are determined.

The value of $t^* = \max(t_1,...t_n)$ is also computed.

Step 4° The numbers of alternatives for which $t_i=t^*$ are determined; it is denoted as l_1. These alternatives are then placed according to their increasing numbers in the positions from 1 to l_1.

Step 5° Alternatives for which $t_i = t^* - \tau_j; \tau_j = 1,...,t^*$ are determined. The number of such alternatives is denoted as l_j. These alternatives are then placed according to their increasing numbers in the positions from $\sum_{w=1}^{j-1} l_w$ to $\sum_{w=1}^{j} l_w$. This procedure is repeated until the positions of all the alternatives are determined.

Step 6° For alternatives placed in positions from $\sum_{w=1}^{j-1} l_w$ to $\sum_{w=1}^{j} l_w$, they are denoted as $O_{i_{j1}},...,O_{i_{jl_j}}$, the submatrix of matrix T of these alternatives is determined and the elements $t_{i_{jz}} = \sum_{v=1}^{l_j} t_{i_{jz} i_{jv}}$ z=1,...,l_j are derived.

Step 7° The value of $t_{ij}^* = \max(t_{ij_1},...,t_{ij_{l_j}})$ is determined.

Step 8° For alternatives $O_{i_{j1}},...,O_{i_{jl_j}}$ the steps 4°, 5° are repeated until the positions of all the alternatives are determined. This procedure terminates when the order of only two alternatives is to be determined. If for these alternatives $r_{i_{jz}i_{jv}} = r_{i_{jv}i_{jz}}$, then the sequence $O_{i_{jz}}, O_{i_{jv}}$ as well as the sequence $O_{i_{jv}}, O_{i_{jz}}$ may occur in the median.

Step 9° The step 8° is repeated for all the groups of alternatives for which l_j>1, $j = s < n$.

Step 10° For each of the steps 4° to 9° one has to verify whether for given order of alternatives $O_{i_1},...O_{i_n}$ the condition $r_{i_t i_y} = r_{i_t i_y}^*$, $n \geq y > t$, t=1,...n-1, holds true. If for alternatives O_{i_t} and O_{i_y} it is not the case, then the whole sequence of alternatives (O_{i_t} ,..., O_{i_y}) is subject to further analysis.

Example 1. Given the set of 7 alternatives $\mathcal{O}=\{O_1, O_2, O_3, O_4, O_5, O_6, O_7\}$ and the set of 11 preference orders $\{P^1,..., P^{11}\}$ given by the experts:

	O_1	O_2	O_3	O_4	O_5	O_6	O_7
P^1	7	4	5	6	2	3	1
P^2	2	5	7	1	3	4	6
P^3	7	6	1	2	4	5	3
P^4	4	7	6	5	1	3	2
P^5	1	2	6	3	4	7	5
P^6	2	3	7	5	1	4	6
P^7	5	1	3	4	7	6	2
P^8	2	7	1	4	5	6	3
P^9	2	3	1	4	6	5	7
P^{10}	2	4	3	6	5	7	1
P^{11}	7	2	6	1	5	4	3

(19)

The numbers given in the k-th row (k = 1,...11) of this matrix define the position of a given alternative as determined by the k-th expert.

The loss coefficients matrix [R] is as follows

$$R = \begin{bmatrix} 0 & \mathbf{8} & 12 & \mathbf{10} & \mathbf{10} & \mathbf{8} & 12 \\ 14 & 0 & \mathbf{10} & \mathbf{10} & 12 & \mathbf{10} & \mathbf{10} \\ \mathbf{10} & 12 & 0 & \mathbf{10} & 12 & \mathbf{10} & 16 \\ 12 & 12 & 12 & 0 & \mathbf{8} & \mathbf{6} & \mathbf{10} \\ 12 & \mathbf{10} & \mathbf{10} & 14 & 0 & \mathbf{6} & 12 \\ 14 & 12 & 12 & 16 & 16 & 0 & 16 \\ \mathbf{10} & 12 & \mathbf{6} & 12 & \mathbf{10} & \mathbf{6} & 0 \end{bmatrix} \tag{20}$$

Summing up the elements of this matrix smaller than K=11 above and below the diagonal, the lower bound is obtained. It is equal to 188.

The T matrix is the following

	O_1	O_2	O_3	O_4	O_5	O_6	O_7	t_i
O_1	0	1	0	1	1	1	0	4
O_2	0	0	1	1	0	1	1	4
O_3	1	0	0	1	0	1	0	3
O_4	0	0	0	0	1	1	1	3
O_5	0	1	1	0	0	1	0	3
O_6	0	0	0	0	0	0	0	0
O_7	1	0	1	0	1	1	0	4

(21)

The T matrix implies the following preference order of the alternatives:

$$\begin{Bmatrix} O_1 \\ O_2 \\ O_7 \end{Bmatrix}, \begin{Bmatrix} O_3 \\ O_4 \\ O_5 \end{Bmatrix}, \{O_6\} \tag{22}$$

$r_{71} < r_{17}$, hence we have $O_7 \succ O_1$
$r_{27} < r_{72}$, hence we have $O_2 \succ O_7$
$r_{12} < r_{21}$, hence we have $O_1 \succ O_2$.

Therefore, the order of alternatives taking the first three positions can be O_2, O_7, O_1 (in this preference order there is the inconsistency $O_2 \succ O_1$; it should be $O_1 \succ O_2$) or O_1, O_2, O_7 (in this preference order there is the inconsistency $O_1 \succ O_7$; it should be $O_7 \succ O_1$).

We have $r_{34} < r_{43}$, hence $O_3 \succ O_4$
$r_{35} < r_{53}$, hence $O_3 \succ O_5$
$r_{45} < r_{54}$, hence $O_4 \succ O_5$.

Therefore the preference order of alternatives to be placed in the positions 4 to 6 is as follows O_3, O_4, O_5.

The final preference order may have the following form:

$$O_1, O_2, O_7, O_3, O_4, O_5, O_6. \tag{23}$$

The R matrix for this preference order is as follows:

	O_1	O_2	O_7	O_3	O_4	O_5	O_6	$\sum_{i,j\in I_p^{(l)}}^{n} r_{ij}$
O_1	0	8	12	12	10	10	8	60
O_2	14	0	10	10	10	12	10	52
O_7	10	12	0	6	12	10	6	34
O_3	10	12	16	0	10	12	10	32
O_4	12	12	10	12	0	8	6	14
O_5	12	10	12	10	14	0	6	6
O_6	14	12	16	12	16	16	0	
								d = 198

(24)

In this preference order there are five inconsistencies:

it should be $O_7 \succ O_1$ – due to this inconsistency the value of the distance is increased by 2
it should be $O_3 \succ O_1$ – the distance is increased by 2
it should be $O_5 \succ O_2$ – the distance is increased by 2
it should be $O_5 \succ O_3$ – the distance is increased by 2
it should be $O_4 \succ O_7$ – the distance is increased by 2.

These inconsistencies cannot be removed.

Hence the distance of the preference order (23) from the set (19) is equal to the lower bound H of the distance increased by the losses implied by the inconsistencies d = 188+(2+2+2+2+2) = 198.

The equivalent preference order is O_7, O_2, O_1, O_3, O_4, O_5, O_6.

4. Median Defined Using the Concept of a Preference Vector

4.1. **Basic notions**

In his book Litvak [11] introduced the concept of a so-called preference vector defined for a preference order. The definition is as follows:

Definition 4 [11]. The preference vector associated with a preference order P^k is defined as follows

$$\pi^k = \{\pi_1^k,...,\pi_n^k\} \tag{25}$$

where π_i^k is equal to the number of alternatives that precede the i-th one in the preference order under consideration.

For such vectors Litvak defined a measure of the distance.

Definition 5 [11]. Given two preference vectors π^{k_1} and π^{k_2}. A measure of the distance between them is defined as follows

$$d(\pi^{k_1},\pi^{k_2}) = \sum_{i=1}^{n} \left|\pi_i^{k_1} - \pi_i^{k_2}\right| \tag{26}$$

It can be proved [11] that a measure defined in such a way satisfies all axioms necessary to determine in a unique way the measure of preference vector closeness. They are as follows:

Axiom 1. $d(\pi^{k_1},\pi^{k_2}) \geq 0$, $d(\pi^{k_1},\pi^{k_2}) = 0$ if and only if π^{k_1} and π^{k_2} vectors are the same.

Axiom 2. $d(\pi^{k_1},\pi^{k_2}) = d(\pi^{k_2},\pi^{k_1})$

Axiom 3. $d(\pi^{k_1},\pi^{k_2}) \leq d(\pi^{k_1},\pi^{k_3}) + d(\pi^{k_3},\pi^{k_2})$

Axiom 4. If π^{k_1} and π^{k_2} vectors differ only in i-th component, then $d(\pi^{k_1},\pi^{k_2}) = \left|\pi_i^{k_1} - \pi_i^{k_2}\right|$.

It is said that the preference vector π^{k_3} is between preference vectors π^{k_1} and π^{k_2} if for each of its components the following inequality is satisfied

$$\min\{\pi_i^{k_1},\pi_i^{k_2}\} \leq \pi_i^{k_3} \leq \max\{\pi_i^{k_1},\pi_i^{k_2}\} .$$

This fact is denoted as $[\pi^{k_1},\pi^{k_2},\pi^{k_3}]$.

Axiom 5. If $[\pi^{k_1},\pi^{k_2},\pi^{k_3}]$, then $d(\pi^{k_1},\pi^{k_2}) = d(\pi^{k_1},\pi^{k_3}) + d(\pi^{k_3},\pi^{k_2})$.

This axiom is a special case of Axiom 3.

Definition 6 [11]. Given a set of preference orders $\{P^{(k)}\} = \{P^1,..., P^K\}$. The distance of some preference order P from this set is given by

$$d(\pi^P, \pi^{(k)}) = \sum_{i=1}^{n} \sum_{k=1}^{K} \left| \pi_i^P - \pi_i^k \right| \tag{27}$$

Let us introduce the following notation

$$\rho_i^{k(j)} = \left| \pi_{i(j)}^P - \pi_i^k \right| \quad i=1,...,n;\ k=1,...K \tag{28}$$

where $\pi_{i(j)}^P$ is the number of alternatives that precede the i-th one in the case when the i-th alternative takes the j-th position in the preference order P.

Summing coefficients $\rho_i^{k(j)}$ over k (k=1,...,K) we obtain $\rho_i^{(j)}$

$$\rho_i^{(j)} = \sum_{k=1}^{K} \rho_i^{k(j)}, \ j = 1, \ldots, n \tag{29}$$

This coefficient represents the aggregated difference among the position of the i-th alternative in the preference order P and its positions in P^k (k=1,...K).

Using this notation one can rewrite the expression (27) as follows

$$d = \sum_{i=1}^{n} \sum_{k=1}^{K} \left| \pi_i^P - \pi_i^k \right| = \sum_{i=1}^{n} \sum_{j=1}^{n} \rho_i^{(j)} x_{ij}, \tag{30}$$

where

$$x_{ij} = \begin{cases} 1 & \text{if the i-th alternative takes the j-th position in a preference order P} \\ 0 & \text{otherwise} \end{cases}$$

Making use of the notion of distance of some preference order P from the set of given preference orders Litvak introduced his definition of a median, being a counterpart of the definition formulated by J. Kemeny.

Definition 7 [11]. A preference order P^M such that

$$M(P^1, ..., P^K) = \arg\min_{\pi} d(\pi, \pi^{(k)}) \tag{31}$$

is called the *median* of a set $\{P^1, \ldots, P^K\}$.

In other words it is a preference order that in the sense of the distance (27) is the "closest" one to all the preference orders of the set $\{P^{(k)}\}$.

4.2. Properties of the median defined with the use of preference vectors

As in the case of Kemeny's median it is assumed that there are no equivalent alternatives in a median, so the number of alternative positions is equal n.

Theorem 6 [6]. The lower bound of the distance (30) is as follows

$$G = \sum_{j=1}^{n} \rho_{min}^{(j)}, \text{ where } \rho_{min}^{(j)} = \min_i [\rho_1^{(j)}, ..., \rho_n^{(j)}] \tag{32}$$

It should be emphasized that $\rho_{min}^{(j)}$ need not correspond to one O_i (i=1,...n) only. If this is the case for all O_i, then the preference order determined by $\rho_{min}^{(j)}$ is the median. Moreover, it should be noted that $\sum_{i=1}^{n}\rho_{i\,min}$ where $\rho_{i\,min}=\min_j[\rho_i^{(j)}],\ j=1,...,n$ determines the lower bound of the distance (30) if there is only one j for which $\rho_i^{(j)}=\rho_{i\,min}$.

In the case under consideration the problem of determining the median can be defined as the following zero-one integer programming problem [7], [11]

$$\sum_{i=1}^{n}\sum_{j=1}^{n}\rho_i^{(j)}x_{ij}\rightarrow\min \quad (33)$$

subject to

$$\sum_{j=1}^{n}x_{ij}=1,\quad i=1,...,n\,,\ \sum_{i=1}^{n}x_{ij}=1,\quad j=1,...,n \tag{34}$$

where the first constraint of (34) means that the i-th alternative should be placed in one position j only and the second one means that in each position j should be placed one alternative only.

4.3. Heuristic algorithm for the median defined with the use of preference vectors

Given the matrix $[\rho]$ with elements $\rho_i^{(j)}$, j = 1, ..., n; i=1, ..., n. A new variable $\delta_i^{(j)}$ is defined as follows

$$\delta_i^{(j)}=\rho_i^{(j)}-\rho_{min}^{(j)}\,,\ j=1,\ldots,n;\ i=1,\ldots,n. \tag{35}$$

Using this variable, one can rewrite the minimization problem (33) as

$$\begin{aligned}&\min_{x_{ij}}\sum_{i=1}^{n}\sum_{j=1}^{n}x_{ij}\rho_i^{(j)}=\min_{x_{ij}}\sum_{i=1}^{n}\sum_{j=1}^{n}x_{ij}(\rho_{min}^{(j)}+\delta_i^{(j)})=\\&\min_{x_{ij}}\sum_{i=1}^{n}\sum_{j=1}^{n}x_{ij}\rho_{min}^{(j)}+\min_{x_{ij}}\sum_{i=1}^{n}\sum_{j=1}^{n}x_{ij}\delta_i^{(j)}\end{aligned} \tag{36}$$

Taking into account that $\sum_{i=1}^{n} x_{ij} = 1$ we have

$$\min_{x_{ij}} \sum_{i=1}^{n}\sum_{j=1}^{n} x_{ij}\rho_i^{(j)} = \sum_{j=1}^{n} \rho_{\min}^{(j)} + \min_{x_{ij}} \sum_{i=1}^{n}\sum_{j=1}^{n} x_{ij}\delta_i^{(j)} \qquad (37)$$

Since $\sum_{j=1}^{n} \rho_{\min}^{(j)}$ is fixed for a given matrix [ρ], then the optimisation problem (33 - 34) reduces to determining the values of x_{ij} (j=1,…,n; i=1,…,n).

A heuristic algorithm described below can be used to solve this problem.

The matrix $[\delta_i^{(j)}]$ is constructed according to (35). A matrix $\Theta_0 = [\tau_a^{(j)}]$ is formed. The rows of this matrix correspond to natural numbers a = 0, 1, 2,… and columns correspond to columns of the matrix [ρ], j = 1,…, n. Elements $\tau_a^{(j)}$ are defined as follows

$$\tau_a^{(j)} = \begin{cases} O_i & \text{when } \delta_i^{(j)} = a \\ \phi & \text{otherwise} \end{cases} \qquad i = 1,\dots,n \qquad (38)$$

Step 1° A set $\mathcal{A}_0$ is formed; it consists of such elements O_i that appear only once in the first row of the Θ_0 matrix (a = 0).

Step 2° A set $\mathcal{A}_1$ is formed; it consists of such elements O_i that appear two or more times in the first row of the Θ_0 matrix (a = 0).

Step 3° Alternatives $O_i \in A_0$ are located in positions corresponding to the columns in which they appear in the matrix Θ_0.

Step 4° To each element of the set $\mathcal{A}_1$ a set of columns in which it appears is assigned: $O_{i_b} \in A_1 \Rightarrow J_{i_b} = \{j_{i_b 1},\dots,j_{i_b l_{i_b}}\}$, where l_{i_b} is the number of columns in which the alternative O_{i_b} appears.

Step 5° Alternatives $O_i \in A_0 \cup A_1$ are eliminated from all the rows of the matrix Θ_0 excluding the first one, i.e. that of a = 0. As a result a new matrix Θ_1 is formed.

Step 6° Among alternatives $O_{i_b} \in A_1$ that for which l_{i_b} is maximal is chosen, $l_{i_{bz}} = \max_{i_b}\{l_{i_b}\}$. It is denoted by $O_{i_{bz}}$ and the corresponding set of columns in which it appears is denoted as $J_{i_{bz}}$.

Let A_1 denotes the set of alternatives which do not appear in the first row of the matrix Θ_1 (a=0), $A_1 = \mathcal{O} - (A_0 \cup A_1)$. The column $j_{l_{bz}} \in I_{l_{bz}}$ in which the alternative $O_{i_{bz}}$ is to be placed is determined in the following way (columns determined in Step 3° are eliminated from the sets J_{i_b}).

$$\delta_{i_{bz}}^{(j_{i_{bz}})} = \max_{j \in J_{i_{bz}}} \min_{O_i \in A_1} \delta_i^{(j)} \qquad \text{where } \delta_{i_{bz}}^{(j)} = \min_{O_i \in A_1, j \in J_{i_{bz}}} \delta_i^{(j)} \tag{39}$$

A solution obtained as a result of solving the problem (39) need not be unique.

In the remaining columns $(J_{i_{bz}} - j_{i_{bz}})$ the alternatives are located according to the rule

$$\min_{O_i \in A_1} \delta_i^{(j)}. \tag{40}$$

When an alternative is located in its position, the sets J_{i_b} and the set A_1 are updated, i.e. the alternatives located in the columns $j_{i_b} \in J_{i_{bz}}$ are eliminated from the matrix Θ_1 as well as from the set A_1 and the corresponding columns j_{i_b} are not taken into account. As a result a new matrix Θ and a new set A are obtained.

If all the operations mentioned in Step 6° can be accomplished in a unique way, then the next $l_{i_{bz}}$ from the decreasing sequence l_{i_b} is chosen and this procedure is repeated for the updated matrix Θ.

It is terminated when in the updated Θ matrix only one alternative appears in rows $a > 0$.

Step 7° Having all the alternatives placed in chosen columns, one has to calculate the distance d according to the formulae (30).

Step 8° Positions of two subsequent alternatives are interchanged and the value of d is calculated. If the distance is decreased, then the positions of these alternatives are interchanged in the preference order determined in Step 6°. This procedure starts from the last alternative in the preference order determined in Step 6° or that for which $\delta_i^{(j)}$ is the largest.

Step 9° If some of the operations of Step 6° is not unique, then Steps 7° and 8° are to be repeated for all the preference orders under considerations.

Example 2. Given the set of 7 alternatives $\mathcal{O} = \{O_1, O_2, O_3, O_4, O_5, O_6, O_7\}$ and the set of 11 preference orders $\{P^1,...,P^{11}\}$ given by the experts:

	O_1	O_2	O_3	O_4	O_5	O_6	O_7
P^1	1	5	2	4	3	4	5
P^2	2	3	3	1	2	1	1
P^3	4	3	2	1	5	3	2
P^4	1	4	1	4	2	3	1
P^5	1	4	3	1	2	3	3
P^6	3	4	3	2	3	1	4
P^7	4	3	1	4	3	5	2
P^8	3	2	1	3	3	3	4
P^9	1	3	3	4	4	5	2
P^{10}	1	2	3	2	4	3	2
P^{11}	2	2	1	2	3	1	1

(41)

The numbers given in the k-th row (k = 1,...11) of this matrix define the position of a given alternative as determined by the k-th expert. If the same number corresponds to different alternatives then it means that they are equivalent.

$$[\rho_i^{(j)}] = \begin{bmatrix} 19 & 18 & 17 & 20 & 27 & 36 & 47 \\ 38 & 27 & 20 & 17 & \mathbf{18} & \mathbf{19} & \mathbf{28} \\ \mathbf{18} & \mathbf{15} & \mathbf{16} & 21 & 28 & 37 & 48 \\ 23 & 18 & 17 & 18 & 23 & 32 & 43 \\ 38 & 27 & \mathbf{16} & \mathbf{15} & \mathbf{18} & 23 & \mathbf{28} \\ 31 & 26 & 21 & 18 & 21 & 28 & 35 \\ 23 & 18 & 21 & 24 & 29 & 34 & 43 \end{bmatrix}$$

$$[\delta_i^{(j)}] = \begin{bmatrix} 1 & 3 & 1 & 5 & 9 & 17 & 19 \\ 20 & 12 & 4 & 2 & 0 & 0 & 0 \\ 0 & 0 & 0 & 6 & 10 & 18 & 20 \\ 5 & 3 & 1 & 3 & 5 & 13 & 15 \\ 20 & 12 & 0 & 0 & 0 & 4 & 0 \\ 13 & 11 & 5 & 3 & 3 & 9 & 7 \\ 5 & 3 & 5 & 9 & 11 & 15 & 15 \end{bmatrix}$$

$$\rho_{min}^{(j)} = \begin{bmatrix} 18 & 15 & 16 & 15 & 18 & 19 & 28 \end{bmatrix}$$

Hence the lower bound G of the distance is equal to 129.

Matrix Θ_0 is as follows:

a=	(1)	(2)	(3)	(4)	(5)	(6)	(7)
0	O_3	O_3	O_3,O_5	O_5	O_2,O_5	O_2	O_2,O_5
1	O_1		O_1,O_4				
2				O_2			
3		O_1,O_4,O_7		O_4,O_6	O_6		
4			O_2			O_5	
5	O_4,O_7		O_6,O_7	O_1	O_4		
6				O_3			
7							O_6
8							
9				O_7	O_1	O_6	
10					O_3		
11		O_6			O_7		
12		O_2,O_5					
13	O_6					O_4	
14							
15						O_7	O_4,O_7
16							
17						O_1	
18						O_3	
19							O_1
20	O_2,O_5						O_3

The steps of the algorithm are the following:

Step 1° $\mathcal{A}_0 = \Phi$

Step 2° $\mathcal{A}_1 = \{O_2, O_3, O_5\}$

Step 3° There are no $O_i \in \mathcal{A}_0$

Step 4° $J_2 = \{5, 6, 7\}$, $l_2 = 3$

$J_3 = \{1, 2, 3\}$, $l_3 = 3$

$J_5 = \{3, 4, 5, 7\}$, $l_5 = 4$

Step 5° Alternatives $O_i \in A_0 \cup A_1$ are deleted from the Θ_0 matrix; the Θ_1 matrix and the set $A_1 = \{O_1, O_4, O_6, O_7\}$ are formed.

Matrix Θ_1 is as follows

a=	(1)	(2)	(3)	(4)	(5)	(6)	(7)
0	O_3	O_3	O_3,O_5	O_5	O_2,O_5	O_2	O_2,O_5
1	O_1		O_1,O_4				
2							
3		O_1,O_4,O_7		O_4,O_6	O_6		
4							
5	O_4,O_7		O_6,O_7	O_1	O_4		
6							
7							O_6
8							
9				O_7	O_1	O_6	
10							
11		O_6			O_7		
12							
13		O_6				O_4	
14							
15						O_7	O_4,O_7
16							
17						O_1	
18							
19							O_1
20							

Step 6° The alternative for which $l_{i_b} = l_{i_{bz}}$ is to be chosen: $i_{bz} = 5$, $l_5 = 4$, hence the position of O_5 is to be determined first. It is accomplished using the rule (39):

$$\max_{j \in J_5} \min_{O_i \in A_i} \delta_i^{(j)} = \max_{j \in J_5} \left\{ \delta_1^{(j)}, \delta_4^{(j)} \right\} = \delta_1^{(7)} = 19 .$$

The alternative O_5 is placed in the 7th position, i.e. $j_5=7$ and the set J_5 is updated

$$J'_5 = J_5 \setminus \{ j_5 \} = J_5 \setminus \{7\}$$

Alternatives $O_i \in A_1$ are to be located in the positions $j_b \in J'_5$ according to the rule (40). We have $\min_{O_i \in A_i, j \in J'_5} \delta_i^{(j)} = \delta_1^{(3)} = \delta_4^{(3)} = 1$.

Hence, in the 3rd position the alternative O_1 as well as O_4 can be located. In the first case the alternative O_4 is placed in the 4th position and O_6 in the 5th position. In the second case we have $\min_{O_i \in A_i, j \in J'_5} \delta_i^{(j)} = \delta_6^{(4)} = \delta_6^{(5)} = 3$.

So O_6 can take the 4th position as well as the 5th one and O_1 can take the 5th one.

As a result the following segments of a searched for preference order are determined:

	Positions	1	2	3	4	5	6	7
Case 1°	Alternatives			O_1	O_4	O_6		O_5
Case 2°	Alternatives			O_4	O_6	O_1		O_5

Now, the set $J'_3 = J_3 \setminus \{3\}$ is to be analyzed. The set $A'_1 = \{O_7\}$. Using the rule (39) we have $\max_{j \in J'_3} \min_{O_i \in A'_i} \delta_i^{(j)} = \max_{j \in J'_3} \{\delta_7^{(j)}\} = \delta_7^{(1)} = 5$.

Hence O_3 is to be placed in the 1st position and O_7 in the 2nd position. So O_2 has to take the 6th position.

The resulting preference orders to be analysed are as follows:

	Positions	1	2	3	4	5	6	7
Case 1°	Alternatives	O_3	O_7	O_1	O_4	O_6	O_2	O_5
Case 2°	Alternatives	O_3	O_7	O_4	O_6	O_1	O_2	O_5

Now, one has to check whether the distance (30) can be decreased by interchanging the positions of subsequent elements.

Case	Positions	1	2	3	4	5	6	7	$\sum_i \delta_i^{(j)}$	d	Alternatives which positions are to be interchanged
1°	Alternatives	O_3	O_7	O_1	O_4	O_6	O_2	O_5		139	
	$\delta_i^{(j)}$	0	3	1	3	3	0	0	10		
2°	Alternatives	O_3	O_7	O_4	O_6	O_1	O_2	O_5		145	
	$\delta_i^{(j)}$	0	3	1	3	9	0	0	16		O_1, O_6
	Alternatives	O_3	O_7	O_4	O_1	O_6	O_2	O_5		141	
	$\delta_i^{(j)}$	0	3	1	5	3	0	0	12		O_4, O_1 – preference order of the Case 1° results

It is evident that for the preference order $O_1, O_7, O_3, O_4, O_6, O_2, O_5$ the distance d is the same as for that given for the Case 1° in Table 3.

5. Ranking of University Teachers

In the spring 2001 a group of students of the University of Information Technology and Management in Warsaw was asked to present their opinions on the teaching skills of their teachers. For this purpose a special questionnaire was prepared. From the whole group of students presenting their judgements two sets were chosen. The first consisted of 17 third year-students and the second of 12 second year-students. Their questionnaires were complete in the sense that they gave their judgement on the entire group of teachers under examination (12 teachers for the third year-students and 9 for the second year-students).

5.1. Judgements given by the third year-students

Student judgements are as follows:

P^1 : $O_5, (O_3, O_4, O_6, O_8, O_9, O_{12}), (O_1, O_{11}), O_2, (O_7, O_{10})$

P^2 : $(O_3, O_4, O_5), (O_6, O_7, O_{11}, O_{12}), (O_2, O_8, O_9), O_1, O_{10}$

P^3 : $(O_2, O_{11}), (O_1, O_3, O_4, O_5, O_8, O_9, O_{10}, O_{12}), (O_6, O_7)$

P^4 : $(O_4, O_5), (O_2, O_3, O_6, O_7, O_8, O_{11}, O_{12}), (O_1, O_9), O_{10}$

P^5 : $(O_3, O_4, O_6), (O_5, O_7, O_8, O_{11}), (O_1, O_9, O_{12}), O_2, O_{10}$

P^6 : $(O_1, O_3, O_4, O_5, O_6), (O_{11}, O_{12}), (O_7, O_9), (O_2, O_8), O_{10}$

P^7 : $(O_1, O_2, O_3, O_4, O_6), (O_5, O_7, O_8, O_{11}, O_{12}), (O_9, O_{10})$

P^8 : $(O_1, O_3, O_4, O_5, O_6, O_{11}), O_{12}, (O_7, O_9), (O_2, O_8), O_{10}$

P^9 : $(O_3, O_4, O_{11}), (O_5, O_6, O_8, O_{12}), (O_1, O_2, O_9, O_{10}), O_7$

P^{10} : $(O_3, O_4, O_6, O_{11}), O_5, (O_1, O_2, O_8, O_{10}, O_{12}), (O_7, O_9),$

P^{11} : $(O_3, O_4, O_5, O_6, O_{11}), (O_1, O_2, O_7, O_8, O_{12}), (O_9, O_{10})$

P^{12} : $(O_3, O_4), (O_1, O_2, O_5, O_6, O_7, O_8, O_9, O_{10}, O_{11}, O_{12})$

P^{13} : $O_4, (O_2, O_3, O_5, O_6, O_7, O_8), (O_9, O_{11}, O_{12}), (O_1, O_{10})$

P^{14} : $O_6, (O_3, O_4, O_7, O_8, O_{11}, O_{12}), (O_2, O_5), (O_1, O_9), O_{10}$

P^{15} : $(O_4, O_5), (O_3, O_7, O_8, O_{11}, O_{12}), (O_1, O_2, O_6, O_9), O_{10}$

P^{16} : $(O_1, O_2, O_3, O_4, O_6, O_{11}, O_{12}), (O_5, O_9), (O_7, O_8), O_{10}$

P^{17} : $(O_3, O_4), (O_1, O_2, O_6, O_{11}, O_{12}), (O_5, (O_7, O_8, O_9), O_{10}$

Brackets are used to indicate that in opinion of a particular expert a given group of alternatives is considered as equivalent. Teachers placed in the first position are considered to be the best from the point of view of the criterion applied.

The corresponding $\left[\delta_i^{(j)}\right]$matrix has the following form

$$\begin{bmatrix} 78 & 59 & 39 & 30 & 24 & 18 & 16 & 28 & 30 & 33 & 55 & 73 \\ 78 & 57 & 37 & 28 & 22 & 16 & 14 & 26 & 28 & 31 & 55 & 73 \\ 5 & \mathbf{0} & \mathbf{0} & 11 & 25 & 39 & 53 & 81 & 91 & 102 & 128 & 146 \\ \mathbf{0} & 1 & 5 & 16 & 30 & 44 & 58 & 86 & 96 & 107 & 133 & 151 \\ 37 & 24 & 12 & 9 & 13 & 19 & 27 & 49 & 59 & 70 & 96 & 114 \\ 27 & 16 & 8 & 11 & 21 & 31 & 41 & 65 & 73 & 82 & 106 & 124 \\ 91 & 64 & 40 & 27 & 21 & 15 & 13 & 25 & 25 & 26 & 44 & 60 \\ 69 & 42 & 20 & 11 & 9 & 7 & 11 & 29 & 33 & 38 & 64 & 82 \\ 112 & 85 & 59 & 42 & 28 & 14 & \mathbf{0} & \mathbf{0} & \mathbf{0} & 1 & 21 & 39 \\ 151 & 124 & 96 & 77 & 61 & 45 & 31 & 31 & 15 & \mathbf{0} & \mathbf{0} & \mathbf{0} \\ 35 & 20 & 6 & 5 & 11 & 17 & 27 & 51 & 61 & 72 & 98 & 116 \\ 54 & 29 & 7 & \mathbf{0} & \mathbf{0} & \mathbf{0} & 8 & 32 & 42 & 53 & 79 & 97 \end{bmatrix}$$

The lower bound of the distance G=357.

The preference order obtained with use of the heuristic algorithm presented in Section 3.3 is as follows

$$O_4, O_3, O_1, O_6, O_{11}, O_{12}, O_5, O_7, O_9, O_2, O_8, O_{10}.$$

For such a preference order the distance (30) is equal to 565.

Interchanging the positions of alternatives O_1, O_2, O_5, O_6, O_7, O_8, O_9, O_{11}, O_{12} the following preference order is obtained

$$O_4, O_3, O_6, O_{11}, O_5, O_{12}, O_8, O_1, O_2, O_9, O_7, O_{10}$$

and the distance (30) is reduced to 495.

5.2. Judgements given by the second year-students

Student judgements are as follows:

P^1 : $O_2, (O_1, O_4, O_5, O_6, O_9), O_3, O_7, O_8$
P^2 : $(O_1, O_2, O_4, O_9), (O_6, O_8), (O_5, O_7), O_3$
P^3 : $(O_1, O_9), (O_2, O_4, O_6, O_8), (O_5, O_7), O_3$
P^4 : $(O_1, O_4), (O_7, O_9), (O_2, O_8), (O_3, O_6), O_5$
P^5 : $O_1, (O_3, O_4, O_5, O_6, O_7, O_9), O_8, O_2$
P^6 : $(O_1, O_2, O_4), (O_6, O_8), (O_5, O_7, O_9), O_3$
P^7 : $(O_5, O_6), O_1, (O_2, O_4, O_9), (O_3, O_8), O_7$
P^8 : $O_1, (O_2, O_3, O_4, O_5, O_6, O_8, O_9), O_7$
P^9 : $O_8, (O_1, O_2, O_4, O_5, O_7, O_9), (O_3, O_6)$
P^{10} : $(O_1, O_4), (O_2, O_6, O_8, O_9), (O_3, O_5, O_7)$
P^{11} : $(O_1, O_4), O_7, (O_2, O_8, O_9), (O_5, O_6), O_3,$
P^{12} : $(O_1, O_4, O_6), (O_3, O_5, O_9), O_2, (O_7, O_8)$

The corresponding $\left[\delta_i^{(j)}\right]$ matrix has the following form

$$\begin{bmatrix} \mathbf{0} & 1 & 6 & 14 & 19 & 30 & 46 & 56 & 64 \\ 26 & 15 & 8 & 6 & 5 & 12 & 24 & 32 & 38 \\ 64 & 47 & 34 & 22 & 9 & 2 & \mathbf{0} & \mathbf{0} & \mathbf{0} \\ 5 & \mathbf{0} & 3 & 9 & 14 & 25 & 41 & 51 & 59 \\ 40 & 25 & 18 & 12 & 5 & 4 & 10 & 18 & 24 \\ 29 & 16 & 9 & 7 & 4 & 9 & 19 & 27 & 35 \\ 54 & 37 & 24 & 16 & 5 & \mathbf{0} & 2 & 6 & 10 \\ 43 & 28 & 15 & 7 & \mathbf{0} & 3 & 11 & 15 & 21 \\ 18 & 5 & \mathbf{0} & \mathbf{0} & 3 & 12 & 28 & 38 & 46 \end{bmatrix}$$

The lower bound of the distance G=170.

The preference order obtained with use of the heuristic algorithm presented in Section 3.3 is as follows $O_1, O_4, O_9, O_2, O_8, O_7, O_5, O_6, O_3$.

For this preference order the distance (30) is equal to 213.

Interchanging the positions of alternatives O_6, O_7, O_8 the following preference order is obtained

$O_1, O_4, O_9, O_2, O_6, O_8, O_5, O_7, O_3$

and the distance is reduced to 199.

6. Concluding Remarks

In the paper two definitions of a median for preference orders are presented and analysed in detail: the classic one due to Kemeny, based on pairwise comparisons and that of Litvak, based on the concept of preference vectors. Calculations performed point out that these medians need not be the same as it is shown in the below tables.

Example 1	Preference order	Distance defined by	
		J.Kemeny	B.G. Litvak
Kemeny's median	$O_1, O_2, O_7, O_3, O_4, O_5, O_6$	198	162
Litvak's median	$O_7, O_1, O_2, O_4, O_5, O_3, O_6$	198	148

It follows from this table that for the set of expert judgements $\{P^{(k)}\}$ (19) preference orders obtained with use of Kemeny as well as Litvak method are different but they are both Kemeny's medians. On the other side preference order obtained with use of Kemeny's method is not the median in the sense of Litvak.

Example 2	Preference order	Distance defined by	
		J.Kemeny	B.G. Litvak
Kemeny's median	$O_1, O_3, O_4, O_7, O_5, O_6, O_2$	175	151
Litvak's median	$O_3, O_7, O_1, O_4, O_6, O_2, O_5$	183	139

As it is shown in the above table, for the set of expert judgements $\{P^{(k)}\}$ (41) preference orders obtained with use of Kemeny as well as Litvak method are different. Preference order obtained with use of Kemeny's method is not the Litvak's median nor preference order obtained with use of Litvak's method is the median in the sense of Kemeny.

It would be of some interest to derive the relationship between the medians under consideration. One of possible approaches to solve this problem can be as follows.

Let us introduce the following notation.

$$J_1^{ik} = \{j : a_{ij}^k = 1\}, \; J_2^{ik} = \{j : a_{ij}^k = 0\}, \; J_3^{ik} = \{j : a_{ij}^k = -1\}, \; i=1,\dots n, \; k=1,\dots K, \quad (42)$$

It is evident that

$$J_1^{ik} \cap J_2^{ik} \cap J_3^{ik} = \phi \text{ and } J_1^{ik} \cup J_2^{ik} \cup J_3^{ik} = \{1,\dots,n\} \quad (43)$$

for all i and k.

Moreover, the following equalities are satisfied

$$j \in J_1^{ik} : \left| a_{ij}^k - 1 \right| = 0; \qquad j \in J_2^{ik} : \left| a_{ij}^k - 1 \right| = 1; \qquad j \in J_3^{ik} : \left| a_{ij}^k - 1 \right| = 2 \tag{44}$$

for all i and k.

Hence, making use of (7), we have for the given i and j

$$r_{ij} = \sum_{k=1}^{K} \left| a_{ij}^k - 1 \right| = \sum_{k=1}^{K} \left[0 \cdot \text{card}(j \in J_1^{ik}) + 1 \cdot \text{card}(j \in J_2^{ik}) + 2 \cdot \text{card}(j \in J_3^{ik}) \right] \tag{45}$$

where the symbol card(·) is used to denote the number of elements of a set.

Therefore, the distance defined by Kemeny d^K (10) can be written as follows

$$d^K = \sum_{\substack{i \\ (i,j) \in I_P^{(1)}}} \sum_{j} r_{ij} = \sum_{\substack{i \\ (i,j) \in I_P^{(1)}}} \sum_{j} \sum_{k=1}^{K} \left[\text{card}(j \in J_2^{ik}) + 2\text{card}(j \in J^{ik}{}_3) \right] \tag{46}$$

The distance d^L based on the definition given by Litvak can be written in the following form

$$d^L = \sum_{i=1}^{n} \sum_{k=1}^{K} \left| \pi_i^P - \pi_i^k \right| = \sum_{i=1}^{n} \sum_{k=1}^{K} \left| \text{card}\, I_{Pi}^{(3)} - \text{card}\, J_3^{ik} \right| \tag{47}$$

where $I_{Pi}^{(3)} = \{ j : a_{ij}^P = -1 \}$. It is evident that $I_{Pi}^{(3)} \subset I_P^{(3)}$.

The derived expressions can be used to determine the relationship between the definitions of median under examination. However, the analysis is very tedious and further studies are necessary to obtain an expression to be used for practical purposes.

References

[1] Aizerman M. and F. Aleskerov. Theory of Choice. North Holland, Amsterdam-New York-Tokyo, 1995.

[2] Black D. The Theory of Committees and Elections. Cambridge University Press, Cambridge 1958.

[3] de Borda J-C. Memoire sur les elections au Scrutin Histoire de l'Academie Royale des Sciences 1781.

[4] Le Breton M. and M. Truchon. A Borda measure for social choice functions. Mathematical Social Sciences, vol. 34, 1997.

[5] de Condorcet M. Essai sur l'Application de l'Analyse à la Probabilité des Rendues à la Pluralité des Voix. Paris 1785.

[6] Bury H., Petriczek G. and D. Wagner. Determining of group opinion with Kemeny's median method (in Polish). In: Preference Modelling and Risk' 99, ed. T. Trzaskalik, Katowice 1999.

[7] Bury H. and D. Wagner. The use of Kemeny's median for group decision making. Integer programming approach. Proceedings of the Sixth International Conference on Methods and Models in Automation and Robotics, Międzyzdroje, 28-31 August 2000.

[8] Kelly J.S. Arrow impossibility theorems. Academic Press, New York, 1978.

[9] Kemeny J. Mathematics without numbers, Daedalus 88, 1959.

[10] Kemeny J. and L. Snell. Mathematical Models in the Social Sciences. Ginn. Boston 1960.

[11] Litvak B.G. Expert Information. Methods of acquisition and analysis (in Russian). Radio and Communication, Moscow 1982.

[12] Michaud P. The true rule of the Marquis de Condorcet. In: Compromise, Negotiation and Group Decision, eds. B.R.Munier, M.F.Shakun, D.Reidel Publishing Company, Dordrecht 1988.

[13] Nurmi H. Comparing voting systems, Dordrecht, D.Reidel, 1987.

[14] Nurmi H. Some techniques of preference profile analysis. Paper presented at NPO meeting on "Power and Fairness", Bad Segeberg, September 3-6, 2000.

[15] Nurmi H. Voting Paradoxes and How to Deal with Them. Springer Verlag, Berlin-Heidelberg-New York, 1992.

[16] Nurmi H. Voting procedures: a summary analysis. British Journal of Political Science, vol.13, 1983.

[17] Saari D.G. Basic Geometry of Voting. Springer Verlag, Berlin Heidelberg, New York, 1995.

[18] Saari D.G. Explaining positional voting paradoxes I. The simple case. NU Center for Mathematical Economy, discussion paper No. 1179, January 1997.

[19] Saari D.G. and V.R. Merlin. A geometric examination of Kemeny's rule. Social Choice and Welfare, vol. 17, 2000.

[20] Slater P. Inconsistencies in a schedule of paired comparisons. Biometrica vol. 48, 1961.

[21] Young H.P. and A. Levenglick. A consistent extension of Condorcet's election principle. SIAM Journal of Applied Mathematics, vol.35, no 2, September 1978.

An Internet-based Group Decision and Consensus Reaching Support System

Sławomir Zadrożny and Janusz Kacprzyk

Systems Research Institute, Polish Academy of Sciences
ul. Newelska 6, 01-447 Warsaw, Poland
and
Warsaw School of Information Technology
ul. Newelska 6, 01-447 Warsaw, Poland
E-mail: kacprzyk@ibspan.waw.pl

Abstract. We present an idea of a group decision making and consensus reaching decision support system and its implementation within the framework of the Internet (WWW) service are presented. The focus is on the support of the consensus reaching process. The preferences are represented using linguistic terms and aggregation schemes employ various soft computing techniques. The point of departure is a multi-criteria evaluation of particular options given by each individual. These are processed to obtain various agreement assessment measures.

Keywords: group decision support system, consensus, fuzzy preferences, linguistic preferences, fuzzy majority, multicriteria choice, Internet, WWW.

1. Introduction

We deal here with a decision support task in the following setting. We assume that there is a set of options characterised by a set of attributes/criteria, and a group of individuals (decision-makers, experts etc.). The individuals discuss the issues under consideration and present their opinions providing *values* for the attributes/criteria of all options, as well as importance weights for the particular criteria/attributes. The former values belong to a set of linguistic terms expressing the compatibility of a particular option (in respect to a given attribute) with an overall goal of the decision problem considered. It may happen that a given individual is not able to assess the aforementioned compatibility or that a given attribute/criterion is not applicable for the option considered. Hence, additional linguistic terms are available for this. Then, our consensus reaching support system automatically generates a preference structure corresponding to a given characterisation of all options in terms of the attributes/criteria considered. More precisely, it is an additive preference structure, as recently proposed in [16].

The next step is the assessment of (an extent of) consensus in the opinion of group members. It may be done on three levels: (1) an attributes/criteria evaluation, (2) a generated preference structure, or (3) implied individual choice sets. In this paper we concentrate on the second type of consensus evaluation, further developing our previous approach (see [4]). The other consensus evaluation schemes are only briefly discussed. Possible hints for a further group discussion, leading to a higher consensus degree, are also briefly mentioned.

We conclude the paper with the presentation of a preliminary implementation of the system. Recent advancements in ICT (Information and Communication Technology) have triggered a spectacular development of the Internet. One of the most popular services available is there the World Wide Web (WWW). The ease of access, use and maintenance of WWW based applications makes them a well-suited solution for group work tasks. These applications are equally attractive in case of spatially distributed groups as well as those connected by local area networks. In the former case, the Internet becomes a medium of communication and, in the latter, the very same technology may be used in the Intranet.

In particular, the WWW based applications may be useful for a decision making support. They provide an attractive and flexible user interface to the core of a decision support system as well as the access to a plethora of information available on the Internet. Thus, the decision-maker may have an immediate access to all data relevant to the problem. In case of group decision making, such systems may easily facilitate an open discussion between group members as well as support various processes within the group in a more structured way.

2. Source Data

Here we assume a classical group decision making setting. There is a *group of individuals* (decision makers, experts, etc.) and a *set of options* being potential solutions to the decision problem under consideration (as we concentrate on the support of the consensus reaching process, it is not important if one or more options may be chosen as a final solution). Moreover, each option is characterised by a *set of criteria/attributes* - in what follows we will employ the term criteria. For the time being, we just assume that the set of criteria is given. Each individual assesses how good a candidate solution an option is with respect to each criterion. These assessments are presented using linguistic labels such as *Very good*, *Good*, *Bad*, etc. A possible *inapplicability* of a criterion for a given option or the *inability* of an individual to rate an option with respect to a given criterion are represented by additional linguistic labels, *NA* and *UNK*, respectively.

Formally, let $X=\{x_1,...,x_M\}$ and $S=\{s_1,...,s_N\}$ denote the set of individuals and options, respectively. Let $C=\{c_1,...,c_K\}$ and $W^m=\{w_1^m,\dots,w_K^m\}$, $w_k\in[0,1]$, be sets of criteria characterising particular options and importance weights of these criteria, respectively. Notice that we assume the same set of criteria for all individuals and options but the importance weights are assigned to particular criteria independently by each individual. Additionally, we assume that $\max_k w_k^m = 1, \forall m$. Let

$$E_m: S\times C\rightarrow LL\cup\{NA, UNK\}$$

be a mapping representing the assessment of particular options as given by individual x_m, where LL=\{*Very good*, *Good*, *Fair*, *Bad*, *Very bad*\}. We assume that there is a linear order $\succ$ defined on the set LL such that $\forall l_1,l_2\in LL$ $l_1\succ l_2$ means that label l_1 denotes a better performance of an option than label l_2. We assume that "*Very good* $\succ$ *Good* $\succ$ *Fair* $\succ$ *Bad* $\succ$ *Very bad*".

The interpretation of *NA* and *UNK* is as given in the previous paragraph. In what follows, we will refer to the elements of the set *LL* as *regular values*.

Example. Let the decision problem consist in the selection of proper candidates for the dean position at a university. The group of individuals is assessing the quality of particular candidates with respect to the following criteria: age, education, experience, self-presentation during an interview, language skills and the quality of scientific publications. We assume that, for an individual, some facts about a candidate may be unknown (e.g., their publications) or may be not applicable (e.g., an interview with some candidates was not conducted). As mentioned, each individual may assign different weights to particular criteria.

Then, the part of source data of the m-th individual may look as follows:

> *S*={Goerge Adams, John Doe, Brian Kelly, John Smith, James Twain, Goerge Welsh}
> *C*={*Age, Education, Experience, Interview, Language skills, Publications*}
> W^m={0.6, 1.0,1.0,0.5,0.8,0.7}
> E_m(John Smith,Age) = Good,
> E_m(John Smith,Education)=Fair,
> E_m(John Smith,Experience)=Fair,
> E_m(John Smith,Interview)=NA,
> E_m(John Smith,Language skills)=Very good,
> E_m(John Smith,Publications)=UNK

3. Preference Structure

The source data mentioned above form the lowest level on which consensus may be sought. An agreement among the individuals on such a low level may be unnecessary. Even if a basic evaluation of particular options, in terms of assigned linguistic labels, may be quite different, still implied preferences on pairs of options may be completely satisfactory. Thus in the next step the system automatically generates for each individual a *preference structure* corresponding in a specific sense to the input data he or she has given.

We adopt here the approach and notation proposed in [16], with some simplifications. A fuzzy preference structure is a triple of binary fuzzy relations (P,I,J) defined on the set $S \times S$. Originally, in [16], it is a sixtuple containing additionally the set of alternatives (options), S, as well as a t-norm and t-conorm used to express required properties of the relations. We will omit here these additional elements as the set of options is here always S, and we assume the Łukasiewicz t-norm and t-conorm (as advocated in [15,16]).

The binary fuzzy relations in the triple correspond to classical notions of strong preference, indifference and incomparability, respectively. Thus, usual properties are required for these relations, including the antisymmetry for P, and the symmetry for I and J. The completeness of the fuzzy preference structure is expressed, due to the use of the Łukasiewicz connectives, as $\mu_P(.,.) + \mu_{P^t}(.,.) + \mu_I(.,.) + \mu_J(.,.) = 1$, where P^t is the transpose of P, i.e., $P^t(s_i,s_j) = P(s_j,s_i)$. The completeness of the preference structure implies the pairwise disjointness, in the sense of the Łukasiewicz conjunction, of all relations P, P^t, I, J.

A fuzzy preference structure is automatically generated according to the following reasoning. Let us assume the following notation (we omit indices referring to the individual as all these computations are done separately for each individual):

$$UC : S \to 2^C \qquad UC(s_i) = \{k : E(s_i, c_k) = UNK\}$$

$$NC : S \to 2^C \qquad NC(s_i) = \{k : E(s_i, c_k) = NA\}$$

$$KC : S \to 2^C ; KC(s_i) = \{k : E(s_i, c_k) \notin \{UNK, NA\}\}$$

The above definitions distinguish some useful crisp subsets of indices of criteria, namely, those taking the value *UNK* (*UC*), the value *NA* (*NC*), and the values different from *UNK* and *NA* (*KC*).

Next, we define for each pair of options four other auxiliary entities:

$$ALLC : S \times S \to [1, K]; \quad ALLC(s_i, s_j) = \sum_{k \in AC} w_k$$

where $AC = \{k : k \notin NC(s_i) \cap NC(s_j)\}$

$$FFC: S \times S \rightarrow [1, K]; \quad FFC(s_i, s_j) = \sum_{k \in FF} w_k;$$

where FF
$$= \{k : k \in KC(s_i) \cap KC(s_j) \wedge E(s_i, c_k) \succ E(s_j, c_k)\}$$

$$FBC: S \times S \rightarrow [1, K]; \quad FBC(s_i, s_j) = \sum_{k \in FB} w_k$$

where FB
$$= \{k : k \in KC(s_i) \cap KC(s_j) \wedge E(s_i, c_k) = E(s_j, c_k)\}$$

In all the above formulas, the set of criteria is meant to be a fuzzy set with membership degrees equal to the importance weights (w_i) of the particular criteria.

Thus, *ALLC* is equal to the cardinality (in the sense of $\sum$Counts) of the fuzzy set of criteria that do not simultaneously take value *NA* for both options. Intuitively, all the three remaining entities - *FFC*, *FSC* and *FBC* - count (in the same sense as *ALLC*) the criteria taking regular values for compared options and favouring the first options from the pair (*FFC*), the second (*FSC*), or not differentiating the options (*FBC*).

Then, the following formulas are used to compute the membership functions of the particular fuzzy preference relations constituting our fuzzy preference structure:

$$\mu_P(s_i, s_j) = FFC(s_i, s_j) / ALLC$$

$$\mu_{P^t}(s_i, s_j) = FSC(s_i, s_j) / ALLC$$

$$\mu_I(s_i, s_j) = FBC(s_i, s_j) / ALLC$$

$$\mu_J(s_i, s_j) = \frac{\sum_{k \in UC(s_i) \cup UC(s_j) \cup NC(s_i) \cup NC(s_j) - NC(s_i) \cap NC(s_j)} w_k}{ALLC}$$

The following reasoning supports the adopted formulas. A degree of strong preference, P, of option s_i over option s_j should be proportional to the fraction of the number of criteria for which option s_i obtained a better score than s_j. Thus, a degree of strong preference is assumed to be equal to this fraction. The calculation of this fraction does not take into account the criteria that were classified by an individual as non-applicable for both options. If it were not arranged that way, such criteria - completely irrelevant for the comparison of that pair of options - would artificially diminish a degree of strong preference (the same applies to the indifference and incomparability relation). Exactly the same idea is applied for the calculation of a membership degree of a pair of options in the transposed strong preference relation. On the other hand, the degree of indifference of a pair of options is assumed to be the fraction of the number of criteria for which the score for both options is equal.

The calculation of a degree of incomparability is slightly more complicated as illustrated in Table 1.

Table 1

	UNK	*NA*	*LL*
UNK	+	+	+
NA	+	-	+
LL	+	+	-

Table 1 shows which criteria are considered as supporting the incomparability of two options. Namely, if for the first option a criterion takes on a value listed in the first column and for the second option a value listed in the first row, and the "+" corresponds to that column and row, then such a combination of values is treated as contributing to the incomparability of options. On the other hand, pairs of values marked with "-" are excluded from the calculations of the incomparability degree. Thus, in fact only two combinations are excluded: (*NA*,*NA*) and a pair of regular values. The former is assumed as in the calculation of all other membership functions of fuzzy relations. The latter is obvious as two regular values may be evidently compared due to the "$\succ$" and identity relations assumed in our model.

This discussion clarifies the interpretation of non-regular values *UNK* and *NA*. The former should be used when an individual cannot assess an option with respect to given criteria due to lack of knowledge as to the appropriate regular value. Whenever *UNK* appears as the value of a criterion, it contributes to the incomparability of a pair of options, no matter what the value of the same criterion for the other option is. On the other hand, *NA* should be used if an individual finds a given criterion to be non-applicable for the option considered. If an individual finds the criterion non-applicable for both the options, then such a criterion is completely excluded from the calculation of membership degrees of preference relations.

It is obvious from the above definitions that $\mu_P(s_i,s_j) + \mu_{P^t}(s_i,s_j) + \mu_I(s_i,s_j) + \mu_J(s_i,s_j) = 1$. Thus, we obtain an *additive fuzzy preference structure* [16]. It means that with each pair of options we have associated four numbers expressing the membership of this pair in particular fuzzy relations constituting our preference structure.

4. Consensus Evaluation

Here we further develop the idea of a consensus reaching support system proposed in [7,4,3]. The main point was there to offer a new, flexible definition of a degree of consensus. From the perspective adopted there and here, the primary goal of a decision process support is to provide a framework for the selection of widely agreed, *consensual*, solutions to the decision problem. The classical concept of consensus meant as full and unanimous agreement, of all individuals involved as to all aspects of the decision problem, is hardly helpful in assisting a group of individuals in arriving at a satisfactory consensus. A flexible concept of consensus proposed earlier may be expressed as "the agreement of *most* of the individuals as to *most* of the decision problem aspects". Thus, basically, the idea was to replace the general quantifier "all", traditionally employed in the definition of consensus, by some flexible, linguistic quantifier as, e.g., "most".

In the previously proposed approach we have concentrated on the evaluation of consensus with respect to the fuzzy preference relation. In this paper, we would like to propose another way of calculating a consensus degree, now in the context of a preference structure. Another goal is to point out some other possible ways of exploiting different forms of preference related data for the evaluation of consensus.

In Sections 2 and 3 we have distinguished the following levels of preference related data:

- weights of the criteria, (1a)
- assessment of each option with respect to all criteria, (1b)
- a fuzzy preference structure (1c)

Usually, the ultimate goal of decision making is the selection of one or more options that may be treated as a best solution for the decision problem under consideration. This selection may take various forms - one option, subset of options, ranking of options etc. - depending on the nature of the decision process, requirements of an "ultimate" decision maker etc.

Anyway, we should add to the list mentioned above, another item, namely:

- a solution (1d)

A solution is meant as proposed by or generated for each individual separately. Usually, we expect as a result of a group decision making process another solution - this time supported by the group as a whole. As already stressed, here we are not dealing with such a solution directly. Anyway, reaching the agreement (consensus) in the group makes such a group solution more obvious and sound.

It is worth noticing that in our approach data belonging to levels (1a) and (1b) are directly provided by each individual, while (1c) and (1d) are automatically generated. In the previous section we have described the rules of transition from (1a) and (1b) to (1c). For various ways of going from (1c) to (1d), see, e.g., [1, 8, 10, 11,12,13, 14].

Previously [7,4,3], we proposed some flexible scheme for consensus evaluation based on the level 3 data, i.e. fuzzy preference relations of all individuals. Here we extend the previous approach to a richer fuzzy preference structure and using some other aggregation technique, while preserving the basic idea of flexible consensus definition. After that, we briefly analyse other possibilities.

In [7,4,3] we started with a fuzzy preference relation R, interpreted as follows:

$$\mu_R(s_i,s_j)=\begin{cases}1 & s_i \text{ definitely preferred to } s_j\\ c\in(0.5,1) & s_i \text{ preferred to } s_j \text{ to some extent}\\ 0.5 & \text{indifference}\\ d\in(0,0.5) & s_j \text{ preferred to } s_i \text{ to some extent}\\ 0 & s_j \text{ definitely preferred to } s_i\end{cases} \tag{2}$$

The reciprocity of this relation was assumed: $\mu_R(s_i,s_j)=1-\mu_R(s_j,s_i), \forall i \neq j$. Then, a flexible concept of consensus was introduced that may be expressed in natural language as:

> "We will say that there is consensus among experts if *most* pairs of *important* experts *agree* as to *most* pairs of *important* options." (3)

This expression may be formalised using Zadeh's calculus of linguistically quantified propositions [18,19]. The *most* may be replaced by another quantifying expression and is modelled by a (fuzzy) *linguistic quantifier*. The notions of *importance* and *agreement* are modelled as fuzzy properties. Then, Zadeh's calculus provides a way to calculate the truth degree of (3). Here we adopt a richer preference structure that supports a more straightforward interpretation, and also makes possible a flexible consensus concept without additional definitions of an agreement degree among the experts.

Basically, we follow the previous approach, i.e. in the first step, for each pair of options we calculate a degree of agreement among *most* of the individuals. Then, we obtain the final consensus degree aggregating the partial consensus indicators computed in the first step, i.e. we calculate if these indicators are high for *most* of the pairs of options. The aggregation operator in both steps is a linguistic quantifier in the sense of Zadeh. Some other approaches to linguistic (fuzzy) quantification may be readily used, see, e.g., [9].

In the first step, there is an agreement among the individuals as to a pair of options if *most* of them support this pair of options due to the same fuzzy preference relation from our fuzzy preference structure. Thus, for a given pair of options we define four fuzzy properties of individuals, XP, XP^t, XI, XJ, basing on their support for particular preference relations. The membership functions of the fuzzy sets representing these fuzzy properties are as follows:

$$\mu_{XP}(x_m) = P(s_i,s_j)$$
$$\mu_{XP^t}(x_m) = P^t(s_i,s_j)$$
$$\mu_{XI}(x_m) = I(s_i,s_j)$$
$$\mu_{XJ}(x_m) = J(s_i,s_j)$$

Then the degree of consensus of *most* individuals as to the pair of options (s_i,s_j) may be expressed as:

$$\text{con}(s_i,s_j) = \max\{\mathrm{Q_m}\, XP(x_m),\ \mathrm{Q_m}\, XP^t(x_m),\ \mathrm{Q_m}\, XI(x_m),\ \mathrm{Q_m}\, XJ(x_m)\} \quad (4)$$

where $\mathrm{Q_m}$ denotes a linguistic quantifier employed, e.g., *most,* and m iterates over the set of individuals.

The truth value of the expression, e.g, $\mathrm{Q_m}\, XP(x_m)$ is calculated according to the original Zadeh's approach:

$$\text{truth}(\mathrm{Q_m}\, XP(x_m)) = \Sigma\text{Count}(XP) \,/\, M$$

In the second step we aggregate the results obtained above, for all pairs of options, thus producing the overall consensus degree:

$$\text{CON} = \mathrm{Q}_{(i,j):i<j}\ \text{con}(s_i,s_j)$$

Observe, that we have to take into account only those pairs of options (s_i,s_j) for which $i<j$ because the relations I and J are symmetric and P and P^t are the transposes of each other.

This concept of consensus is *constructive*. In the first step, we acknowledge consensus among the experts only if *most* of them support one of the preference relations. It may happen that their preferences are in perfect agreement but are more or less equally distributed among all four relations. Then, the computed consensus degree will be low. This may be somehow alleviated by replacing formula (4) by the following one:

$$\text{con}(s_i,s_j) = \max\{\ \mathrm{Q_m}\, XP(x_m) \vee \mathrm{Q_m}\, XI(x_m),\ \mathrm{Q_m}\, XP^t(x_m) \vee \mathrm{Q_m}\, XI(x_m),\ \mathrm{Q_m}\, XJ(x_m)\ \} \quad (5)$$

Then the resulting consensus degree will be usually higher (especially in case of the Łukasiewicz disjunction) but still the concept does not completely loose its constructive character.

The question of incomparability requires a more careful consideration. Basically, we can employ the same approach, possibly with some modifications, for data on all remaining levels mentioned in (1a), (1b) and (1d). The resulting consensus degree concepts may be applicable in different contexts. For example, requiring a high consensus at the level of criteria weights, we focus on the very definition of the decision problem. If the weights chosen by particular individuals are essentially different (there is no consensus about them), then it may mean that the understanding of the decision problem among individuals is also rather distinct. Reaching consensus at this level may stimulate a discussion leading to a higher agreement on all other levels.

5. Implementation

The Internet, or more precisely the WWW environment, seems to be close to an ideal platform for the implementation of a group decision support system of the proposed type. The idea of browsing as a universal paradigm of human-computer interaction is very attractive and may prove to be also effective and efficient.

Basically, we have a simple client software (a WWW browser), focused on sending a document request to a WWW server selected by the user and displaying the document received applying some limited formatting commands. This fairly simple protocol turned out to be very efficient and attractive. Soon it started serving not only the purposes of a scientist but also a growing population of "ordinary" users. WWW has become a subject of commercial interest and activity what triggered a fast development of new applications and tools. Capabilities of a modern WWW-related software are much more sophisticated than it was a couple of years ago. Unfortunately, this happened, to some extent, at the expense of standardisation. Anyway, we decided to undertake our implementation using newest developments in this area, such as Dynamic HTML and XML. Before briefly presenting our pilot implementation, we will discuss how modern WWW-based software engineering meets the needs of decision support systems considered in this paper.

There is a vast literature related to the postulated features of (group) decision support systems. We can mention some of them after, e.g., Freyenfeld [5]:

1. an interactive way of data processing and presentation supporting individual (decision maker) involved,
2. user-friendliness,
3. a natural way of dialogue,
4. a filtering of information

The first two requirements are strictly of a software engineering nature. Here WWW provides a rich enough environment to accomplish those goals. The third feature may be connected with the idea of human consistency that is usually argued as a characteristic for fuzzy logic based approaches. Thus, in fact we are combining advanced software engineering solutions with sophisticated modelling and information processing methods. For example, the use of linguistic labels during the assessment of options should be much more comfortable for an "average" user.

There are also some requirements specific to group decision support systems. For example, the Internet may be a perfect communication media preserving the *anonymity* of individuals. It is well known from the literature that the participation in a decision making (or advising) process may imply a severe stress. Members of the group are afraid to express their real opinions in front of other people. They quite often quickly adopt positions of stronger personalities what, obviously, may have nothing to do with the merit of the decision problem. The Internet makes it possible to organise a session by "grouping" people being physically in very distant places and completely anonymous to each other. Huber [6] put stress on the use of software *in the session*. Again, a WWW based solution fits this requirement perfectly. A possible reason for breaking the session could be a need for more information on the topic. The use of an Internet based solution makes it possible to gather needed information from some millions of various sources of information available over the WWW as well as through a fast e-mail contact with other experts. All that may be done in the background without any discontinuation of the current session. In fact, on more and more occasions this is just the way we proceed now looking for some information!

DeSanctis and Gallupe [2] introduced three levels of group decision support systems, depending on their control of decision processes occurring in the group. The level 1 systems are to enhance these processes through a support of the exchange of information among the group members. The level 2 systems additionally provide some modelling tools and, finally, the level 3 systems should control the whole decision making session according to some flexible strategies. In fact, the Internet technology provides means for the implementation of features required by all three levels of the group decision-making systems. The integration of Web tools, e-mail agents, directory services or data transfer protocols makes this environment the richest information exchange forum. Moreover, it is easily accessible from virtually any desktop computer system. The system level 2 requirements may be also easily addressed as the WWW servers are no longer just file servers but full-fledged application servers.

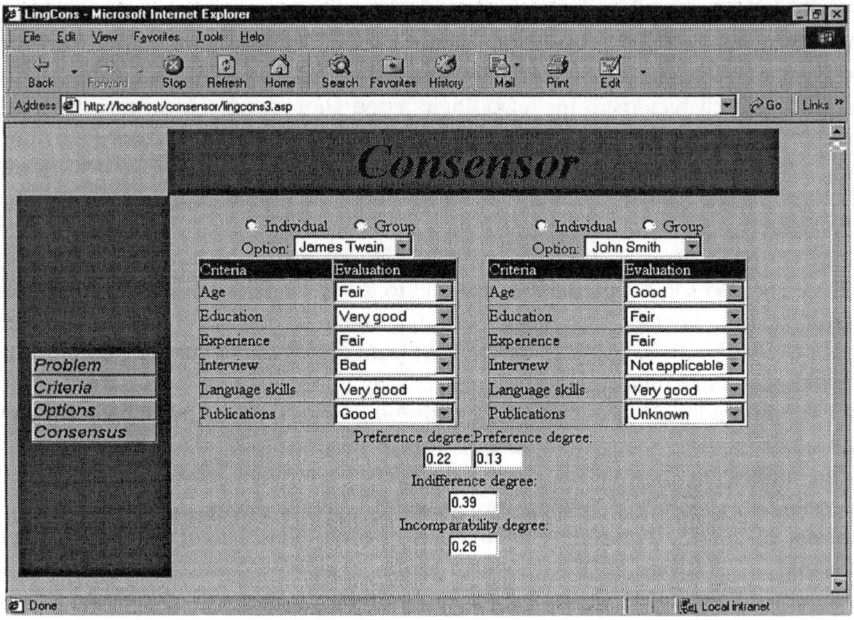

Exhibit 1. Assessment and comparison of options

The pilot implementation of the proposed system makes it possible to:

- set the weights of particular criteria,
- assess the quality of each option according to each criteria, and finally,
- compare opinions of the individuals including the consensus degree evaluation.

Most of the functionality is implemented on the client side using a WWW browser. We employed the Internet Explorer 5 supporting such recent advancements in Web technology as Dynamic HTML or XML. The server side of the application, based on the Internet Information Server, consists of a number of Active Server Pages files. Preferences and problem description data are represented in XML to provide for a high flexibility. The resulting user interface is illustrated in Exhibit 1.

6. Concluding Remarks

We presented an idea of an Internet based group decision support system aimed mainly at supporting the consensus reaching process. We tried to couple a flexible software design with human consistent, sound theoretical foundations. The former was addressed by the use of modern tools and techniques available for system development in the WWW based environment. The latter relied on the application of recent preference modelling techniques elaborated in the framework of fuzzy sets and fuzzy logic.

There are many possible extensions and enhancements to the proposed system. From the software engineering perspective, it would be advantageous to develop a standard, XML based representation of preferences on all levels involved, i.e. linguistic labels, criteria importance weights, particular fuzzy relations constituting the preference structure, etc.

There are some extensions needed to support the completeness of the proposed group decision support system. These include, for example, a support for choosing a set of criteria to be used when assessing particular options. Another possible extension would be to offer more flexible a transition from a criteria based assessment of options to a fuzzy preference structure. Such a more flexible transition could, e.g., take into account compensatory interrelations among the criteria.

References

[1] Barrett CR, Pattanaik PK and M. Salles, "On choosing rationally when preferences are fuzzy", Fuzzy Sets and Systems 1990; 34: 197-212.

[2] DeSanctis G., Gallupe R.B., "A foundation for the study of group decision support systems", Management Sciences 33, 1987.

[3] Fedrizzi M., Kacprzyk J., Owsiński J.W. and S. Zadrożny, "Consensus reaching via a GDSS with fuzzy majority and clustering of preference profiles", Annals of Operations Research 1994; 51: 127-139.

[4] Fedrizzi M, Kacprzyk J and S. Zadrożny, "An interactive multi-user decision support system for consensus reaching processes using fuzzy logic with linguistic quantifiers", Decision Support Systems 1988; 4: 313-327.

[5] Freyenfeld W.A, Decision Support Systems, NCC Publications, Manchester, England, 1984.

[6] Huber G., Issues in the design of GDSS. Management Information Systems Quarterly 7, 1984, 195-204.

[7] Kacprzyk J and M. Fedrizzi, "A 'soft` measure of consensus in the setting of partial (fuzzy) preferences", European Journal of Operational Research 1988; 34: 316 - 325.

[8] Kitainik L., "Fuzzy Decision Procedures with Binary Relations: Towards a Unified Theory", Kluwer Academic Publishers, Boston/Dordrecht/London, 1993.

[9] Liu Y. And E.E. Kerre, "An overview of fuzzy quantifiers. (I). Interpretations", Fuzzy Sets and Systems 95, 1998, 1-21.

[10] Nurmi H., "Approaches to collective decision making with fuzzy preference relations". Fuzzy Sets and Systems 6, 1981, 249-259.

[11] Nurmi H. and J. Kacprzyk, "On fuzzy tournaments and their solution concepts in group decision making", EJOR 51, 1991, 223 - 232.

[12] Orlovsky S.A.: "Decision-making with a fuzzy preference relation", Fuzzy Sets and Systems 1, 1978, 155-167.

[13] Ovchinnikov S.V. and V.M. Ozernoy, "Identifying noninferior decision alternatives based on fuzzy binary relations", in: J. Kacprzyk and M. Roubens (Eds.): Non-Conventional Preference Relations in Decision Making. Springer Verlag, Berlin, 1988, 82-95.

[14] Świtalski Z., "Choice functions associated with fuzzy preference relations.", in: J. Kacprzyk and M. Roubens (Eds.): Non - conventional Preference Relations in Decision Making, Berlin: Springer-Verlag, 1988.

[15] Ovchinnikov S., "Modelling valued preference relations", in J. Kacprzyk and M. Fedrizzi (Eds.): Multiperson Decision Making Models Using Fuzzy Sets and Possibility Theory, Kluwer Academic Publishers, 1990, 64-70.

[16] Van de Walle B., B. De Baets and E. Kerre, "A plea for the use of Łukasiewicz triplets in fuzzy preference structures. Part 1: General argumentation", Fuzzy Sets and Systems 97, 1998, 349-359.

[17] Zadrożny S, "An approach to the consensus reaching support in fuzzy environment", in J. Kacprzyk, H. Nurmi and M. Fedrizzi (Eds.): Consensus under Fuzziness. Kluwer, Boston, 1996

[18] Zadeh LA., "A computational approach to fuzzy quantifiers in natural languages", Comp. and Maths. with Appls. 1983; 9: 149-184.

[19] Zadeh LA., "A computational theory of dispositions", International Journal of Intelligent Systems 1987; 2: 39-64.

Limpio: A DSS for the Urban Waste Collection Problem

J.I. Peláez[*] , M.T. Lamata[**] and F.J. Fuentes[***]

[*] Dpto. Lenguajes y Ciencias de la Computación
Universidad de Málaga
Malaga, España
e-mail: jignacio@lcc.uma.es.

[**] Dpto. Ciencias de la Computación e Inteligencia Artificial
Universidad de Granada
Granada, España
e-mail: Mlamata@goliat.ugr.es.

[***] Director, Andalucía Fcc. Medio Ambiente, Residuos Industriales
e-mail: FJFuentesF@fcc.es

Abstract: Waste collection is a problem which has been extensively researched in recent decades. Studies have highlighted problems such as the initial volume of information needed to run the systems, the direction of traffic in the city, or the measures needed to establish how valid a solution is. These problems have hindered the use of decision support systems by companies of the sector. This paper presents a Decision Support Systems which eliminates the aforementioned problems in urban waste collection.

Keywords: Genetic Algorithm, Vehicle Routing Problems, Clustering, DSS.

1. Introduction

L. Adelman defined Decision Support Systems (DSS) as an interactive computer programs which use analytical methods, such as decision analysis, optimisation algorithms, scheduling algorithms, etc, to help the user formulate alternatives, analyse their impacts, and interpret and select different options [10], [11], [12].

Basically, as shown in figure 1, the DSS consists of three elements. The relation between these elements, which Sprague and Carlson call the DDM paradigm, comprises dialog, data and modelling. This paradigm states that a good DSS must have a good balance between the components that make up the system (see Figure 1): the data base management system (DBMS) which carries out the management of the data stored in the system; the model base management system (MBMS), which manages all the functions and models which the system uses; and the dialog generation and management system (DGMS), which manages the interface between the user and the rest of the system.

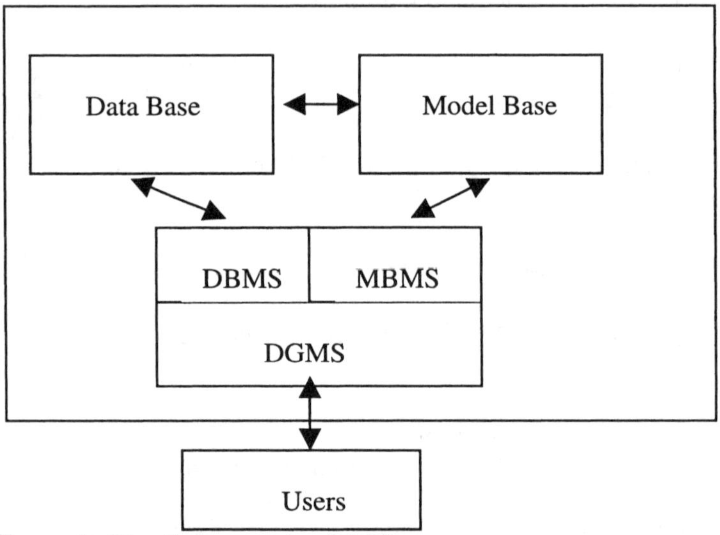

Figure 1. The Components of DSS

During recent years, DSS development has been constantly increasing, incorporating such diverse areas as aircraft control with the Airplan DSS; the evaluation of the Earth's natural resources with Terra-visión, or traffic aid with the DSS Carguide. An area in which DSS development is of great importance is that of the urban waste collection problem (UWCP) [3], [4], [5]. The existence of systems which help the experts to choose the collection routes can imply not only a saving of millions of Euros but also improve the quality of service.

The urban waste collection problem (UWCP) is a optimisation problem where the sectoring and routing problems are present. The problem consists of a set of geographically distributed nodes connected to each other. The direction of these connections is that of the traffic. The aim is to find clusters with the shortest distances, adjacent nodes and greatest load; routes of minimal distance and the number of times which the vehicle visits the same streets.

According to our studies, the principal characteristics needed for the systems to be useful for a company are:

1. The system must be operative. The volume of information to run and maintain it must be minimal.
2. The distance and loads must be similar for the different clusters, and should also take into account the direction of traffic.
3. The routes distance must be minimal.
4. The number of times which the vehicle visits the same streets must be minimal.
5. Finally, the systems must incorporate some strategic planning tools to create runs and examine the effects of change.

In this paper we present Limpio, a DSS for urban waste collection. The paper is laid out as follows: in section 2, we present Limpio; in section 3, we show real examples; and finally, we present the conclusions.

2. Decision Support System Limpio

A way of considering the components of a DSS and the relationships between these parts is to use the dialog, data and model paradigm. In this conceptualisation, there is the dialog (D) between the user and the system, the data (D) which supports the system, and the models (M) which provide the analysis capabilities.

Limpio has been designed for the UWCP, and comprises a dialog component, a data component and a model component.

2.1. The Dialog Component

This is the combination or set of options, action language and presentation language taken at the same time, to implement the data base.

Figure 2 shows the interface of Limpio. It is highly visual, so the user can see the necessary information at all times and can therefore adapt to changes which could help decision making. This interface consists of a menu system which allows the user to: configure the size of the city, adding or eliminating new zones (configuracion); choose the type and capacity of the vehicles (Nuevo camion); choose the type of rubbish containers (puntos de recogida); choose the vehicles which collect these containers, the capacity of the containers and a time penalisation which depends on the capacity of the container (nuevo contenedor); go to different zones of the city (ir a); introduce a another waste container (nodes) or eliminate with the mouse (marcar); connect nodes (enlazar); calculate the routes (simulacion); and finally, show the capacity and nodes for each route (ruta).

The action language allows the user to change the capacity of the vehicles, nodes and the direction of traffic, to choose the capacity for each cluster and the distance for each route with the mouse, GPS or keyboard.

And finally, the presentation language used in Limpio is a graphic representation of a digital plan which shows the routes to the users solving the urban waste collection problem, and enables them to print out copies to distribute to those working in the waste collection service.

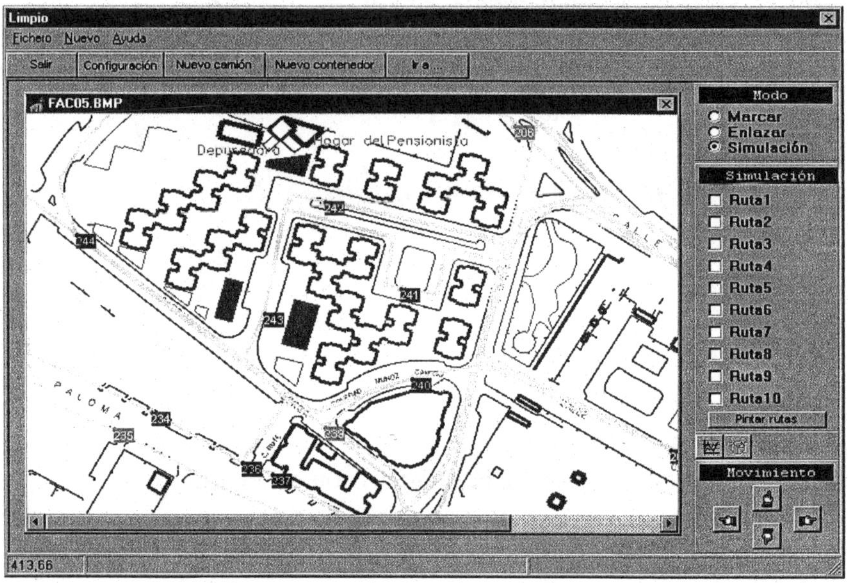

Figure 2. The Dialog Component of DSS Limpio

2.2. The Data Component

Data plays an important role in the DSS, and is the input for the processing of the models. The user enters the data using the action language, so basic information (capacity, type and site of each node, direction of traffic, etc.), as well as his own knowledge of the city (the traffic situation, the order of collection) can be entered. Therefore, the system can evolve with the experts' knowledge

The major criticism which companies have regarding the data components is the volume of information that the user has to input into the system for its running and maintenance. This volume makes the systems inoperative. This information refers to the number of collection points and their connections. For example in Malaga city there are about 10,000 rubbish collection points to which the direction of the streets has to be added. To solve this problem, Limpio uses a GPS navigation system to find out the direction of the traffic, the collection points and the streets where traffic can circulate. This information can periodically be automatically input into the system. This information is stored in a data base to be used by the model component at a later date. We can see the data base table in figure 5 a. b c.

2.3. The Model Component

The model component provides the analysis capabilities for a DSS. Limpio has a connection module, a clustering module, a routing module and an evaluation module.

- **Connection Module**

A second criticism which urban waste collection companies have regarding the model component is the direction of traffic. This problem occurs when nodes which are visibly near, in practice are far away, due to the direction of the traffic. To solve this, we propose the use of work levels.

Work levels. We say that nodes C_i and C_{i+1} belong to level *n*, if the shortest route between C_i and C_{i+1} has *n* intermediate nodes
The work level eliminates the problem because the cluster and route module use it in this process. Furthermore, the work level causes a simplification of the search space. Our studies show that the solution is in work level < 5.

Limpio has a connection module for calculating the connection between nodes of different levels.

- **Sectorisation Module**

This module is responsible for grouping the different collection points while respecting the restrictions of capacity, distance covered and direction of traffic. This module has been carried out using a genetic algorithm, as this technique has proved very effective in dealing with NP-complete problems and can provide an available solution at any time.

- **Individual**

In the UWCP each individual represents a feasible solution. An individual is made up of a set of chromosomes, each one of these representing a sector.

- **Evolutionary process**

The evolutionary process used in the genetic algorithm is a "steady-state".

- **Fitness**

The proposed fitness was based on the expansive competitive learning rule[7]. and is composed of two elements: (1) the first one seeks solutions where the sectors' distance from their centre of gravity is minimum; (2) the second one attempts to find a global minimum by looking for solutions in which the centres of gravity are sufficiently distant from each other.

The first fitness is calculated as follows:

$$Fitness1 = \frac{\sum_{i=1}^{n}\left(\sum_{j=1}^{mi} DE\left(CM_i, C_{ij}\right)\right)}{n}$$

where: DE is the Euclidean distance, CM_i is the "centre of gravity" for chromosome *i*, C_{ij} is the coordenates of the nodes j for the cromosomes i, m_i is the number of nodes of chromosome *i*, and *n* the number or cromosomes.

The second fitness is calulated as follows:

$$Fitness2 = \frac{\sum_{i=1}^{n}\left(\overline{CM} - CM_i\right)^2}{n}$$

and finally, the global fitness is calculated as:

$$Fitness = \alpha \cdot Fitness1 - \beta \cdot Fitness2$$

where $\alpha, \beta \in [0,1]$.

- **Selection operator**

This operator selects the individuals to be crossed, using the roulette method.

- **Crossing operator**

In figure 3 we show the crossing operator. The individual *son* is generated by father and mother individuals, which are selected by selection operator. The functions "Remove_Repeats" and "Introduce_pendientes" carry out their operations considering the work levels restrictions.

```
ALGORITHM Crossing  (Var son: T_Individual; father,m other : T_individual);
Var chromo: T_chromosome;
Begin
     For i = 1  TO Total_Cromos do
     Begin
     chromo=RAND(CHROMOSOME(father,i), CHROMOSOME(mother, i));
     CHROMOSOME (son, i) = chromo;
          REMOVE_ REPEATS (son);
          INTRODUCE_ PENDIENTES  (son),
End
End
```

Figure 3. Crossing operator

- **Mutation operator**

This operator removes links that do not satisfy the work level restriction.

- **The Routing module**

This module is responsible for finding the best route within each of the clusters. Our research shows that companies do not want an approximate route, they need the best solution in function of the information provided. The traditional algorithm uses maximum information, for example Floyd's distance matrix, to resolve routing problems. Therefore, many of these cannot be used as it would take thousands of years.

In our proposed routing algorithm we use minimal information to achieve routing. In the case of a solution not being found, new information is generated until the problem has been solved.

The algorithm that we propose in this paper uses the Backtracking technique to find the best solution. At first sight, this technique could seem inadequate due to its complexity, but together with the zero work level and some suitable branch conditions, excellent results have been obtained.

Our studies show that using the zero work level (systems operative) a node becomes adjacent to two nodes (on average), which minimises the expansion tree which generates the thorough search. If to this restriction we add some branch conditions in which a node will be acceptable if:

1. It has not been visited more than the permitted number of times. Initially this value is 1.
2. There are no nodes whose parents have been visited.
3. There are no nodes whose children have been visited.

4. The ratio between the distance covered to this node and its equivalent (regarding its position on the route) in an already obtained route is < Δ.

means that the best route can be found in real time. In real examples, and with 400 maximum nodes we obtained the best solutions in time < 30 second time.

In figure 4, we show the routing algorithm.

```
ALGORITHM Ruting ( Var M: Distance_Level; Var Route: Type_ Route);

Begin
      Backtraking (M, Route);
       If Not Solution (Route) Then
       Begin
           New_Information (M);
           Backtraking (M, Route);
       End;
End;
```

Figure 4. Routing Module

- **The Evaluation Module**

A problem we find once the clusters and routes have been designed is how to evaluate the suitability of the solutions. To solve this problem, Limpio includes and evaluation module which provides the user with the measures necessary to help reach a decision.

- Nodes by time unit.
- Waste by time unit.
- Constant K

$$k = \frac{N^o\_Nodes\_Sectors \cdot Total\_Kilos}{distance \cdot N^o\_Total\_Nodes}$$

2.4. General Vision of Limpio

In the figure 5. a, b and c, we show the information flow in Limpio.

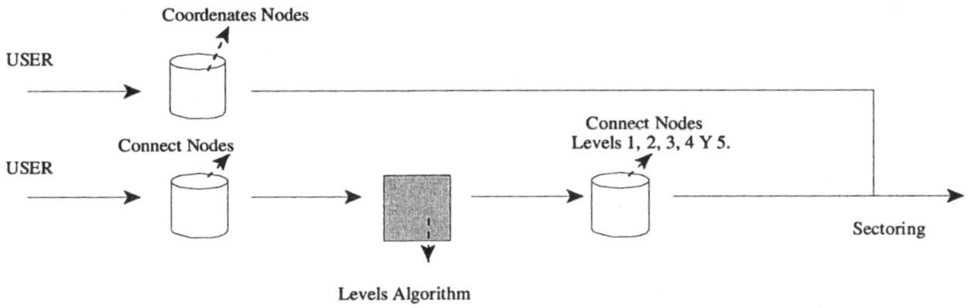

Figure 5.a. Information Flow in Limpio

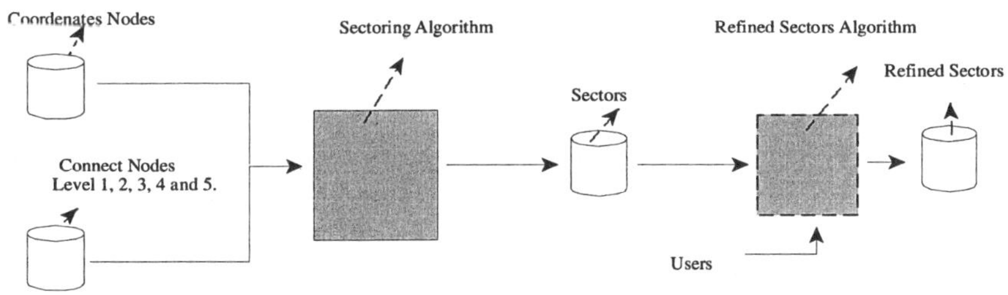

Figure 5.b. Information Flow in Limpio

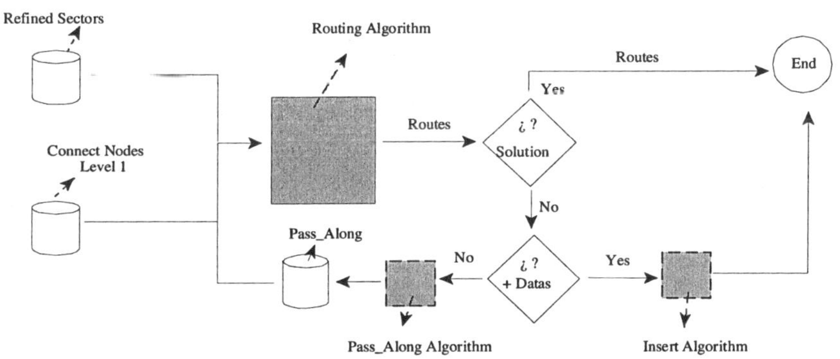

Figure 5.c. Information Flow in Limpio

3. Computational Results

Below we show the results obtained with real data after applying Limpio. This test was carried out on a total of 1739 nodes in the city of Malaga. In figure 6.a and b, we show the nodes before and after applying the sector module, while in table 1 we show the data for the evaluation module. These results show that the solutions are adequate because the ratios are similar. Therefore, they are of much interest to the companies of the sector.

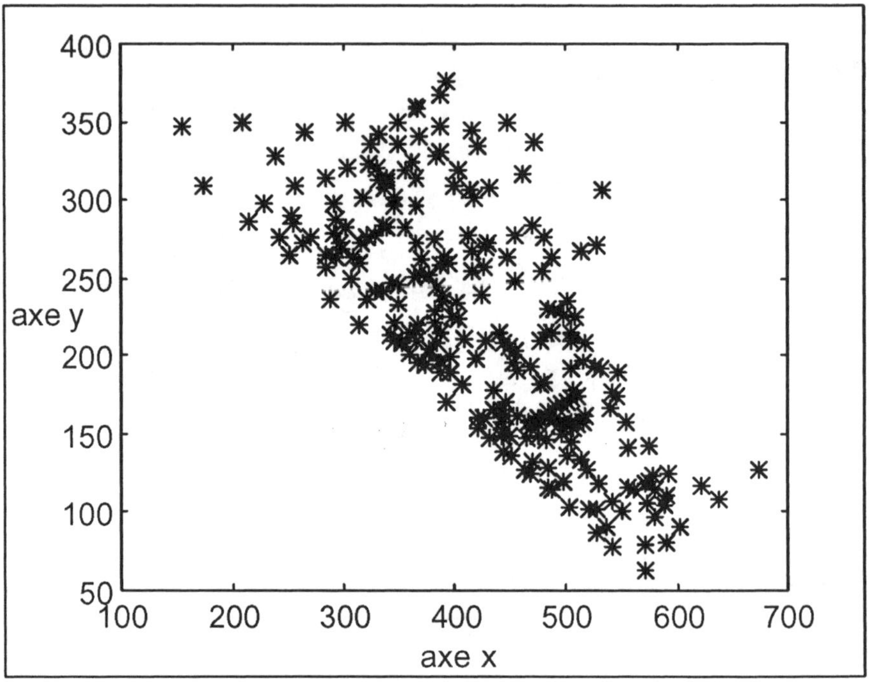

Figure 6. a. Initial Nodes

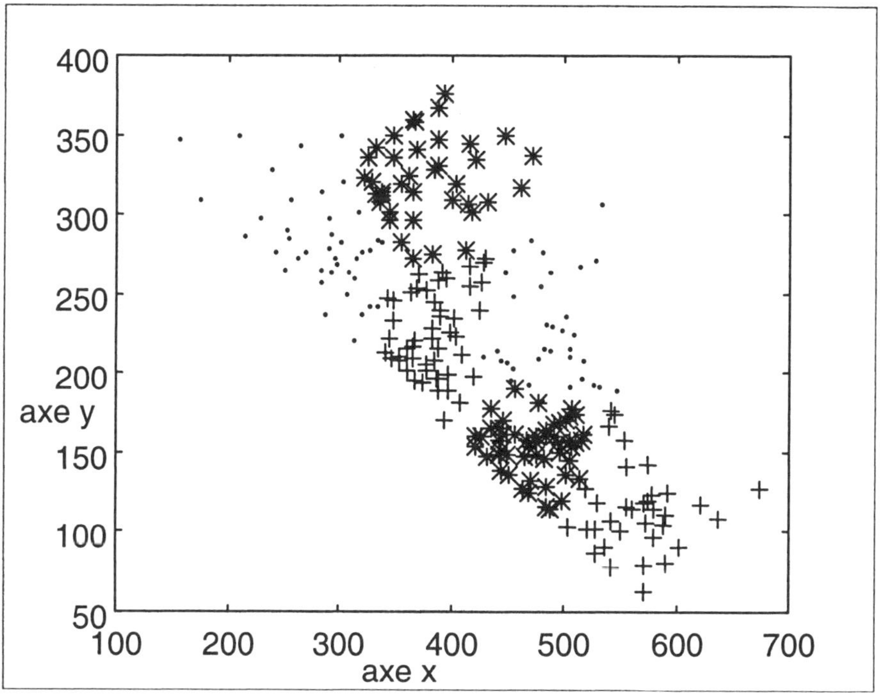

Figure 6. b. Clusters that are obtained after applying the genetic algorithm. Values α and β used in fitness 1 and fitness 2 are equal to 0.6 and 0.4 respectively, and the work level is set to three.

Table 1. Nodes / Hour and Kilos / Hours for the clusters in figure 5. These results show fair routes.

Route	Nodes	Nodes / Hour	Kg / Hour
1	296	60. 65	4696.47
2	286	57. 21	4145.14
3	287	60. 42	4452.12
4	292	53. 24	4320.00
5	295	63. 34	4701.82
6	283	61. 54	4867.56

4. Conclusions

In this paper we have presented the DSS Limpio for the collection of urban waste. This system provides the user with an interface which allows him to interact with the system, offering solutions which allow him to choose the best way to carry out urban waste collection. The system also allows the user to input his own knowledge and obtain solutions in real time, which in turn enables him to predict and evaluate new situations.

Moreover, Limpio has of a model component which eliminates the criticisms made by the companies of the sector i.e. the volume of information needed for its setting up and maintenance; the direction of traffic in the cities; and the measures needed to evaluate the obtained solutions. By means of a connection module which uses work levels, a sectorization and routing module which by using the work levels obtains efficient solutions, and finally an evaluation module which allows the different situations to be evaluated.

Acknowledgements

Research supported in part under project P1398-1305.

References

1. Assad, A. and B. Golden, “Vehicle Routing: Methods and Studies”, North Holland, Amsterdam, 1988.

2. Gilbert Laporte. “The Vehicle Routing Problem: An overview of exact and approsimate algorithms”. European Journal of Operational Research, 59, 1992.

3. Gilbert Laporte, Yves Nobert, Martin Desrocher. “Optimal Routing under Capacity and Distance Restrictions”. Operations Research. 1984.

4. Lamata, M.T. Peláez, J.I. “A Genetic Algorithm for Decision of the Number Sectors of Urban Waste Collection”. Seventh IFSA World Congress, Prague 1997.

5. Lamata, M.T. Peláez, J.I. Sierra J.C. Bravo J. A Sectoring Genetic Algorithm for the Urban Waste Collection Problem. Lecture Notes in Computes Science. Springer. 1999. ISSN 0302-9743.

6. Miller D. "A Matching Based Exact Algorithm for Capacitated Vehicle Routing Problems". Orsa Journal on Computing. Vol 7. No 1. 1995.

7. Oliver, I. M,D,J. Smith and J.R.C. D. Holland, "A study of permutation crossover operators on the travelling salesman problem". Proc. Of the second Int. Conf. On Genetic Algorithms, Lawrence Erlbaum Assoc. Pp. 224-230, 1987.

8. Pérez J. " Aprendizaje Competitivo para la Cuantificación Vectorial". CAEPIA'97. Málaga. Spain.

9. Sam R. Thangiak, Kendall E. Nygard, Paul L. Juell, "Gideon: Agenetic Algorithm System for Vehicle routing with time windows". IEEE. 1991.

10. Sprague, R. H. and E. D. Carlson. Building Effective Decision Support Systems Englewood Clifts, N.J.: Prentice-Hall, Inc.: 1982, ISBN: 0-13-086215-0.

11. Sprague, R. H. and H. J. Watson (editors). Decision Support Systems : Putting Theory into Practice. 3rd ed. Englewood Clifts, N.J. : Prentice Hall, 1993. 437 p. ISBN 0-13-036229-8.

12. Sprague, R. H. and Hugh J. Watson (editors). Decision Support for Management. Englewood Clifts, N.J.: Prentice-Hall, Inc.: 1996, ISBN: 0-13-396268-7.

A Decision Support System for Air Quality Control Based on Soft Computing Methods

Piotr Holnicki and Andrzej Kałuszko

Systems Research Institute
Polish Academy of Sciences
ul. Newelska 6
01-447 Warsaw, Poland
E-mails: {holnicki, kaluszko}@ibspan.waw.pl

Abstract: The problem considered is related to air quality protection on a regional scale. The approach refers to optimal allocation of financial means for emission reduction in a given set of power and heating plants. The implementation considered is sulfur-oriented. The problem is formally stated as cost-constrained minimization of environmental damage function by the optimal choice of desulfurization technologies, within the predefined set of the controlled plants. The receptor-oriented objective function utilizes air pollution forecast preprocessed by a regional scale dispersion model. An integer-type optimization problem is solved by two methods. The first method utilizes a heuristic algorithm designed for solving this specific problem, which directly finds discrete solution. Another approach is based on the classical gradient optimization algorithm and gives continuous, technologically not applicable solution. Then the continuous solution is transformed to the discrete form by enumeration of some discrete cases. Both algorithms has been implemented and tested on the real data for selected region. The case study relates to the set of major power plants in Silesia Region (Poland) and the basic desulfurization technologies, which are to be allocated. The test calculations allows us to evaluate accuracy of the heuristic method as well as applicability of both approaches for supporting decisions concerning optimal strategies of emission abatement on a regional scale.

1. Introduction

The paper addresses the problem of regional strategy of environmental quality protection and computer methods, which allow to implement respective decision support tools. The main task deals with the regional-scale strategy for air quality control, mainly due to sulfur oxide pollution. The objective is to allocate emission reduction technologies to emission sources in such a way that certain environmental quality index is minimized under the constraint on total cost of the operation.

Regional-scale abatement strategy depends on the criteria upon which the environmental damage is evaluated. The straightforward approach is based on the emission reduction in all the plants by the fixed percentage or proportionally to the current emission intensity. This solution, however, is not the most efficient from environmental and economic point of view. Other strategies can also be formulated [1,2,7], where the final environmental impact is considered as the main criterion. Moreover, the problem of cost-effectiveness of emission control can be taken into account. This motivates formulation of the problem in terms of optimization techniques.

The task of defining optimal allocation of emission abatement technologies is formulated as integer optimization problem. Exactly one of technologies available must be assigned to each emission source. To solve this problem, two algorithms have been implemented. One of them is a purely heuristic method that directly finds a discrete solution. Another approach utilizes one of continuous, gradient optimization algorithms, respectively adapted to the specific problem. In the paper two methods are presented and compared on the real-data case study.

Test computations were performed on the set of major power plants located in Upper Silesia Region, where optimal strategy of reduction of SO_2 concentration is considered. Calculations have been performed for real emission values and meteorological scenarios. Results characterize efficiency and accuracy of the discussed methods.

2. Statement of the Problem

Assume that there are N controlled SO_2 emission sources in a region Ω. Moreover, we have M technologies for emission reduction in our disposal. Effectiveness and the unit cost (both for investment and operational costs) characterize each of them. Our goal is to allocate emission reduction technologies to all the sources in such a way, that the value of certain environmental damage index (the objective function) will be minimized subject to constraints on investment and operational costs, in given period T.

Let us denote:

$\Omega = L_x \times L_y$ -- rectangle area under consideration,

N -- number of controlled sources,
M -- number of available desulfurization technologies,
C -- constraint on total (investment and operational), year averaged costs,
$\vec{u} = [u_1, u_2, ..., u_N]$ -- emission vector of controlled sources,
$\vec{e} = [e_1, e_2, ..., e_M]$ -- effectiveness vector of desulfurization technologies,
$F = \{f_{ij}\},\ 1 \le i \le N,\ 1 \le j \le M$ -- matrix of abatement cost per unit emission,
$X = \{x_{ij}\},\ 1 \le i \le N,\ 1 \le j \le M$ -- "0-1" matrix of technology assignment to the sources (decision variable matrix).

Definition of the environmental criterion, which is to be minimized, depends on the objectives of the control strategy considered. We define here a global environmental cost function of the following form:

$$J(c) = \frac{1}{2} \int_{\Omega} w(x,y)[\max(0, c(x,y) - c_{ad})]^2 \, d\Omega\,, \qquad (1)$$

where:

$w(x,y)$ -- area sensitivity (weight) function,
c_{ad} -- admissible level of SO_2 concentration.

The concentration forecast used in (1) is calculated as

$$c(x,y) = c_0(x,y) + \sum_{i=1}^{N} A_i(x,y) \cdot u_i\,, \qquad (x,y) \in \Omega, \qquad (2)$$

where:

$c_0(x,y)$ -- background concentration (impact of uncontrolled sources),
$A_i(x,y)$ -- transfer matrix (relation emission → concentration) of the i-th source.

The unit transfer matrix $A_i(x,y)$ -- represents here the contribution of the i-th source, referred to the unit emission intensity. All the matrices $A_i(x,y)$; ($i = 1,...,N$), for the controlled sources are preprocessed by the respective forecasting model. In a similar way, the background pollution field $c_0(x,y)$ is computed for uncontrolled, background emissions, including the inflow from the neighboring regions. The current emission intensity of the i-th source depends on the initial emission value -- u_i^0 and efficiency of the abatement technology applied, according to the formula

$$u_i = u_i^0 \sum_{j=1}^{N} (1-e_j) \cdot x_{ij}, \quad \sum_{j=1}^{N} x_{ij} = 1, \quad x_{ij} \in \{0,1\}, \quad 1 \le i \le N, \tag{3}$$

where:

u_i -- current emission intensity of the i-th source,

u_i^0 -- initial emission intensity of the i-th source.

Cost of emission abatement in each source consists of two components: investment cost and operational cost. Both investment and operational costs depend on the specific abatement technology and on the parameters of the energy installation where this technology is to be applied. Here a simplified approach is utilized, where the investment cost of the j-th abatement technology applied in the i-th emission source is calculated as annual cost, averaged over the entire amortization period. Thus, the total emission abatement cost per year, considered as a sum of desulfurization costs in the respective plants, is used to formulate the financial constraint

$$\sum_{i=1}^{N} F_i = \sum_{i=1}^{N} u_i^0 \sum_{j=1}^{M} f_{ij} x_{ij} = \sum_{i=1}^{N} u_i^0 \sum_{j=1}^{M} (f_{ij}^1 + f_{ij}^2) x_{ij} \le C, \tag{4}$$

where F_i is the annual total cost of the abatement strategy assigned to the i-th source, while the cost coefficients (cost per unit emission) related to the j-th technology applied to the i-th source are as follows:

f_{ij} -- averaged annual total cost,

f_{ij}^1 -- averaged annual investment cost,

f_{ij}^2 -- averaged annual operational cost.

Now we can formulate the following problem of allocation of emission reduction technologies to emission sources.

ALLOCATION PROBLEM: *Determine the set of emission reduction technologies*

$$X^* = \{x_{ij}^* \in \{0,1\}: \sum_{j=1}^{M} x_{ij}^* = 1, \quad 1 \le i \le N, \quad 1 \le j \le M\},$$

such that the environmental cost function (1) s minimized

$$J(c(X^*)) \Rightarrow \min,$$

subject to the total cost constraint

$$\sum_{i=1}^{N} F_i \leq C \,.$$

3. Implementation of the Optimization Algorithms

In this section three algorithms of solving the problem stated above are presented. The first one is a heuristic algorithm (*heuristic method*) directly solving the discrete programming task formulated in Section 1.

Another approach discussed here (*continuous method*) is based on formulation the main task as the respective continuous problem, where the decision variables are real numbers $x_{ij} \in \langle 0,1 \rangle$, ($i = 1,\ldots,N;\ \ j = 1,\ldots,M$). That means, combination of several technologies allocated to a source is an accepted solution. Despite technologically unrealistic, such a continuous solution can be used as a reference base for evaluation of accuracy of the heuristic, discrete programming algorithm.

On the other hand, the solution in a form of real numbers can be utilized as the starting point for searching by enumeration the neighboring discrete solution. Such an approach is used in the implementation of the third algorithm presented in the sequel and referred to as *combined method*..

3.1. A heuristic discrete-programming algorithm

The flow diagram of the *heuristic algorithm* is shown on Figure 1. In order to compare solutions, we construct for each x_{ij} -- i.e. for each source i and each technology j -- an efficiency factor r_{ij}, defined as follows:

$$r_{ij} = \alpha[u_i^0 \cdot e_j / f_{ij}] + (1-\alpha)[\int_{\Omega} A_i(x,y) \cdot e_j / f_{ij}], \qquad \alpha \in \langle 0,1 \rangle,$$

where α is the weight factor ranging from 0 to 1, reflecting the influence of emission reduction versus concentration reduction (via transfer matrix $A_i(x,y)$).

The main loop of the algorithm consists of the following steps:

1) Set the value of α to 0.

2) Construct the list of all x_{ij}, where i is the number of the source and j is the number of technology, ordering the list in descending mode due to the value of r_{ij} for x_{ij}.
   Set the value of the temporary variable C_{tmp} to 0. C_{tmp} stores the sum of products $f_{ij} \cdot u_i^0$ for x_{ij} present in the solution $\{x_{ij}\}$.
3) Take the first x_{ij}^* on the list , i.e. the x_{ij}^* with the greatest r_{ij} .
4) Check, if x_{ij}^* can be added the solution $\{x_{ij}\}$, i.e. check, if the sum $C_{tmp} + f_{ij}^* \cdot u_i^o$ is greater than the cost constraint C. If so, proceed to step 6. Otherwise proceed to step 5.
5) Update the solution $\{x_{ij}\}$ by adding x_{ij}^* .
   Update C_{tmp} by adding $f_{ij}^* \cdot u_i^o$.
   Remove x_{ij}^* from the list, and all x_{ij} concerning the same source.
   Proceed to step 7.
6) Remove x_{ij}^* from the list.
   Proceed to step 7.
7) Check, if the list is empty. If not, go to step 3. Otherwise proceed to step 8.
8) Check, if the solution obtained in the current iteration is better than the best one known yet. If so, set the current solution as the best one. Proceed to step 9.
9) Increase α by some value (in our case 0.1).
10) Check, if the value of α is equal 1. If so, stop calculations. Otherwise, go to step 2.

3.2. An algorithm based on the continuous optimization

The continuous optimization task is solved by a modified version of the method of linearization proposed by Pshenitchny [6]. General problem formulation consists in the minimization of the objective function

$$f_0(\vec{x}) \to \min \tag{5}$$

subject to the constraints

$$f_j(\vec{x}) < 0, \quad j = 1,\ldots,m, \qquad \vec{x} \in R^n. \tag{5a}$$

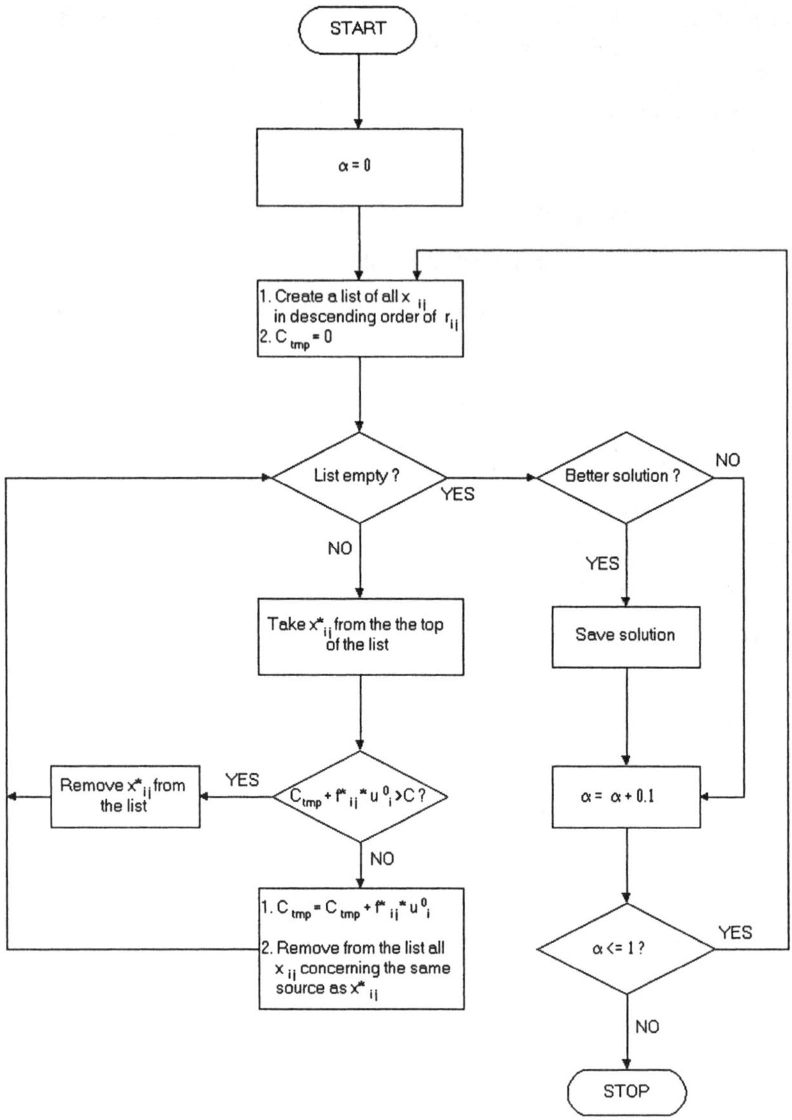

Figure 1. Flow-diagram of the heuristic algorithm

The implementation of the method utilizes the following functions:

$$F(\vec{x}) \overset{df}{=} \max_{1 \le j \le m} f_j(\vec{x}), \qquad \Phi_N(\vec{x}) \overset{df}{=} f_0(\vec{x}) + N(\vec{x})\, F(\vec{x}),$$

where the function $N(\vec{x})$ depends on the dual variables of the main optimization problem and is defined in details in [3]. Moreover the set of the active constraints with the tolerance $\delta > 0$ is defined as follows:

$$M_{\delta}(\vec{x}) \stackrel{df}{=} \left\{ j\text{: } j \geq 1 \quad \text{and} \quad f_j(\vec{x}) \geq F(\vec{x}) - \delta \right\}$$

The consecutive steps of the algorithm are as follows:

1) Set the initial point $\vec{x}_0$ and the parameters $\delta, \varepsilon, \theta > 0$; $0 < \eta < 1$. Put $i = 0$.
2) Calculate $F(\vec{x}_i)$ and $M_{\delta}(\vec{x}_i)$.
3) Solve the quadratic programming problem to find the descent direction $\vec{p}$
4)

$$\frac{1}{2}\|\vec{p}\|^2 + \nabla f_0(\vec{x}_i) \cdot \vec{p} \to \min \tag{6}$$

subject to the constraints

$$f_j(\vec{x}) \cdot \vec{p} + f_j(x_i) \leq 0, \qquad j \in M_{\delta}(\vec{x}_i) \qquad \vec{p} \in R^n. \tag{6a}$$

5) If $\|p\|^2 \leq \varepsilon$ -- *END* of the algorithm; $\vec{x}_i$ is the minimum point.
6) If the solution of (6)--(6a) does not exist, put $\delta = \delta / 2$ and go step 2).
7) Calculate $N(\vec{x}_i)$ (details can be found in [3]).
8) Find the minimum number $k = 0,1,\ldots$ such that

$$\Phi_N\left(\vec{x}_i + \left(\frac{1}{2}\right)^k \vec{p}\right) \leq \Phi_N(x_i) - \left(\frac{1}{2}\right)^k \eta \|p\|^2 \tag{7}$$

9) Substitute $\vec{x}_{i+1} = x_i + \left(\frac{1}{2}\right)^k \vec{p}$ and go to step 2).

To formulate the *continuous method* for the *ALLOCATION PROBLEM* stated in Section 2, we assume the continuity of the decision variables x_{ij}, $(i = 1,\ldots,N;\, j = 1,\ldots,M)$. That means, the solution is a set of real values $x_{ij} \in \langle 0,1 \rangle$. The task generates therefore certain artificial, continuous solution, which can be used as a reference base for the heuristic algorithm discussed above.

The objective function to be minimized is applied in the form (1) along with the constraint (2), which relates the environmental impact of the source (SO_2 concentration) with the actual emission intensity.

Some additional constraint was imposed on the original problem to get the final values of the decision variables close to the integer solution. To this end, the constraints (3) where modified as follows:

$$u_i = u_i^0 \sum_{j=1}^{N} (1-e_j)\cdot u_i, \quad \sum_{j=1}^{M} x_{ij} = 1, \quad x_{ij} \in \langle 0,1 \rangle, \quad 1 \le i \le N; \;\; 1 < j < M, \tag{8}$$

$$x_{ij} = y^2{}_{ij}, \qquad 1 \le i \le N; \;\; 1 < j < M.$$

We apply the algorithm (5)--(7) to the continuous optimization problem (1), (2), (4) and (8) in terms of the auxiliary decision variable $y_{ij} \in \langle 0,1 \rangle$. Due to (8), gradient of the objective function (1) has the following form

$$\frac{\partial J}{\partial y_{ij}} = \frac{\partial J}{\partial d}\cdot\frac{\partial c}{\partial u_i}\cdot\frac{\partial u_i}{\partial y_{ij}} = 2u_i^0(1-e_j)\int_\Omega w(x,y)(\max(0,c(x,y)-c_{ad})A_i(x,y)d\Omega \tag{9}$$

$$(i=1,...,N;\;\; j=1,...,M)$$

where:

$$\frac{\partial J}{\partial d} = \int_\Omega w(x,y)(\max(0,c(x,y)-c_{ad})d\Omega, \tag{9a}$$

$$\frac{\partial c}{\partial u_i} = A_i(x,y), \qquad (i=1,\dots,N), \tag{9b}$$

$$\frac{\partial u_i}{\partial y_{ij}} = 2u_i^0\cdot(1-e_j), \qquad (i=1,\dots,N;\;\; j=1,\dots,M). \tag{9c}$$

Solution of this problem will be directly applied for evaluating accuracy of the previous, heuristic algorithm. Moreover, it will be utilized for searching the closed integer-form solution by enumeration method (*combined method*). Results of the test computations and comparison of two methods are presented in the next section.

4. Case Study Analysis

Three optimization algorithms discussed in Section 3 were applied in the real-data case for selection of desulfurization technologies in the major power plants of the industrial Upper Silesia Region. The region is characterized by high concentration of heavy industry and the energy sector installations.

The domain considered is a rectangle area 110 km x 76km. In this area 20 major power plants were selected and considered as the controlled sources. Moreover, certain number of medium and small industrial sources constitutes the background emission field.

In the example presented, 8 desulfurization technologies are taken into account (5 basic technologies and 3 combined). The technologies and the respective emission reduction efficiencies are as follows:

1. "do nothing" technology $(e_1 = 0)$,
2. low-sulfur fuel $(e_2 \cong 30)$,
3. dry desulfurization method $(e_3 \cong 35)$,
4. low-sulfur fuel + dry desulfurization method $(e_4 \cong 545)$,
5. half-dry desulfurization method $(e_5 \cong 75)$,
6. low-sulfur fuel + half-dry desulfurization method $(e_6 \cong 825)$,
7. MOWAP method $(e_7 \cong 85)$,
8. low-sulfur fuel + MOWAP method $(e_8 \cong 895)$.

The annual unit concentration maps for the controlled sources (the transfer matrices $A_i(x, y), \quad i = 1, \ldots, N$) are preprocessed off-line by the regional-scale forecasting model REGFOR3 defined in [4,5]. This is a dynamical, single-layer model based on the meteorological input data for the period of simulation.

The same technique is used for generating the background concentration field for intermediate, pointwise and area sources. Computations were performed for one representative year, where a sequence of meteorological data with 12-hrs time resolution was applied. Calculations were repeated for several levels of the total cost constraint. In Tables 1--4 some selected results are presented for cost constraints 150 mill $/yr. and 250 mill $/yr., respectively.

Figure 2 presents the map of the initial SO_2 concentration in the region as well as location of the main emission sources. The initial emission intensities of 20 selected sources are presented in Table 1. This table also presents the solution obtained by continuous optimization method for the cost constraint 150 mill $/yr.

Relative share of the abatement technologies selected by the algorithm for specific sources are shown in the consecutive rows of Table 1. Columns refer to 8 abatement technologies, according to their increasing efficiencies. It can be seen that, due to the specific form of the algorithm and the additional constraint imposed, most of the solutions is of integer form.

On the other hand, solutions found for sources no. 6, 13--15, 17,19 are fuzzy and suggests technologically unrealistic combination of several technologies. The last column shows the emissions of the sources, related to selected reduction strategies.

Table 2 shows the results for the same cost limit, but generated by heuristic and combined methods, respectively. Solution get by the heuristic method is shown in the first part of Table 2. In this case, general strategy suggested is similar to that obtained by continuous algorithm, but significant differences appear in some specific sources. The rate of reduction of the environmental cost function is much worse for heuristic method comparing to those of the other two algorithms.

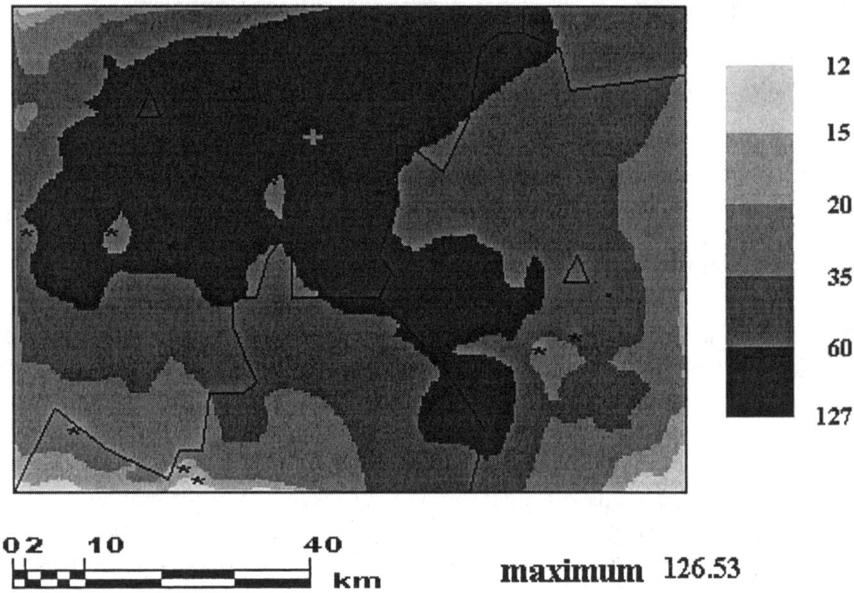

Figure 2. Initial SO_2 concentration in the region [$\mu g / m^3$].

As explained in the previous section, the combined algorithm utilizes the optimal solution of the continuous method as the starting point. Then all the fuzzy solutions x_{ij}, with the value not equal 0 nor 1, are enumerated. In the enumeration process the values of those x_{ij} are set to 0 or 1. The result usually gives a slightly worse rate of cost function reduction comparing to that of continuous method but is much better than heuristic solution.

Table 1. Solutions obtained by continuous optimization and cost constraint 150 mill $/yr.

source		solution by optimization method environmental cost reduction = **0.1175** abatement cost = **150.00** mill $/yr.								
	initial	abatement technology								final
No	emiss.	1	2	3	4	5	6	7	8	emiss.
1	303.20	0.00	0.00	0.00	0.00	0.00	0.00	**1.00**	0.00	45.48
2	225.30	**1.00**	0.00	0.00	0.00	0.00	0.00	0.00	0.00	225.30
3	104.00	0.00	0.00	0.00	**1.00**	0.00	0.00	0.00	0.00	47.32
4	91.80	0.00	0.00	0.00	**1.00**	0.00	0.00	0.00	0.00	41.77
5	90.10	0.00	0.00	0.00	0.00	0.00	**1.00**	0.00	0.00	40.99
6	78.00	**0.55**	**0.45**	0.00	0.00	0.00	0.00	0.00	0.00	54.60
7	65.00	0.00	**1.00**	0.00	0.00	0.00	0.00	0.00	0.00	45.51
8	52.00	**1.00**	0.00	0.00	0.00	0.00	0.00	0.00	0.00	52.00
9	52.00	0.00	**1.00**	0.00	0.00	0.00	0.00	0.00	0.00	36.40
10	45.10	0.00	**1.00**	0.00	0.00	0.00	0.00	0.00	0.00	31.57
11	34.70	0.00	0.00	0.00	0.00	0.00	0.00	**1.00**	0.00	5.21
12	33.80	0.00	0.00	0.00	0.00	0.00	0.00	0.00	**1.00**	3.55
13	29.90	0.00	0.00	0.00	0.00	0.00	**0.05**	**0.94**	**0.01**	4.51
14	25.10	0.00	0.00	0.00	0.00	0.00	0.00	**0.16**	**0.84**	2.82
15	26.00	0.00	0.00	0.00	**0.91**	0.00	**0.09**	0.00	0.00	11.15
16	18.70	0.00	0.00	0.00	**1.00**	0.00	0.00	0.00	0.00	8.51
17	16.90	**0.51**	**0.49**	0.00	0.00	0.00	0.00	0.00	0.00	14.40
18	15.10	0.00	0.00	0.00	**1.00**	0.00	0.00	0.00	0.00	6.87
19	12.30	**0.43**	**0.26**	**0.31**	0.00	0.00	0.00	0.00	0.00	10.00
20	11.60	0.00	**0.01**	0.00	**0.99**	0.00	0.00	0.00	0.00	5.30

Table 2. Integer-type solutions obtained by two approaches and cost constraint 150 mill $/yr.

sou rce		solution by heuristic method env. cost reduction = **0.125** abatement cost = **149.731** mill $/yr.									solution by optimization method env. cost reduction = **0.118** abatement cost = **149.768** mill $/yr.								
	initial	abatement technology								final	abatement technology								final
No	emiss.	1	2	3	4	5	6	7	8	emiss.	1	2	3	1	5	6	7	8	emiss.
1	303.20	0	0	0	0	0	0	**1**	0	45.48	0	0	0	0	0	0	**1**	0	45.48
2	225.30	**1**	0	0	0	0	0	0	0	225.30	**1**	0	0	0	0	0	0	0	225.30
3	104.00	0	0	0	**1**	0	0	0	0	47.32	0	0	0	**1**	0	0	0	0	47.32
4	91.80	0	0	0	0	**1**	0	0	0	22.95	0	0	0	**1**	0	0	0	0	41.77
5	90.10	0	0	0	0	0	0	0	**1**	9.46	0	0	0	0	0	**1**	0	0	15.77
6	78.00	**1**	0	0	0	0	0	0	0	78.00	0	**1**	0	0	0	0	0	0	54.60
7	65.00	**1**	0	0	0	0	0	0	0	65.00	0	**1**	0	0	0	0	0	0	45.50
8	52.00	**1**	0	0	0	0	0	0	0	52.00	**1**	0	0	0	0	0	0	0	52.00
9	52.00	0	0	0	**1**	0	0	0	0	23.66	0	**1**	0	0	0	0	0	0	36.40
10	45.10	**1**	0	0	0	0	0	0	0	45.10	0	**1**	0	0	0	0	0	0	31.57
11	34.70	**1**	0	0	0	0	0	0	0	34.70	0	0	0	0	0	0	**1**	0	5.21
12	33.80	0	0	0	0	0	0	0	**1**	3.55	0	0	0	0	0	0	0	**1**	3.55
13	29.90	0	0	0	0	0	**1**	0	0	5.23	0	0	0	0	0	0	**1**	0	4.49
14	25.10	0	0	0	0	0	0	**1**	0	3.77	0	0	0	0	0	0	**1**	0	3.77
15	26.00	0	0	0	**1**	0	0	0	0	11.83	0	0	0	**1**	0	0	0	0	11.83
16	18.70	0	0	0	**1**	0	0	0	0	8.51	0	0	0	**1**	0	0	0	0	8.51
17	16.90	**1**	0	0	0	0	0	0	0	16.90	0	**1**	0	0	0	0	0	0	11.83
18	15.10	0	0	0	**1**	0	0	0	0	6.87	0	0	0	**1**	0	0	0	0	6.87
19	12.30	**1**	0	0	0	0	0	0	0	12.30	**1**	0	0	0	0	0	0	0	12.30
20	11.60	0	**1**	0	0	0	0	0	0	8.12	0	0	0	**1**	0	0	0	0	5.28

Tables 3 and 4 show analogous solutions of allocation problem obtained for 250 mill $/yr. Figures 3 and 4 present respective maps of SO_2 concentration for the optimal emission reduction strategy (combined method) and the financial constraints 150 mill $/yr and 250 mill $/yr., respectively.

Table 3. Solutions obtained by continuous optimization and cost constraint 250 mill $/yr.

source No	initial emiss.	solution by optimization method environmental cost reduction = **0.0420** abatement cost = **250.00** $/yr.								
		abatement technology								final emiss.
		1	2	3	4	5	6	7	8	
1	303.20	0.00	0.00	0.00	0.00	0.00	0.00	**1.00**	0.00	45.48
2	225.30	0.00	0.00	**1.00**	0.00	0.00	0.00	0.00	0.00	146.45
3	104.00	0.00	0.00	0.00	0.00	0.00	0.00	0.00	**1.00**	10.92
4	91.80	0.00	0.00	0.00	0.00	0.00	0.00	0.00	**1.00**	9.64
5	90.10	0.00	0.00	0.00	0.00	0.00	**0.12**	0.00	**0.88**	10.23
6	78.00	0.00	0.00	0.00	**1.00**	0.00	0.00	0.00	0.00	35.53
7	65.00	0.00	0.00	0.00	**1.00**	0.00	0.00	0.00	0.00	29.58
8	52.00	0.00	0.00	**1.00**	**0.00**	0.00	0.00	0.00	0.00	33.80
9	52.00	0.00	0.00	0.00	**1.00**	0.00	0.00	0.00	0.00	23.66
10	45.10	0.00	0.00	0.00	**1.00**	0.00	0.00	0.00	0.00	20.52
11	34.70	0.00	0.00	0.00	0.00	0.00	0.00	**1.00**	0.00	5.21
12	33.80	0.00	0.00	0.00	0.00	0.00	0.00	0.00	**1.00**	3.55
13	29.90	0.00	0.00	0.00	0.00	0.00	0.00	0.00	**1.00**	3.14
14	25.10	0.00	0.00	0.00	0.00	0.00	0.00	0.00	**1.00**	2.64
15	26.00	0.00	0.00	0.00	0.00	0.00	0.00	0.00	**1.00**	2.73
16	18.70	0.00	0.00	0.00	**0.67**	**0.33**	0.00	0.00	0.00	7.25
17	16.90	0.00	**0.99**	0.00	**0.01**	0.00	0.00	0.00	0.00	11.77
18	15.10	0.00	0.00	0.00	0.00	0.00	**0.70**	0.00	**0.30**	2.33
19	12.30	0.00	0.00	0.00	**1.00**	0.00	0.00	0.00	0.00	5.60
20	11.60	0.00	0.00	0.00	**0.99**	0.00	**0.01**	0.00	0.00	5.24

Table 4. Integer-type solutions obtained by two approaches and cost constraint 250 mill \$/yr.

source		solution by heuristic method env. cost reduction = **0.0456** abatement cost = **249.881** mill \$/yr.									solution by optimization method env. cost reduction = **0.0423** abatement cost = **249.181** mill \$/yr.								
No	initial emiss.	abatement technology								final emiss.	abatement technology								final emiss.
		1	2	3	4	5	6	7	8		1	2	3	1	5	6	7	8	
1	303.20	0	0	0	0	0	0	0	**1**	31.84	0	0	0	0	0	0	**1**	0	45.48
2	225.30	0	0	0	**1**	0	0	0	0	102.51	0	0	**1**	0	0	0	0	0	146.45
3	104.00	0	0	0	0	0	0	0	**1**	10.92	0	0	0	0	0	0	0	**1**	10.92
4	91.80	0	0	0	0	0	0	0	**1**	9.64	0	0	0	0	0	0	0	**1**	9.64
5	90.10	0	0	0	0	0	0	0	**1**	9.46	0	0	0	0	0	0	0	**1**	9.46
6	78.00	0	**1**	0	0	0	0	0	0	54.60	0	0	0	**1**	0	0	0	0	35.49
7	65.00	0	0	**1**	0	0	0	0	0	42.25	0	0	0	**1**	0	0	0	0	29.58
8	52.00	**1**	0	0	0	0	0	0	0	52.00	0	0	**1**	0	0	0	0	0	33.80
9	52.00	0	0	0	**1**	0	0	0	0	23.66	0	0	0	**1**	0	0	0	0	23.66
10	45.10	0	0	**1**	0	0	0	0	0	29.32	0	0	0	**1**	0	0	0	0	20.52
11	34.70	0	**1**	0	0	0	0	0	0	24.29	0	0	0	0	0	0	**1**	0	5.21
12	33.80	0	0	0	0	0	0	0	**1**	3.55	0	0	0	0	0	0	0	**1**	3.55
13	29.90	0	0	0	0	0	0	0	**1**	3.14	0	0	0	0	0	0	0	**1**	3.14
14	25.10	0	0	0	0	0	0	0	**1**	2.64	0	0	0	0	0	0	0	**1**	2.65
15	26.00	0	0	0	0	0	0	0	**1**	2.73	0	0	0	0	0	0	0	**1**	2.73
16	18.70	0	0	0	0	0	**1**	0	0	3.27	0	0	0	**1**	0	0	0	0	8.51
17	16.90	**1**	0	0	0	0	0	0	0	6.90	0	**1**	0	0	0	0	0	0	11.83
18	15.10	0	0	0	0	**1**	0	0	0	3.78	0	0	0	0	0	**1**	0	0	2.64
19	12.30	0	0	**1**	0	0	0	0	0	7.99	0	0	0	**1**	0	0	0	0	5.60
20	11.60	0	0	0	**1**	0	0	0	0	5.28	0	0	0	**1**	0	0	0	0	5.28

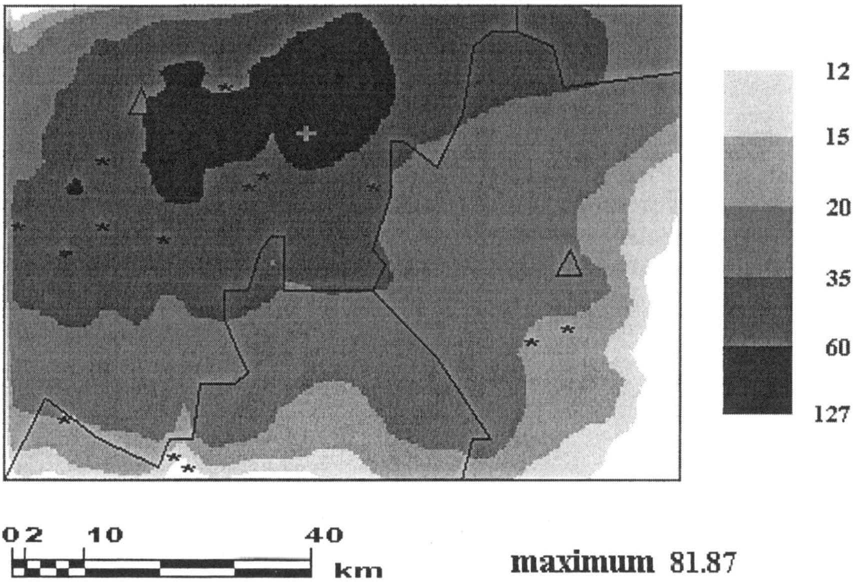

Figure 3. SO_2 concentration [$\mu g / m^3$] map for abatement technology constraint 150 mill \$/yr.

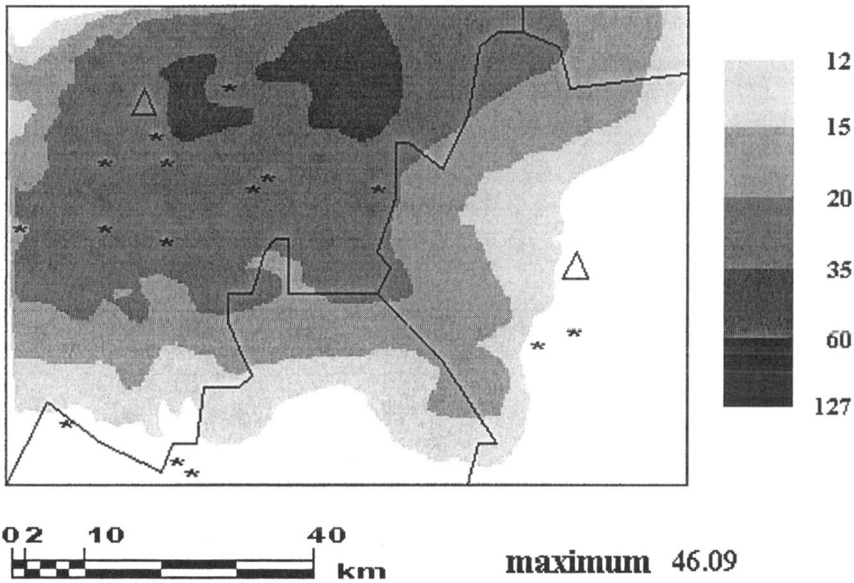

Figure 4. SO_2 concentration [$\mu g / m^3$] map for abatement technology constraint 250 mill \$/yr.

5. Concluding Remarks

Results presented in the previous section for two levels of cost limit show that general allocation strategies are similar for three algorithms discussed. Reduction of the objective function is the best for continuous method and definitely the worse for heuristic one. On the other hand, the absolute values of that index obtained by three methods only slightly differ from each other. This is the result of very flat shape of criterion function (1) in the neighborhood of the optimum point.

The methods discussed here were also tested on 50 different, randomly generated data sets, with the same cost constraint. The comparison of the achieved results is shown on Figure 5.

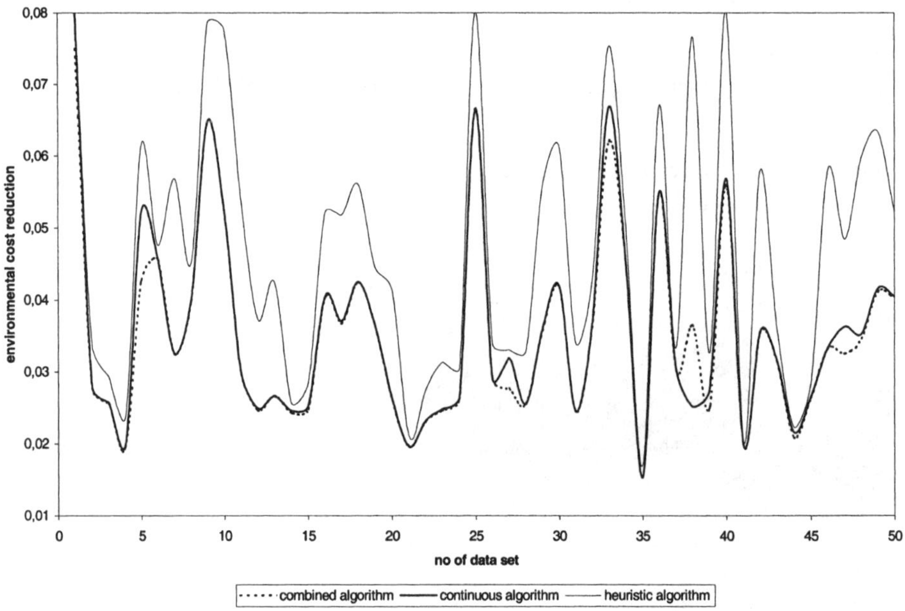

Figure 5. Comparison of the efficiency of different algorithms

As one can observe, the heuristic method gives the worst results (reduction of the environmental objective function), although it is the fastest method. The difference in solution quality calculated by the continuous method and the combined method varies from case to case. It depends on the stop criterion of the continuous algorithm, which is the compromise between computing time and the final accuracy of the solution. In cases, where the continuous solution is computed with high accuracy condition, the solution obtained by the combined method must be no better than the continuous one. On the contrary, where the continuous algorithm stops early, the better solution can be obtained by enumerating some possible solution, what is done by the combined method.

Bibliography

1. Alcamo J., Hordijk L., Shaw R. (1990) *The Rains Model of Acidification: Cost-effective Sulfur Emission Reduction Under Uncertainty.* IIASA, WP-94-119, Laxenburg.

2. Dernwent R.G. (1988) A better way to control air pollution. *Nature* 331, February, str. 575 - 578.

3. Holnicki P., .Żochowski A. (1990) *Selected Methods of Air Quality Analysis* (in Polish). PWN, Warszawa.

4. Holnicki P., Kałuszko A., Żochowski A. (1994) A microcomputer implementation of air quality forecasting system for urban scale. *Microcomputer Applications*, 13, No. 2, str. 76 -- 84.

5. Holnicki P., Nahorski Z., Żochowski A. (2000) *Modeling of Environmental Processes* (in Polish).. WSISiZ Publ., Warszawa.

6. Pshenitschny B.N.: (1983) *Method of Linearization* (in Russian). Nauka, Moscow

7. Zannetti P. (1990) *The Air Pollution Modeling*. Computational Mechanics Publ., Southampton.

Multicriteria Genetic Tuning for the Optimization and Control of HVAC Systems

Rafael Alcalá[1], José Manuel Benítez[2], Jorge Casillas[2], Juan Luis Castro[2], Oscar Cordón[2], Antonio González[2], Francisco Herrera[2], and Raúl Pérez[2]

[1] University of Jaén, Dept. of Computer Science, E-23071 Jaén, Spain, email: alcala@ujaen.es

[2] University of Granada, Dept. of Computer Science and A.I., E-18071 Granada, Spain, email: {jmbs,casillas,castro,ocordon,a.gonzalez,herrera,fgr}@decsai.ugr.es

Abstract. This work presents the use of genetic algorithms for the optimization and control of Heating, Ventilating and Air Conditioning (HVAC) systems developing smartly tuned fuzzy logic controllers for energy efficiency and overall performance of these systems.

An optimum operation of the HVAC systems is a necessary condition for minimizing energy consumptions and optimizing indoor comfort in buildings. This problem has some specific restrictions that make it very particular and complex because of the large time requirements existing due to the need of considering multiple criteria (which enlarges the solution search space) and to the long computation time models require to assess the accuracy of each individual.

To solve these problems, three efficient genetic tuning strategies, considering different multicriteria approaches, have been presented and tested in two real test sites (buildings) obtaining satisfactory results.

1 Introduction

Heating, Ventilating, and Air Conditioning (HVAC) systems are equipments usually implemented for maintaining satisfactory comfort conditions in buildings. The energy consumption as well as indoor comfort aspects of ventilated and air conditioned buildings are highly dependent on the design, performance and control of their HVAC systems and equipments.

Therefore, the use of appropriate automatic control strategies, as Fuzzy Logic Controllers (FLCs) [13], for HVAC systems decision support could result in important energy savings when compared to manual control [1]. In this way, some works apply FLCs to solve simple problems such as thermal regulation, maintaining a temperature setpoint [3,19,28].

However, in current systems, various criteria are considered and optimized independently one from another: variable air flows are used for indoor air quality control, controlled air temperature is used for thermal comfort management, and temperature set points are modified for energy consumption control. No global strategy for a coupled and integrated management of all these criteria has been yet efficiently implemented. Moreover, control systems in buildings are often designed and tuned using rules of thumb not always compatible with the controlled equipment requirements, energy performance and users expectations and demand.

Therefore, an optimum operation of these systems is a necessary condition for minimizing energy consumptions and optimizing indoor comfort. To do so, this contribution aims to develop optimum fuzzy controllers dedicated to the control of HVAC systems with regard to the energy performance and indoor comfort requirements, with the following main innovations:

- The use of FLCs: it will enable the implementation of multicriteria control strategies incorporating expert knowledge.
- The development of smart setting and tuning techniques for these controllers: it will enable a rational operation and improved performance of the fuzzy logic controllers –by fitting the Data Base (DB) parameters of previously obtained Knowledge Bases (KBs) provided by experts– and is a necessary condition for implementing complex control techniques.

Genetic algorithm (GAs) [21] have been recognized to be possibly well-suited to multicriteria optimization [6,11,16,41,45]. However, since the controller accuracy is assessed by means of simulations which usually take a long time, neither of them can be used satisfactorily because they do not properly address this restriction. Therefore, in order to solve these problems, efficient multicriteria genetic tuning approaches considering them should be developed [2]. To do so, three efficient genetic tuning strategies considering different multicriteria approaches have been developed and tested in two real test sites (buildings). Accurate models of the controlled buildings have been provided by experts.

This chapter is organized as follows. In Sect. 2 the HVAC systems are presented, showing the importance of these kinds of systems in the building sector. Section 3 explains how the FLCs can be applied to HVAC systems. In Sect. 4, the HVAC systems tuning restrictions are introduced, proposing some specific solutions to them. Section 5 presents three efficient genetic tuning techniques considering different multicriteria approaches. Section 6 shows the experiments performed in the two test sites. To sum up, in Sect. 7, some concluding remarks are pointed out, showing how this methodology could be applied to other systems and progressively implemented at industrial level.

2 Heating, Ventilating, and Air Conditioning Systems

In EU countries, primary energy consumption in buildings represents about 40% of total energy consumption and it has grown from 1974 over 13% overall. This energy consumption is highly dependent on weather conditions. Moreover, depending on the countries, more than a half of this energy is used for indoor climate conditions. On a technological point of view, it is estimated that the consideration of specific technologies like Building Energy Management Systems (BEMSs) can save up to 20% of the energy consumption of the building sector, i.e., 8% of the overall Community consumption. BEMSs are generally applied only to the control of active systems, i.e., Heating, Ventilating, and Air Conditioning (HVAC) systems.

HVAC systems are equipments usually implemented for maintaining satisfactory comfort conditions in buildings. The energy consumption as well as indoor comfort aspects of ventilated and air conditioned buildings are highly dependent on the design, performance and control of their HVAC systems and equipments. Therefore, the use of appropriate automatic control strategies, as FLCs, for HVAC systems control could result in important energy savings when compared to manual control.

An HVAC system is comprised of all the components of the appliance used to condition the interior air of a building. The HVAC system is needed to provide the occupants with a comfortable and productive working environment which satisfies their physiological needs. Temperature and relative humidity are essential factors in meeting physiological requirements. When temperature is above or below the comfort range, the environment disrupts person's metabolic processes and disturbs his activities.

Therefore, an HVAC system is essential to a building in order to keep occupants comfortable. A well-designed operated, and maintained HVAC system is essential for a habitable and functional building environment. Outdated, inappropriate, or misapplied systems result in comfort complaints, indoor air quality issues, control problems, and exorbitant utility costs. Moreover, many HVAC systems do not maintain an uniform temperature throughout the structure because those systems employ unsophisticated control algorithms. In a modern intelligent building, a sophisticated control system should provide excellent environmental control [3].

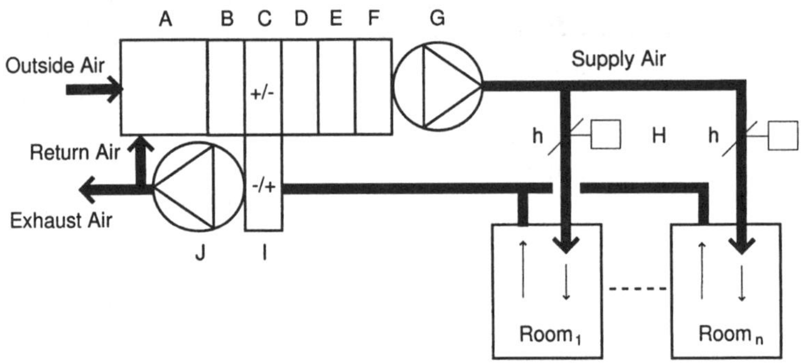

Fig. 1. Generic structure of an office building HVAC system

In Fig. 1, a typical office building HVAC system is presented. This HVAC system would comprise the following components to be able to raise and lower the temperature and relative humidity of the Supply Air (SA):

A. This module mixes the Return Air (RA) and the Outside Air (OA) to provide SA, and also closes OA-damper and opens RA-damper when fan stops.
B. It is a filter to reduce the OA emissions to SA.
C. The preheater/heat recovery unit preheats SA and recovers energy from the Exhaust Air EA.
D. A humidifier raising the relative humidity in winter.
E. This is a cooler to reduce the SA temperature and/or humidity.
F. An after-heater unit to raise the SA temperature after humidifier or to raise the SA temperature after latent cooling (dehumidifier).
G. The SA fan.
H. The dampers to demand controlled SA flow to rooms.
I. It is a heat recovery unit for energy recovery from EA.
J. The EA fan.

There are no statistical data collected on types and sizes of HVAC systems delivered to each type of building in different European countries. Therefore, to provide an HVAC system compatible with the ambiance is a task of the BEMS designer depending on its own experience. In the following section we will see how FLCs can be applied to the control of HVAC Systems.

3 Fuzzy Control of HVAC Systems

Nowadays, there are a lot of real-world applications of FLCs like intelligent suspension systems, mobile robot navigation, wind energy converter control, air conditioning controllers, video and photograph camera autofocus and imaging stabilizer, anti-sway control for cranes, and many industrial automation applications.

An FLC [13,33,34] is a kind of Fuzzy Rule-Based System which is composed of a *KB* that comprises the information used by the expert operator in the form of linguistic control rules, a *Fuzzification Interface*, that transforms the crisp values of the input variables into fuzzy sets that will be used in the fuzzy inference process, an *Inference System* that uses the fuzzy values from the fuzzification interface and the information from the KB performing the reasoning process, and the *Defuzzification Interface*, which takes the fuzzy action from the inference process and translates it into crisp values for the control variables. The KB is comprised of two components: the DB and the RB. The DB contains the definitions of the linguistic labels, that is, the membership functions for the fuzzy sets. The RB is a collection of fuzzy control rules, comprised by the linguistic labels, representing the expert knowledge of the controlled system.

In the specific case of HVAC Systems, some works apply FLCs to solve simple problems such as thermal regulation, maintaining a temperature set-point [3,19,28]. However, in this work various different criteria must be considered in order to reduce the energy consumption maintaining a desired comfort level. In building automation, the objective of a global controller would be to maintain the indoor environment within the desired (or stipulated) limits. In our case, to maintain environmental conditions within the comfort zone and to control the Indoor Air Quality (IAQ). Furthermore, other important objectives could be required, e.g, energy savings, system stability, etc. In any case, numerous factors have to be considered in order to achieve these objectives. It makes the system being controlled very complex and present a strong non linearity. In these cases, FLCs are very robust tools which would enable the implementation of multiple criteria control strategies incorporating expert knowledge.

As it is known, the design of an FLC is focused on the following parameters and characteristics:

- *Control and controlled parameter selection.* Controlled parameters are variables which are affected by the action of a controlled device receiving signals from a controller, whilst control parameters are variables which may be used as inputs or outputs for a control strategy.
- *The composition of the FLC KB*, that is, the set of fuzzy control rules forming the RB, and the set of linguistic terms in the fuzzy partitions of the input and output spaces forming the DB.
- *FLC architecture and operators*, i.e., the rule type and architecture of the FLC, the membership function type, the conjunctive operator *and*, the implication function, the defuzzification mode, the characteristic value and the control crisp value.

In this way, after the BEMS designer defines the system to be controlled (building and HVAC specification), the construction of the correspondent FLC can be performed. This task can be subdivided in the following subtasks:

1. Knowledge extraction method selection.
2. Identification of the controlled and the control parameters.
3. Identification of global indices for assessment of the indoor building environment.
4. Description of number and architecture of fuzzy controllers.
5. KB derivation method selection.
6. Selection of the inference system operators.
7. KB derivation.

3.1 Knowledge Base Derivation

As said, the KB encodes the expert known knowledge of the controlled system. Therefore, it depends on the concrete application making the accuracy of the designed FLC directly depends on its composition. There are four modes of derivation of fuzzy control rules, that are not mutually exclusive [32]. These modes are the following:

a *Expert experience and control engineering knowledge*: It is the most widely used, being effective when the human operator is able to linguistically express the control rules he uses to control the system. Since they present an adequate form to represent expert knowledge, these rules are usually of Mamdani type.
b *Modeling of the operator's control actions*: The control action is formed making a model of the operator actions without interviewing him.
c *Based on the fuzzy model of a process*: It is based on developing a fuzzy model of the system and constructing the fuzzy rules of the KB from it. This approach is similar to that traditionally used in Control Theory. Hence, structure and parameter identification are needed [36].
d *Based on learning and self-organization*: This method is based on the ability for creating and modifying the fuzzy control rules in order to improve the controller performance by means of automatic methods.

In these kinds of problems (HVAC System controller design), the KB is usually constructed by using the first method, i.e., based on the operator's experience. However, FLCs sometimes fail to obtain satisfactory results with the initial rule set drawn from the expert's experience [28]. This is because of:

- the gathering and structuring of expertise is not easy,
- the setting up of the knowledge base is an extensive task, and
- although a lot of knowledge is generic, the structure of the system to which it will apply varies substantially.

Moreover, in our case the system being controlled is too complex and optimal controllers are required. Therefore, this approach needs to modify the initial KB to obtain an optimal controller. To do so, a tuning on the semantic of an FLC previously obtained from human experience could be performed by modification of the DB components.

In this work, FLCs will be obtained from human experience to subsequently be tuned by the application of automatic tuning techniques. Thus, the learning method is a combination of the first and fourth derivation modes.

On the other hand, to evaluate the FLCs performance, physical modelization of the controlled buildings and equipments is needed. These models will be developed by BEMS designers using building simulation tools, and they will have to be able to account for all the parameters considered in control. The models will be validated using experimental data corresponding to the real sites being simulated. Many data corresponding to various operation conditions and heat or cooling load will be prepared and compared with simulations.

Thus, we will have the chance to evaluate the FLCs designed in the simulated system with the desired environmental conditions. In the same way, these system models can be used by the experts to validate the initial KB before the tuning process. On the other hand, it is of major importance to assess the fitness function in tuning.

3.2 Control and Controlled Parameters

Control and controlled parameters have to be chosen regarding the control strategy being implemented, the technical feasibility of the measurements as well as economic considerations. Fortunately, the BEMS designer is usually able to determine these parameters.

However, our intention is to develop both controllers and tuning strategies. This requires the use of explicit parameters (directly used as fuzzy controller's inputs or outputs) as well as implicit parameters used in the fitness function developed in order to evaluate the performance of each controller.

To identify the FLC's variables various parameters (control or explicit parameters) may be considered depending on the HVAC system, sensors and actuators. We propose the following parameters:

- *Predicted Mean Vote (PMV) index for thermal comfort*: Instead of only using air temperature as a thermal comfort index, we could consider the more global PMV index selected by international standard ISO 7730 (incorporating relative humidity and mean radiant temperature).
- *Difference between supply and room temperatures*: possible disturbances can be related to the difference between supply and mean air temperature. When ventilation systems are used for air conditioning, such a criterion can be important.
- *CO_2 concentration*: IAQ was found to be critical. As CO_2 concentration is a reliable index of the pollution emitted by occupants, it can be selected as IAQ index. It is therefore supposed that the building and HVAC system have been properly designed and that occupants actually are the main source of pollution.
- *Outdoor temperature* also needed to be accounted for, since during mid-season periods (or even mornings in summer periods) its cooling (or heating) potential through ventilation can be important and can reduce the necessity to apply mechanical cooling (or heating).
- *HVAC system actuators*: It directly depends on the concrete HVAC system, e.g., valve positions, operating modes, fan speeds, etc.

To identify global indices for assessment of the indoor building environment, various parameters (controlled or implicit parameters) may be measured depending on the objectives of the control strategy. In these kinds of problems, these parameters could be selected among:

- Thermal comfort parameters: indoor climate control is one of the most important goal of intelligent buildings. Among indoor climate characteristics, thermal comfort is of major importance. This might include both global and local comfort parameters.
- IAQ parameters: IAQ is also of major concern in modern buildings. It is controlled either at the design stage by reducing possible pollutants in the room and during operation thanks to the ventilation system. As our

work is dedicated to HVAC systems, IAQ is also an important parameter to account for.

- Energy consumption: If appropriate IAQ and thermal comfort levels have to be guaranteed in offices, this has to be achieved at a minimum energy cost. Therefore, energy consumption parameters would need to be incorporated.
- HVAC system status: A stable operation of the controlled equipments is necessary in order to increase life cycle and thus reduce the maintenance cost. Information of the status of the equipments at the decision time step or on a longer period must thus be considered.
- Outdoor climate parameters: Indoor conditions are influenced by outdoor conditions (air temperature, solar radiation, wind). Moreover, in an air distribution HVAC system, the power required to raise or lower the supply temperature is a function of outdoor temperature and humidity. Some of these parameters would thus need to be selected.

The selection of these parameters is a task concerned to the BEMS designer as well.

3.3 Architecture and Operators

Architecture and inference operators are factors that have a significant influence on the FLC behavior. The influence of several of these factors is analyzed in [8,30], taking as a basis several control applications.

As we have already seen, there are different alternatives to select these factors. In this section, we will propose one of them attending to their advantages and weaknesses in some aspects of the KB derivation process. We will strive to apply operators as simple as possible without loss in the system accuracy. If so, these operators will be easier to implement and faster to compute.

A distributed hierarchical architecture [18,44], which allows to divide the control tasks, is proposed for our FLC. Using the expert knowledge of the system to partition the controller allows adequate control with much fewer rules. Moreover, with this approach the control tuning becomes easier since the modification of one parameter influences a smaller number of rules.

In adition, it is recommended that three controllers (rather than a single one) be developed for each testing site. The reason for this lies in important climate variations all over the year and variable expectations from occupants according to season. Therefore, one controller per season will be developed considering fall and spring as the same kind of season. These controllers could be switched according to dates or by mixing the three controllers including a new meta-level in the hierarchical FLC.

On the other hand, the remaining factors to be considered are the following: rule type, type of membership functions, conjunctive operator, implication function, defuzzification mode, characteristic value and control crisp value. The selection of all of them is presented below.

We use the Mamdani-type rules because they provide a natural framework to include expert knowledge in the form of linguistic rules which is of major importance in our problem. In the same way, we use the triangular membership functions instead of the trapezoidal or the gaussian ones —being the former two linear functions and the latter a non-linear function—. Since we expect a KB derivation from experts, linear functions are more intuitive and easier to manage. Moreover, as all of them achieve similar results [12], we will use triangular membership functions, which are simpler. Their formula is:

$$\mu_{A_i}(x) = \begin{cases} \dfrac{x-a}{b-a}, & \text{if } a \leq x < b \\ \dfrac{c-x}{c-b}, & \text{if } b \leq x \leq c \\ 0, & \text{otherwise} \end{cases} .$$

Among all the associative functions, t-norms are the more suitable to be used to define the connective *and* [8]. Two basic t-norms have been considered: minimum ($Min(x,y) = min(x,y)$) and product ($\Pi(x,y) = x \cdot y$). Minimum operator achieves co-operative rules while product operator achieves competitive rules. Since we have recommended triangular membership functions and a good co-operation among rules is interesting in this case, the minimum operator is proposed. On the other hand, from the results reported in [8], we recommend the use of the minimum t-norm (Mamdani implication) also as implication operator (rule connective) because it yielded the best behavior among the 41 implication operators tested.

We use the FITA (First Infer, Then Aggregate) scheme considering the rule antecedent matching, h_i, because the defuzzification method working in this mode is more robust and easier to compute than those using the FATI (First Aggregate, Then Infer) scheme [8]. As characteristic value and control crisp value, we propose the Weighted Mean Of Maxima (WEMOM), since according to the results reported in [8], it renders the best accuracy among the 17 different defuzzification methods tested.

4 Multicriteria Genetic Tuning of FLCs for HVAC Systems

Tuning approaches are usually based on the availability of a predefined RB and a preliminary set of membership functions associated to the fuzzy partitions, DB. Their main aim is to find a better set of parameters by only changing the DB components, thus making optimum the FLC behavior. In this way, FLCs could be obtained from human experience or learning methods to subsequently be tuned by the application of automatic tuning techniques. We have followed the same approach but, in our case, the problem has two specific restrictions which make it very particular and complex:

- The evaluation is based on multiple criteria (energy consumption, occupants thermal comfort, indoor air quality, peak load electrical demand, ...). This fact adds complexity to the search because we must obtain the best trade-off among the different criteria.
- The controller accuracy is assessed by means of simulations which usually take a long time. This causes the run time of the algorithms to be extremely long.

In this case, numerous factors have to be considered in order to address these restrictions. It makes the system being controlled very complex and presents a strong non linearity. Therefore, an efficient operation of the automatic tuning techniques is a necessary condition in order to achieve good results.

GAs are general-purpose global search techniques that use principles inspired by natural population genetics to evolve solutions to problems. The basic principles of the GAs were first laid down rigorously by Holland [27] and are well described in many texts such as [21,35]. The basic idea is to maintain a population of knowledge structures that evolves over time through a process of competition and controlled variation. Each structure in the population represents a candidate solution to the specific problem and has an associated *fitness* to determine which structures are used to form new ones in the process of competition.

GAs present flexibility to work with different FLC architectures and have a good capability to include expert knowledge [9]. Furthermore, the ability to handle complex problems, involving features such as discontinuities, multimodality, disjoint feasible spaces and noisy function evaluations, reinforces the potential effectiveness of GAs in multicriteria search and optimization. For these reasons, GAs have been recognized to be possibly well-suited to multicriteria optimization [6,11,16,41,45].

Although there are many genetic tuning approaches [5,7,22,24,29], neither of them can be used satisfactorily because they do not properly address the said restrictions. Therefore, in order to solve these two problems, efficient genetic tuning approaches considering both restrictions should be developed.

4.1 Tuning Restrictions

As we have indicated, the two important restrictions we want to solve are the need of considering multiple criteria (which enlarges the solution search space) and the long computation time models require to assess the accuracy of each individual.

The **first restriction** will be solved by using multicriteria optimization techniques that allow us to work with fitness functions comprised by competitive objectives. In these cases, we could obtain not only an optimal solution, but a possible solution set. Depending on the number of solutions obtained, we can distinguish between two multicriteria approaches:

- Multicriteria aggregation-based methods: All classical methods scalarize the objective vector reducing it to a scalar optimization problem. Probably, the simplest of all these classical techniques is the method of objective weighting. In this case, multiple criteria functions are combined into one overall objective function by means of a vector of weights. This technique has much sensitivity and dependency toward weights. However, when trustworthy weights are available, this approach reduce the search space providing the adequate direction into the solution space and its use is highly recommended. Therefore, the main question to consider this approach is: Have we trusted weights to estimate the importance of each objective?
- Multicriteria non aggregation-based methods or multiobjective methods: In a typical multicriteria optimization problem, there is a set of solutions that are superior to the rest in the search space when all the objectives are considered. These solutions are known as non-dominated solutions (pareto set), whilst the remaining solutions are known as dominated solutions. None of the solutions in the non-dominated set is absolutely better than the other ones.
  Mathematically, the concept of *Pareto-optimality* or *non-dominance* is defined as follows. Let us consider, without loss of generality, a multicriteria function $f(x) = (f_1(x), f_2(x), \ldots, f_n(x))$ to be minimized with m parameters, $x = (x_1, x_2, \ldots, x_m) \in X$, and n objectives. A decision vector $a \in X$ dominates $b \in X$ (noted as $a \succ b$) if, and only if:

$$\forall i \in \{1, 2, \ldots, n\},\ f_i(a) \leq f_i(b) \quad \wedge$$
$$\exists j \in \{1, 2, \ldots, n\} \ \mid\ f_j(a) < f_j(b)\ .$$

  Any vector that is not dominated by any other is said to be pareto-optimal or non-dominated.

In order to solve the **second restriction**, the use of efficient tuning methods is necessary. There are some approaches that increase the convergence speed of GAs:

- An objective weighting technique would reduce the search space providing the adequate direction into the pareto when trustworthy weights are used.
- A steady-state GA [42], that consists of selecting two of the best individuals in the population and combining them to obtain new offspring. This approach improves the convergence and simultaneously decreases the number of evaluations.
- In order to reduce the GAs search space, an integer coding could be used. This one uses discrete parameter domains forcing to take values from a finite value set [20] (see Fig. 2). The cardinality of this set must be rich enough in order to allow the tuning process to achieve accurate results, but small enough so that small changes provoke significant variations.

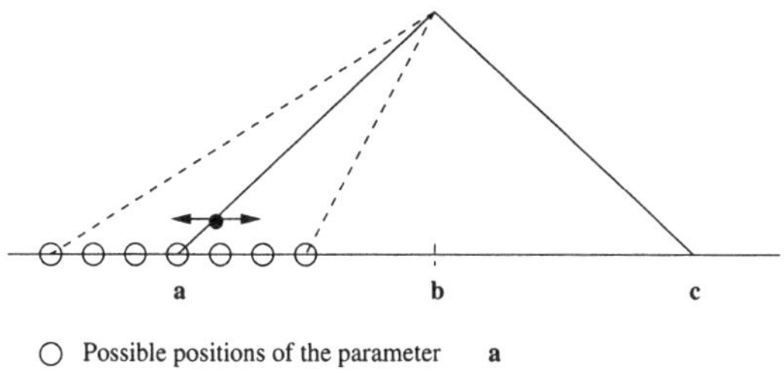

Fig. 2. Integer coding to tune fuzzy sets

- Reducing the population size, the number of evaluations is significantly decreased. However, this size must be large enough in order to maintain the diversity in the genetic population.

Several efficient tuning methods with different characteristics will be proposed based on the combination of multicriteria techniques with these different approaches. In order to do so, the different multicriteria genetic approaches will be introduced in the following subsection.

4.2 Multicriteria Genetic Optimization

Generally, multicriteria GAs only differ from the rest of GAs in the fitness function and/or in the selection operator. The evolutionary approaches in multicriteria optimization can be classified in three groups [16]: plain aggregating approaches, population-based non-pareto approaches, and pareto-based approaches.

Plain Aggregating Approaches

As conventional GAs require scalar fitness information to work on, a scalarisation of the objective vectors is always necessary. In most problems, where no global criterion directly emerges from the problem formulation, objectives are often artificially combined, or aggregated, into a scalar function according to some understanding of the problem, and then the GA is applied. Practically, all the classical aggregation approaches can be used with GAs. Some approach of this kind has been reported in the literature: weighted sum [39], distance functions [43], etc.

Optimizing a combination of the objectives has the advantage of producing a single compromise solution, requiring no further interaction with the decision-maker. The problem is that, if the optimal solution can not be accepted, new runs of the optimizer may be required until a suitable solution is found. However, in the case of objective weighting, when trustworthy weights are available this problem disappears.

Population-Based Non-Pareto Approaches

This approach allows to exploit the special characteristic of GAs. The use of an population of individuals offers the possibility to treat non-commensurable objectives separately and to search for multiple non-dominated solutions concurrently in a single GA run.

Now, a non-dominated set of individuals is obtained instead of obtaining only one of them. In order to do this, the selection operator is changed. Generally, the best individuals according to each objective are selected (in many occasions, an order according to its importance is followed) and then these partial results are combined to obtain the new population.

Different proposals based on this idea can be found in [17,23,31,37].

Pareto-Based Approaches

The population-based non-pareto approaches attempt to promote the generation of multiple non-dominated solutions. However, none makes direct use of the actual definition of Pareto-optimality. As the best solution cannot be selected among the non-dominated solutions set, the approaches should assign equal probability of reproduction to all of them. In GAs based on the concept of pareto-optimality, to calculate the reproduction probability of each individual, the solutions are compared by means of the dominance relation ($\succ$). Some approaches of these kinds can be found in [15,38].

Although the Pareto-based ranking correctly assigns all the non-dominated individuals the same fitness, it does not guarantee that the Pareto set be uniformly sampled. When it is presented with multiple equivalent optima, finite populations tend to converge to only one of them, due to stochastic errors in the selection process. This phenomenon is known as *genetic drift* [10].

Since preservation of diversity is crucial in the field of multiobjective optimisation, several multiobjective GAs have incorporated the *niche* and *specie* concepts for the purpose of favouring such behaviour [15,38].

5 Three Different Multicriteria Genetic Tuning Strategies

Thinking on these three different multicriteria approaches, we must consider an important aspect in the selection of the best techniques and methods to accomplish the tuning process: Have we trusted weights to estimate the importance of each objective? Depending on whether we can obtain them or not, the recommended strategy will be different. Nevertheless, we will be able to use several tuning processes with and without weights if we want to compare the results or if we are not sure of the weight reliability. In this way, combining the multicriteria and the said efficient tuning approaches, three strategies have been developed (Table 1 shows a summary of the three developed methods):

- Taking into account the existence of trusted weights and in order to benefit from them, we propose a simple steady-state GA with the classical real coding [26] and with a fitness function based on objective weighting that considers them. It will be called, Weighted Multi-Criterion Steady-State Genetic Algorithm (WMC-SSGA) [1,2]. Furthermore, the use of fuzzy goals for dynamically adapting the search direction in the space of solutions will be considered. It will make the method robust and more independent from the weight selection for the fitness function.
- If we do not have trustworthy weights, we must search for advanced multiobjective techniques with adequate characteristics to obtain the desired convergence, e.g., the use of the integer coding. In this way, two different multiobjective approaches based in the two well-known algorithms presented in [15] and [38] are proposed. On the one hand, the so called Multi-Objective Steady-State Genetic Algorithm (MO-SSGA) [2] presents likely premature convergence getting a quick search speed at the expense of decreasing the diversity. This algorithm is complemented with the use of a steady-state approach. On the other hand, the so called Non-Dominated Sorting Genetic Algorithm (NDS-GA) obtains a good distribution over the non-dominated individuals set at the expense of the search speed by using a generation-based GA. Therefore, while the first one is quicker obtaining good solutions, the second one is theoretically more robust and sure.

Table 1. Summary on Tuning Strategies

Method	When	Multicriteria approach	Features
One	TW exist	Aggregation	SSGA + RC
Two	TW not exist	Pareto-based	SSGA + IC
Three	TW not exist & robust results are required	Pareto-based	GA + IC + FSh

[TW] Trusted Weights. [RC/IC] Real/Integer Coding. [FSh] Fitness Sharing.

There is an important aspect that the proposed methods address in the same way, the definition of the variation intervals for each gene. In the following subsection, this common characteristic is introduced. After this, the three proposed methods will be widely explained.

5.1 Dynamic Variation Intervals

In order to be meaningful, each chromosome (a complete DB) must maintain their genes (the DB definition points) within their respective variation intervals. These intervals are usually computed and fixed from the initial solution

—DB provided by experts—. However, in our case, these intervals are dynamically adapted from the best individual for each GA iteration, avoiding the restrictions of fixing them from the beginning of the GA run [2]. Thus, once these intervals have been calculated, the genes out of range are randomly generated within them.

Let (a^i_j, b^i_j, c^i_j) be the definition points of the j-th membership function label of the i-th variable. In a strong fuzzy partition (those in which the membership degree within the variable domain is kept to 1.0) the vertex of each label (b^i_j) coincides with the nearest extreme points of its neighbor labels, $c^i_{j-1} = b^i_j = a^i_{j+1}$. In this case, only the vertex of the labels has to be considered and the same variation interval can be defined for coincident points. Thus, the variation intervals are usually defined by the middle points between the correspondent vertex and the vertex of the previous and the next label.

In our case, a more flexible approach is considered and the vertex of the labels does not have to coincide with the nearest extreme points of its neighbor labels (see Figure 3). However, considering these three points as a simple set for each label $B_j = \{c^i_{j-1}, b^i_j, a^i_{j+1}\}$ and taking into account that they have the same variation interval, the same approach can be followed. In this way, the middle point between two sets can be computed considering the maximum point of the first set and the minimum point of the second set. Therefore, to calculate the left extreme of the variation interval for a concrete definition point $x \in B_j$, we should consider the maximum point of B_{j-1} (l^1_x) and the minimum point of the corresponding set B_j (l^2_x). And for the corresponding right extreme, we should consider the maximum point of B_j (r^1_x) and the minimum point of B_{j+1} (r^2_x).

Finally, taking into account that $a^i_j \in B_{j-1}$, $b^i_j \in B_j$ and $c^i_j \in B_{j+1}$, the variation intervals of each definition point of the j-th label membership function of the i-th variable, (a^i_j, b^i_j, c^i_j), are calculated from the initial or best individual as,

$$\{l^1_{a^i_j}, l^2_{a^i_j}\} = \{\max(c^i_{j-3}, b^i_{j-2}, a^i_{j-1}), \min(c^i_{j-2}, b^i_{j-1}, a^i_j)\}$$
$$\{r^1_{a^i_j}, r^2_{a^i_j}\} = \{\max(c^i_{j-2}, b^i_{j-1}, a^i_j), \min(c^i_{j-1}, b^i_j, a^i_{j+1})\}$$
$$[L_{a^i_j}, R_{a^i_j}] = [l^2_{a^i_j} - \frac{l^2_{a^i_j} - l^1_{a^i_j}}{2}, r^1_{a^i_j} + \frac{r^2_{a^i_j} - r^1_{a^i_j}}{2}] \ ,$$

$$\{l^1_{b^i_j}, l^2_{b^i_j}\} = \{\max(c^i_{j-2}, b^i_{j-1}, a^i_j), \min(c^i_{j-1}, b^i_j, a^i_{j+1})\}$$
$$\{r^1_{b^i_j}, r^2_{b^i_j}\} = \{\max(c^i_{j-1}, b^i_j, a^i_{j+1}), \min(c^i_j, b^i_{j+1}, a^i_{j+2})\}$$
$$[L_{b^i_j}, R_{b^i_j}] = [l^2_{b^i_j} - \frac{l^2_{b^i_j} - l^1_{b^i_j}}{2}, r^1_{b^i_j} + \frac{r^2_{b^i_j} - r^1_{b^i_j}}{2}] \ ,$$

$$\{l^1_{c^i_j}, l^2_{c^i_j}\} = \{\max(c^i_{j-1}, b^i_j, a^i_{j+1}), \min(c^i_j, b^i_{j+1}, a^i_{j+2})\}$$
$$\{r^1_{c^i_j}, r^2_{c^i_j}\} = \{\max(c^i_j, b^i_{j+1}, a^i_{j+2}), \min(c^i_{j+1}, b^i_{j+2}, a^i_{j+3})\}$$
$$[L_{c^i_j}, R_{c^i_j}] = [l^2_{c^i_j} - \frac{l^2_{c^i_j} - l^1_{c^i_j}}{2}, r^1_{c^i_j} + \frac{r^2_{c^i_j} - r^1_{c^i_j}}{2}] \ ,$$

Notice that the associated variation intervals of the corresponding extreme values, a^i_j and c^i_j, are calculated exactly as the intervals for b^i_{j-1} and b^i_{j+1}, respectively.

Figure 3 graphically depicts the variation intervals for the i-th variable following the proposed approach. We have considered that the vertex of the labels at the edges of the variables' domain must coincide with the extreme points. These labels will be symmetrical with respect to their vertexes.

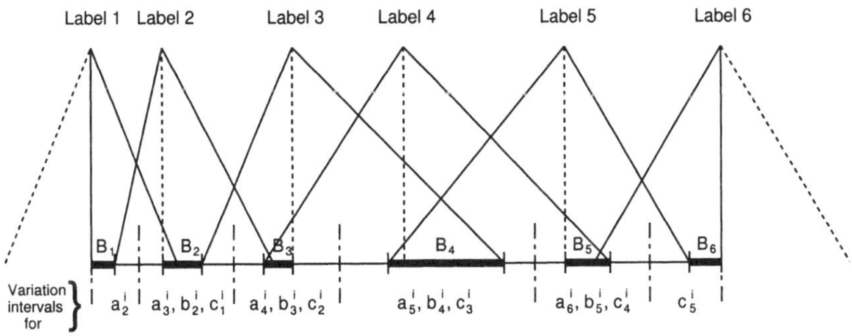

Fig. 3. Variation intervals of the i-th variable

5.2 Weighted Multi-Criterion Steady-State Genetic Algorithm

WMC-SSGA [1,2] consists of a GA based on the well-known steady-state approach [42]. Its fitness function is based on objective weighting. However, in order to make the method robust and more independent from the weight selection for the fitness function, the use of fuzzy goals for dynamically adapting the search direction in the space of solutions will be considered.

Coding scheme: WMC-SSGA uses a real coding scheme [26]. A solution is directly encoded in a chromosome by joining the definition points (a^i_j, b^i_j, c^i_j) of the l_i labels of each one of the m variables composing the DB. For example:

$$C_i = (a^i_1, b^i_1, c^i_1, \ldots, a^i_{l_i}, b^i_{l_i}, c^i_{l_i}), \ \ i = 1, \ldots, m \ ,$$
$$C = C_1 C_2 \ldots C_m \ .$$

Initial gene pool: To make use of the existing knowledge, the DB previously obtained from expert knowledge is included in the population as an initial solution. The remaining individuals are randomly generated maintaining their genes within their respective variation intervals. These intervals are computed from the initial solution (see Sect. 5.1).

Evaluating the chromosome: The fitness function is based on objective weighting. However, it has been modified in order to consider the use of fuzzy goals for dynamically adapting the search direction in the space of solutions, decreasing the improvement possibility of those objectives which approach their goals in the first place. Thus, a function modifier parameter, $\delta_i(x)$, is used to penalize each objective (taking values over 1.0) whenever its value gets worse with respect to the initial solution or to decrement the importance of each individual fitness value whenever it comes to its respective goal (taking values close to 0.0). Moreover, a penalization rate has been included in $\delta_i(x)$, allowing the user to set up priorities in the objectives. This penalization rate, p_i, for each objective is a real number from 0.7 to practically 1, although the user specifies this penalization from 0 to 1 (less and more priority, respectively), which is more interpretable. Therefore, the global fitness is evaluated as:

$$F = \sum_{i=1}^{n} w_i \cdot \delta_i(C_i) \cdot C_i \ ,$$

with C_i being the considered criteria (objectives) for each specific problem and w_i being the corresponding weighting coefficients.

Two cases can be presented in the corresponding individual according to the value of the goal, g_i, and the value of the initial solution, i_i. Depending on these values, two different δ functions will be applied.

- The first case is when the value of g_i is lesser than the value of i_i, presenting the following behavior (see Fig. 4).

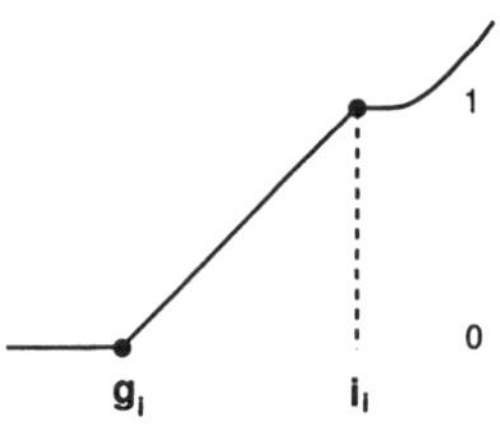

$$\delta_i(x) = \begin{cases} 0, & \text{if } x \leq g_i \\ \dfrac{x - g_i}{i_i - g_i}, & \text{if } g_i < x < i_i \\ \dfrac{x - i_i}{x - x \cdot p_i} + 1, & \text{if } i_i \leq x \ . \end{cases}$$

Fig. 4. $\delta_i(x)$ when $g_i \leq i_i$

In this case, the objective is not considered if the goal is met and penalized if the initial results are worsen.

- The second case happens when the initial value, i_i, is lesser than the goal value, g_i (see Fig. 5).

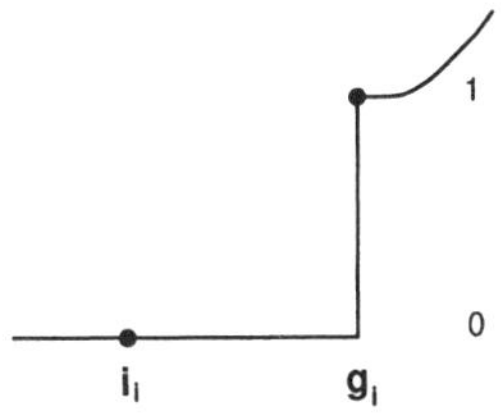

$$\delta_i(x) = \begin{cases} 0, & \text{if } x < g_i \\ \dfrac{x - g_i}{x - x \cdot p_i} + 1, & \text{if } g_i \leq x \end{cases} .$$

Fig. 5. $\delta_i(x)$ when $g_i > i_i$

Now, the initial results can be worsen while the goal is met, and it is penalized otherwise.

Notice that the penalization function allows the search to slightly worsen the goal, improving other objectives to subsequently met the goal again.

Genetic operators: The **selection** is based on the Baker's stochastic universal sampling [4] (by only selecting two individuals). WMC-SSGA also follows the interval adapting scheme explained in Sect. 5.1.

Since WMC-SSGA uses the real coding scheme, the crossover and mutation operators have been selected according to this aspect: the Max-Min-Arithmetical crossover [25] and Michalewicz's non-uniform mutation [35].

Using the max-min-arithmetical **crossover**, if $C_v^t = (c_1, \ldots, c_k, \ldots, c_H)$ and $C_w^t = (c'_1, \ldots, c'_k, \ldots, c'_H)$ are going to be crossed, the next four offspring are obtained:

$$\begin{aligned} C_1^{t+1} &= aC_w^t + (1-a)C_v^t \\ C_2^{t+1} &= aC_v^t + (1-a)C_w^t \\ C_3^{t+1} &\text{ with } c_{3k}^{t+1} = \min\{c_k, c'_k\} \\ C_4^{t+1} &\text{ with } c_{4k}^{t+1} = \max\{c_k, c'_k\} \ , \end{aligned}$$

with a being a constant parameter chosen by experts, and H being the number of genes.

In the case of the Michalewicz's non-uniform **mutation**, a gene c_k, with a variation interval $[L_{c_k}, R_{c_k}]$, can be mutated as $c'_k = c_k + \Delta(t, R_{c_k} - c_k)$ with probability 0.5, or as $c'_k = c_k - \Delta(t, c_k - L_{c_k})$, in other case. With t being the current generation, function $\delta(t, y)$ returns a value in the range $[0, y]$ such that the probability of $\delta(t, y)$ being close to 0 increases as the number of generations increases. This function is formulated as $\delta(t, y) = y(1 - r^{(1-\frac{t}{T})^b})$, with r being a random number in $[0, 1]$, T the total number of generations, and b being selected by the user to determine the dependency with t.

Thus, once the mutation operator is applied over the four offspring obtained from the crossover operator, the resulting descendents are the two best of these four individuals.

Restart approach: Finally, to get away from local optima, this algorithm uses a restart approach [14]. Thus, when the population of solutions converges to very similar results (practically the same fitness value in the population), the entire population but the best individual is randomly generated within the variation intervals. It allows the algorithm to perform a better exploration in the search space and to avoid getting stuck at local optima.

5.3 Multi-Objective Steady-State Genetic Algorithm

MO-SSGA [2] consist on an integer coded steady-state GA for multiobjective optimization. This algorithm presents likely premature convergence getting a quick search speed at the expense of decreasing the diversity. Its fitness function is based on the multiobjective approach presented in [15]. However a new scheme to accommodate goal attainment have been proposed.

Coding scheme: Once again, a solution is encoded in a chromosome by joining the representation (definition points) of the l_i labels of each one of the m variables composing the DB. However, in this case, an integer coding scheme is used to represent the possible solutions. This one uses discrete parameter domains for each one of the m variables, forcing to take values from a finite value set. The cardinality of these domains, G_i, is determined by the global granularity parameter G —chosen by experts— and directly depends on the corresponding number of labels l_i:

$$G_i = G * l_i + 1, \quad i = 1, \ldots, m.$$

Let L^i and R^i be the left and right extremes of the i-th variable domain. The corresponding possible values for this variable are uniformly distributed from L^i to R^i. Therefore, the distance between a point and the next one will be,

$$d_i = \frac{R^i - L^i}{G_i - 1}.$$

In this way, a mapping between an integer ordered set (from 0 to $G_i - 1$) and the set B_i of the corresponding real discrete values could be established as follows:

$$D : \{0, 1, \ldots, G_i - 1\} \longrightarrow B_i,$$
$$D(x^i) = L^i + x^i * d_i, \ \forall\, x^i \in \{0, 1, \ldots, G_i - 1\}.$$

The integer coding scheme consists on coding the possible solutions by using these integer ordered sets to represent the corresponding real discrete values —which can be easily obtained by using $D(x)$—. This representation eases the application of the genetic operators.

Initial gene pool: The initial pool is comprised of one individual containing the DB previously obtained from expert knowledge and the remaining ones generated at random maintaining their genes within their respective variation intervals. The intervals in which will switch around each gene are established from the initial solution as explained in Sect. 5.1. However, since this method is based on integer coding, once the variation intervals are computed as in the real coding, the upper and lower limits defining these intervals and the initial solution values must be encoded by rounding to the closer corresponding integer in the coding scheme as follows:

$$C : [L^i, R^i] \longrightarrow \{0, 1, \ldots, G_i - 1\},$$
$$C(x^i) = round\left(\frac{x^i - L^i}{\frac{R^i - L^i}{G_i - 1}}\right), \ \forall\ x^i \in [L^i, R^i].$$

After this, the initial gene pool can be generated within these variation intervals following the integer coding.

Evaluating the chromosome: A rank-based fitness assignment method for Multiple Objective GA (MOGA) was developed by Fonseca and Fleming in [15]. With this purpose, a rank is assigned to each individual of the population. These individuals are sorted to be selected according to rank. MO-SSGA is based on this approach. Thus, the position in the individuals' ranking can be given by:

If *non-dominated* **then** $rank(x_i) = 1$
Else $rank(x_i) = 1 + (dominants\ of\ x_i)$

The traditional assignment of fitness according to rank may be extended as follows:

1. Sort population according to rank.
2. Assign fitness to individuals interpolating from the best (rank 1) to the worst (rank $n^* \leq N$) in the usual way, according to some function, usually linear but not necessarily.
3. Average the fitness of individuals with the same rank, so that all of them will be sampled at the same rate. Note that this procedure keeps the global population fitness constant while maintaining appropriate selective pressure, as defined by the function used.

A scheme to accommodate goal attainment was proposed in [15]. It consists on modifying the ranking scheme for selection. In this way, the ranking in MO-SSGA was extended to accommodate goal information by altering the way in which individuals are compared with one another. However, considering the original approach interesting individuals were lost. Therefore, a new scheme has been implemented replacing the *concept of dominance* by the *concept of preferable.* Following this new scheme, individuals that meet more goals are more important than individuals that meet less even if they do not dominate to these latter ones. Thus, $y_a = (y_{a,1}, \ldots, y_{a,q})$ is preferable to $y_b = (y_{b,1}, \ldots, y_{b,q})$:

- If y_a meets more number of goals than y_b, or,
- If y_a meets exactly the same goals than y_b, $k+1, \ldots, q$, and y_a dominates to y_b in $k+1, \ldots, q$ objectives.

Genetic operators: The chromosome **selection** is based in the Baker's stochastic universal sampling [4]. However, in this case only two individuals are selected. These two individuals replace the to worst in the population.

The genetic operators are a two points crossover and a mix mutation. Using the two points **crossover**, two points are randomly generated and the genes between these points are exchanged to obtain two offsprings. After this, the mix **mutation** is applied over these two offspring. It consists on randomly using one of the following mutation operators:

- A random mutation, by randomly selecting an integer value within the variation interval of the corresponding gene.
- The Thrift mutation [40], by changing the gene one level either up or down.

In the same way that WMC-SSGA, this method follows the same interval adapting scheme explained in Sect. 5.1. However, since the integer coding is considered, the best solution in each iteration must be decoded —by using $D(x)$— to compute the variation intervals as in the real coding scheme. After this, the upper and lower limits defining these intervals must be encoded by using $C(x)$.

5.4 Non-Dominated Sorting Genetic Algorithm

This method is based on the non-dominated sorting of the population and the niche method in the parameters values of each individual presented in [38]. In this case, the method consists on an integer coded generation-based GA, obtaining a good distribution over the non-dominated individuals set at the expense of the search speed. Therefore, this method is theoretically more robust and sure, but much more slower.

The same coding scheme, initialization and genetic operators in MO-SSGA have been considered in NDS-GA. Therefore, the integer coding, the two points crossover and the mix mutation have been implemented. Furthermore, this method follows the interval adapting scheme explained in Sect. 5.1 in the same way that MO-SSGA.

The main differences with MO-SSGA are the generational approach and the multiobjective approach considered. In this case, a percentage of the population is selected for reproduction, and once the new population is generated the mutation operator is applied with a predetermined probability. The population ranking in NDS-GA is computed as follows.

Evaluating the chromosome: The population is progressively grouped in fronts and for all the individuals in a front is given an equal ranking value that must be lesser than the value assigned to the previous front. Thus, the non-dominated individuals present in the population are picked out to constitute the first non-dominated front and assigned a large dummy fitness value. To maintain diversity in the population, these classified individuals are then shared with their dummy fitness values. Afterwards, these non-dominated individuals are temporary ignored and the rest of the population is processed in the same way to find the second front. The new assigned dummy fitness value must be kept smaller than the minimum shared dummy fitness value of the previous front. This process is continued until the entire population is classified into several fronts.

When each individual has its sharing dummy fitness value assigned, the algorithm continues using a stochastic remainder proportionate selection. In this case, the Baker selection [4] is used. This would result in quick convergence of the population toward non-dominated regions, but sharing helps to distribute it over these regions. The used sharing function is calculated in the following way:

$$Sh(d_{ij}) = \begin{cases} 1 - \left(\dfrac{d_{ij}}{\sigma_{share}}\right)^2, & \text{si } d_{ij} < \sigma_{share} \\ 0, & \text{otherwise} \end{cases},$$

where d_{ij} is the phenotypic distance between two individuals i and j in the current front, and σ_{share} is the maximum phenotypic distance allowed between any two individuals to become members of a niche (some guidelines to set σ_{share} appear in [10]). A parameter niche count is calculated adding the above sharing function values for all individuals in the current front. Finally, the shared fitness value of each individual is calculated dividing its dummy fitness value by its niche count.

The same scheme presented in MO-SSGA to accommodate goal attainment is considered, replacing the *concept of dominance* by the *concept of preferable.*

6 Experimental Results and Analysis

In order to evaluate the goodness of the proposed tuning techniques, several experiments have been carried out within the framework of the JOULE-THERMIE programme under the GENESYS [1] project. Two real test sites (buildings) were available for the experiments. The first one is provided by both Centre National de la Recherche Scientifique and the Ecole Nationale des

[1] GENESYS Project: Fuzzy controllers and smart tuning techniques for energy efficiency and overall performance of HVAC systems in buildings, European Commission, Directorate-General XII for Energy (contract JOE-CT98-0090).

Travaux Publics de l'Etat (CNRS ENTPE) from France, whilst the second belongs to a French private enterprise whose name must remain anonymous. From now on, the latter will be called ATC test site —from Anonymous Test Cell—.

In both cases, the main objective was the energy performance but maintaining the required indoor comfort levels. Therefore, we should consider the development of a fitness function aiming to characterize the performance of each tested controller towards thermal comfort, indoor air quality, energy consumption and system stability criteria. In this way, the objective was to **minimize** the following five criteria:

C_1 Upper thermal comfort limit: $if\ PMV > 0.5, C_1 = C_1 + (PMV - 0.5)$.
C_2 Lower thermal comfort limit: $if\ PMV < -0.5, C_2 = C_2 + (-PMV - 0.5)$.
C_3 IAQ requirement: $if\ CO_2\ conc. > 800ppm, C_3 = C_3 + (CO_2 - 800)$.
C_4 Energy consumption: $C_4 = C_4 +$ Power at time t.
C_5 System stability: $C_5 = C_5 +$ System change from time t to $(t - 1)$.

To assess the proposed tuning techniques for fitness computation, accurate models of these controlled buildings (as well as the corresponding initial FLCs) were provided by experts for each season, considering fall and spring as the same kind of season. These models require long computation times, which makes more complex the tuning process. The results obtained were very satisfactory, specially for the ATC Summer model. However, due to the large number of results, we will work only with a cross-section of the models, the CNRS ENTPE Spring and Fall model, the CNRS ENTPE Summer model, and the ATC Summer model. In this section, the experiments performed with the said models are presented.

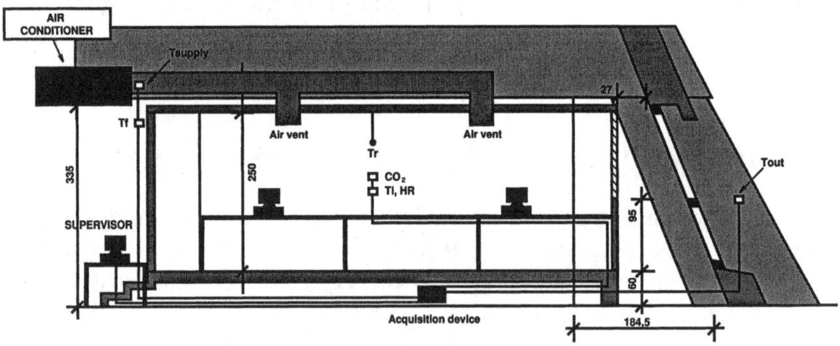

Fig. 6. Representation of the CNRS ENTPE test cells

6.1 The CNRS ENTPE and ATC Test Sites

The first task was to develop the thermal models of the two test sites that would be used in the complete learning process. These test sites have different characteristics, specially regarding the composition of their HVAC system. The main characteristics of these sites are the followings:

- CNRS ENTPE test site: Two single zone twin cells with low thermal mass located in a large hall whose climatic conditions can be controlled. The climatic control of the large hall temperature make it possible to create artificial climate with at least $8°C$ amplitude per day (e.g. from 23 to $31°C$ in summer conditions). The HVAC system is based on an air supply ventilation system with a maximum air flow rate of $2000m^3/h$ (test cells volume is $80m^3$), with direct expansion cooling and an electric coil controlled through a triac. Three fan speeds make it possible to slightly control supplied air flow rates (Fig. 6 illustrates these test cells).
- ATC test site: Also located in France, this test environment consists of two adjacent twin cells. Around these test cells walls, an artificial climate can be created at any time (winter conditions can be simulated in summer and vice-versa). These test cells are medium weight constructions. The HVAC system tested is a fan coil unit supplied by a reverse-cycle heat pump, and a variable fan speed mechanical extract for ventilation.

These test cells were equipped with all sensors required according to the selected control and controlled parameter.

The main achievement was the development of a full monozone building model. This model was built from scratch within the Matlab-Simulink environment, being developed as a general purpose model which could be used for any other conditions, projects or applications in the future. However, in order to improve its performance, it was later customized to suit each testing facility (different test sites and seasons). This customization (such as including HVAC systems models) might be slightly changed in the future in order to account for further experiments and calibration.

The thermal simulation was based on finite-differences methods for the conduction model. The maximum value for the time-step of the simulation was calculated using the stability condition according to the discretization scheme. Simulation time step could be reduced to 60 seconds for these test cells. Due to the relatively small thickness and large thermal conductive of windows, the heat conduction model for the windows was considered constant. Convective heat exchanges were based on constant heat convection coefficients. Radiant temperature is calculated as a function of surface temperature, weighted by their relative area. The HVAC system models were based on manufacturers data and modules developed in the frame of *IEA task 22* provided by the Royal Technical Institute of Stockholm.

The fitness function and fuzzy inference algorithm were also added within these models. Data were available and used for models calibration. The main problems in the calibration concerned the modelization of the HVAC equipments as well as solar radiation effects on internal heat gains.

For each of the two testing sites, a different hierarchical FLC architecture was selected, regardless of the season considered in each case. They are very slightly different in their structure but all of them include at least PMV, CO_2 concentration, previous HVAC system status and outdoor temperature. In addition, the architecture developed for the ATC FLC included measures of thermal discomfort, IAQ discomfort and energy consumption for a 30 minutes to 1 hour period prior to the control decision. Both, the control parameters finally selected (see Sect. 3.2) and the FLC architecture for the ATC Summer FLC are presented in Fig. 7 as example. Each module in the figure represents a simple fuzzy controller and as we mentioned in Sect. 3.3, the *minimum t-norm* playing the role of the implication and conjunctive operators, and the WEMOM deffuzification strategy were used.

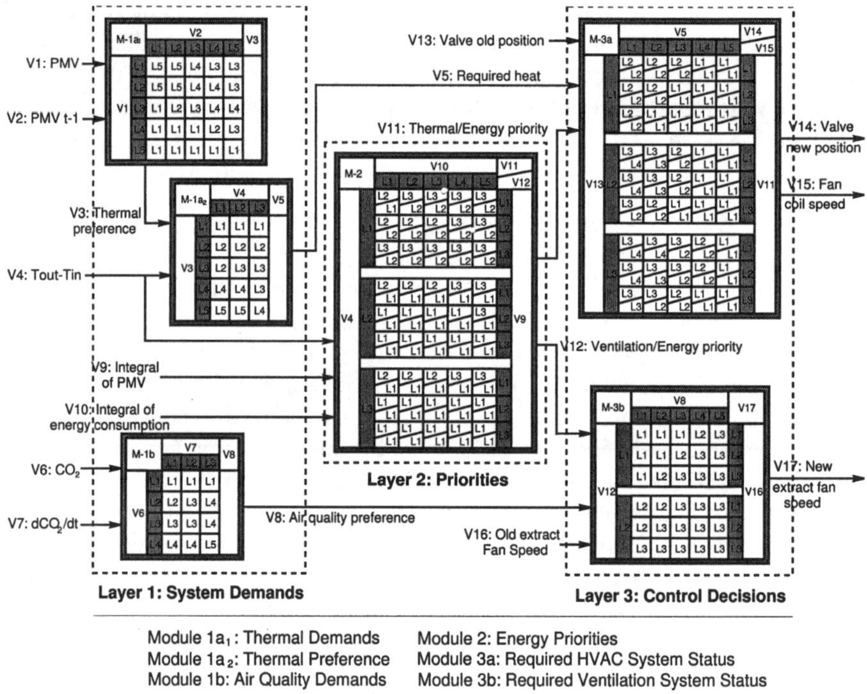

Fig. 7. Initial rule base and generic structure of the ATC Summer FLC

Initial KBs were obtained from BEMS designers for each model and season. For instance, Figs. 7 and 8 show the initial RB and DB of the ATC Summer FLC. This initial RB is fixed for all the tuning process. As initial DB, we considered symmetrical fuzzy partitions of triangular-shaped membership functions for each one of the m variables. These membership functions were labeled from $L1$ to Ll_i, with l_i being the number of membership functions of the i-th variable. The decision table (initial RB) for each simple FLC (module) for the ATC Summer FLC is represented in Figure 7 in terms of this labels. Therefore, each cell of the table represents a fuzzy subspace and contains its associated output consequent(s), i.e., the correspondent label(s).

Thinking on the WMC-SSGA method, another important outcome was to assign appropriate weights to each criterion of the fitness function. The basic idea in this weight definition was to find financial equivalents for all of them. Such equivalences are difficult to define and there is a lack of confident data on this topic. Whereas, energy consumption cost is easy to set, comfort criteria are more difficult. Recent studies have shown that an 18% improvement in people's satisfaction about indoor climate corresponds to a 3% productivity improvement for office workers. Based on typical salaries and due to the fact that PMV and CO_2 concentrations are related to people's satisfaction, such equivalences can be defined.

The same strategy can be applied to the systems stability criterion, life-cycle of various systems being related to number of operations. Based on this, weights can be obtained for each specific building (test site). Thus, trusted weights for both test cells were obtained:

- For CNRS ENTPE test site the chosen values were: $w_1 = 0.0083022$, $w_2 = 0.0083022$, $w_3 = 0.00000456662$, $w_4 = 0.0000017832$ and $w_5 = 0.000761667$.
- And for the ATC test site: $w_1 = 0.0041511$, $w_2 = 0.0041511$, $w_3 = 0.00000228333$, $w_4 = 0.0000017832$ and $w_5 = 0.000761667$.

6.2 Experiments Developed on Simulated Systems

Three different models were implemented, the CNRS ENTPE Spring and Fall model, the CNRS ENTPE Summer model, and the ATC Summer model. The FLCs obtained from the proposed techniques will be compared to the performance of the initial expert FLC and to the performance of a classic On-Off controller for all of these models (the goals and improvements will be computed with respect to this classical controller).

The tuning strategies were assessed with simulations of 10 days with the corresponding climatic conditions. The results obtained by the tuning methods for each model are presented in the following subsections.

CNRS ENTPE Spring and Fall Model

The results obtained by the proposed methods with the CNRS ENTPE Spring and Fall model are shown in Table 2 together with the goal values (g_i) imposed to them, where % stands for the improvement rate with respect to the On-Off controller for each criterion. In this case, WMC-SSGA was run two times, first from the initial DB and after from the best DB obtained in the previous run. Each run had 500 iterations. MO-SSGA was run 500 iterations, but restarting the method from a previously tuned DB did not substantially improve the results by which it made no sense more runs with this technique. Something similar occurred with NDS-GA, which was run for 150 iterations.

Since the time required for each model evaluation was approximately 200 seconds, the estimated run times were, four days for 500 iterations in WMC-SSGA, two days for 500 iterations in MO-SSGA and eight days for 150 iterations in NDS-GA (computed as product of the number of evaluations per generation, the evaluation time and the number of generations).

The intention from experts was to achieve up to 15% energy saving (C_4) with system stability (C_5) at least equal to On-Off stability (2730) and PMVlow criteria (C_1) no more than 10% higher than for On-Off (PMVlow < 105). Following the experts intention, the values imposed to the methods were the following ones (see Table 2): $g_1 = 0$, $g_2 = 108$, $g_3 = 0$, $g_4 = 19363$ and $g_5 = 2800$, with penalization rates of $p_1 = 0.0$, $p_2 = 0.0$, $p_3 = 0.0$, $p_4 = 0.5$, and $p_5 = 0.7$ for WMC-SSGA, and granularity $G = 40$ for MO-SSGA and NDS-GA.

Table 2. Results obtained with the CNRS ENTPE Spring and Fall model

ATC	PMV>0.5		PMV<-0.5		CO_2		Energy		Stability	
MODEL	C_1	%	C_2	%	C_3	%	C_4	%	C_5	%
ON-OFF	0	–	95.7	–	0	–	22780	–	2730	–
INIT.	0	–	100.1	-4.7	0	–	21857	4.05	1340	50.9
GOALS (g_i)	0	–	108.0	–	0	–	19363	–	2800	–
WMC-1	0	–	100.3	-4.9	0	–	20044	12.01	2557	6.3
WMC-2	0	–	100.3	-4.9	0	–	20065	11.92	2527	7.4
WMC-3	0	–	103.7	-8.4	0	–	19700	13.52	2960	-8.4
WMC-4	0	–	104.0	-8.7	0	–	19484	14.47	3270	-19.8
MO-1	0	–	100.3	-4.8	0	–	20488	10.06	2730	0.0
MO-2	0	–	100.3	-4.8	0	–	20489	10.06	2770	-1.5
NDS-1	0	–	101.8	-6.4	0	–	20587	9.63	2790	-2.2
NDS-2	0	–	101.9	-6.5	0	–	20594	9.60	2620	4.0

From Table 2, and taking into account the requested goals, experts considered as the best solution the first obtained by WMC-SSGA, that practically meets the energy goal with a 12% of improvement compared to On-Off, and completely meets the remaining ones. On the other hand, the third solution from WMC-SSGA with only an 8% of loss in stability gets notorious improvements in energy. It shows that even in the case of considering an objective-weighting fitness function, diverse individuals could be obtained. Moreover, all these individuals increase the global fitness in more than 10% showing that all of them are very acceptable solutions.

The results obtained from MO-SSGA and NDS-GA are exceeded in every respect by the results from WMC-SSGA. However, they present more or less 10% of improvement in energy and fitness having a good behavior. Moreover, solutions from MO-SSGA requires much less computation time.

CNRS ENTPE Summer Model

As in the Spring and Fall model, the computation time was also spent in its largest share by WMC-SSGA because from the very first it presented a better behavior than the rest ones. In this case, WMC-SSGA was run three times from the best DB obtained in the previous run. Each run had 500 iterations. Once more time MO-SSGA was only run 500 iterations and NDS-GA was run only 150 generations. The time required for each model evaluation was 220 seconds approximately. Therefore, the computation times were similar to the mid-season model ones.

The intention from experts was to reduce PMVupp criteria (C_1) to 0 and to maintain HVAC stability (C_5) as close as possible to the On-Off controller (1160), with energy (C_4) not larger than 10000. Following the experts intention, the goal values imposed to WMC-SSGA, MO-SSGA and NDS-GA were the following ones (see Table 3):

- WMC-SSGA: $g_1 = 0$, $g_2 = 13.7$, $g_3 = 0$, $g_4 = 9000$ and $g_5 = 1477$, with penalization rates of $p_1 = 1$, $p_2 = 1$, $p_3 = 1$, $p_4 = 0.9$, and $p_5 = 0.99$, respectively.
- MO-SSGA and NDS-GA: $g_1 = 0$, $g_2 = 13.7$, $g_3 = 0$, $g_4 = 9823$ and $g_5 = 1200$, with granularity $G = 40$.

In view of the results shown in Table 3, all the goals but the stability were practically met. In this case, the solution presenting the best stability value (-25.1%) is the first from WMC-SSGA, by which it was considered the best one by the experts. However, this solution does not meet the PMV goal, thus making the fourth solution a good alternative. In any case, values in stability were improved in a 100% from the initial FLC results, and all the remaining goals have been practically met, what is a very good result for this tuning method.

The results obtained from MO-SSGA and NDS-GA present very high stability values respect the On-Off controller results. In this case, to guide the search thorough the pareto space is more difficult than for WMC-SSGA and the results are not exactly what were expected. However, stability was

Table 3. Results obtained with the CNRS ENTPE Summer model

ATC	PMV>0.5		PMV<-0.5		CO_2		Energy		Stability	
MODEL	C_1	%	C_2	%	C_3	%	C_4	%	C_5	%
ON-OFF	0.00	–	13.84	–	0	–	11557	–	1160	–
INIT.	4.50	-450	13.70	1.01	0	–	9148	20.85	2579	-122.3
WMC (g_i)	0.00	–	13.70	–	0	–	9000	–	1477	–
MO/NDS (g_i)	0.00	–	13.70	–	0	–	9823	–	1200	–
WMC-1	2.35	-234	13.55	2.04	0	–	9799	15.21	1451	-25.1
WMC-2	0.05	-4.7	13.77	0.49	0	–	9823	14.99	1486	-28.1
WMC-3	0.03	-2.5	13.77	0.49	0	–	9827	14.96	1476	-27.2
WMC-4	0.03	-2.5	13.77	0.49	0	–	9811	15.09	1476	-27.2
MO-1	0.00	0.0	13.04	5.72	0	–	9636	16.62	1766	-52.2
MO-2	0.00	0.0	13.06	5.64	0	–	9624	16.73	1783	-53.7
NDS-1	0.00	0.0	13.54	2.17	0	–	9795	15.25	1635	-41.0
NDS-2	0.00	0.0	13.54	2.17	0	–	9800	15.20	1635	-41.0

improved in about a 50% from the initial FLC results for both methods, and the computation time for MO-SSGA is much lesser than the required for WMC-SSGA.

It is noticeable that energy savings were about 15% for all the solutions, being this the main objective in the project. Moreover, the improvement of the fitness function was about 13% which show a good general behavior of the obtained FLCs.

ATC Summer Model

The tuned DBs presented in Table 4 for the Summer ATC model correspond to three individuals from the population at iteration 500 with WMC-SSGA. As in the previous models, the remaining results were obtained at iteration 500 with MO-SSGA and generation 150 with NDS-GA. The time required for each model evaluation is 215 seconds approximately. Therefore, once again the algorithms were in the known times.

The intention from experts was to try to have 15% energy saving (C_4) together with a global improvement of the system behavior compared to On-Off control. Comfort parameters could be slightly increased if necessary (no more than 1.0 for criteria C_1 and C_2). Following the experts intention, the goal values imposed to WMC-SSGA, MO-SSGA and NDS-GA were the following ones:

- WMC-SSGA: $g_1 = 1$, $g_2 = 1$, $g_3 = 7$, $g_4 = 2000000$ and $g_5 = 1000$, with penalization rates of $p_1 = 1$, $p_2 = 1$, $p_3 = 1$, $p_4 = 0.9$, and $p_5 = 0.97$, respectively.
- MO-SSGA and NDS-GA: $g_1 = 1$, $g_2 = 1$, $g_3 = 8$, $g_4 = 2700000$ and $g_5 = 1100$, with granularity $G = 40$.

Notice that these goals imposed to the algorithms are higher than the ones initially required since the initial goals were easily met.

Table 4. Results obtained with the ATC Summer model

ATC	PMV>0.5		PMV<-0.5		CO_2		Energy		Stability	
MODEL	C_1	%	C_2	%	C_3	%	C_4	%	C_5	%
ON-OFF	0.0	–	0	–	0	–	3206400	–	1136	–
INIT.	0.0	–	0	–	0	–	2901686	9.50	1505	-32.48
WMC (g_i)	1.0	–	1	–	7	–	2000000	–	1000	–
MO/NDS (g_i)	1.0	–	1	–	8	–	2700000	–	1100	–
WMC-1	0.0	–	0	–	0	–	2575949	19.66	1115	1.85
WMC-2	0.0	–	0	–	0	–	2587326	19.31	1077	5.19
WMC-3	0.0	–	0	–	0	–	2596875	19.01	1051	7.48
MO-1	0.1	-10	0	–	0	–	2697449	15.87	1543	-35.83
MO-2	0.5	-51	0	–	0	–	2836160	11.55	1053	7.31
NDS-1	0.3	-32	0	–	0	–	2703631	15.68	1095	3.61
NDS-2	0.3	-28	0	–	0	–	2693120	16.01	1156	-1.76

In this case, the expert goal has been easily met by WMC-SSGA. Moreover, the solutions present a desirable diversity that allow us to select different and interesting FLCs.

From the results in Table 4, experts selected the third DB from WMC-SSGA as the most promising one. In this case, the solutions obtained from this method present improvement ratios of about 20% in energy and 5% in stability. Again, the results obtained with MO-SSGA and NDS-GA were clearly worse than the ones obtained with WMC-SSGA. However, in the case of MO-SSGA acceptable solutions could be quickly obtained.

Figure 8 represents the initial and the final DBs for the ATC FLC taking as final DB the third solution from WMC-SSGA in Table 4. It shows that small variations in the membership function parameters provoke large improvements in the FLC behavior.

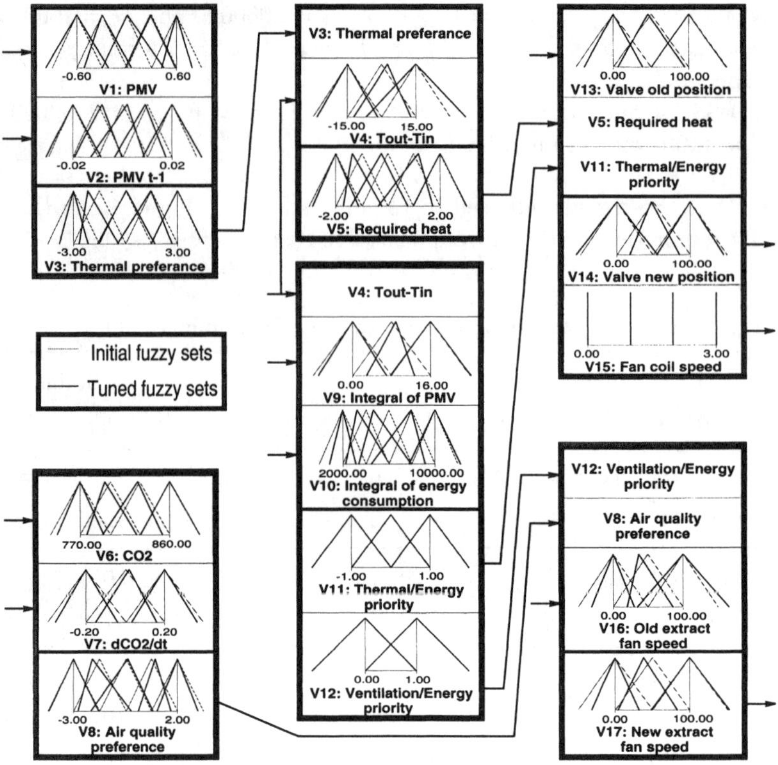

Fig. 8. Initial and tuned DB of the ATC summer FLC

6.3 Methods Analysis

A similar behavior for each method can be observed in all the different models, presenting WMC-SSGA the best results over the remaining methods and practically meeting all the requested goals. On the other hand, we also have seen as MO-SSGA generates acceptable solutions with less computation time than WMC-SSGA. In any case, the computation time for WMC-SSGA is reasonable.

About MO-SSGA and NDS-GA they present similar results but NDS-GA is far slower than MO-SSGA, by which MO-SSGA would be preferable and it could be a very interesting technique in order to quickly obtain acceptable solutions.

All the proposed techniques have yielded much better results than the classical On-Off controller, showing the good behavior FLCs can achieve on these kinds of complex multicriteria problems.

From this experimental study, we could say that WMC-SSGA is the best strategy for this problem. However, we must think that it is possible thanks to properly guide the search by using trusted weights, while MO-SSGA and

NDS-GA perform a robust search. Therefore, in the case in which trusted weights can not be provided probably MO-SSGA would be the best strategy.

The good results obtained by WMC-SSGA can be attributed to the use of a method of objective weighting that can directly guide to the best solution, to the use of fuzzy goals for dynamically adapting the search direction in the space of solutions, and to the restart approach getting away from local optima. In the following, a convergence analysis on WMC-SSGA will be made in order to see the way in which these factors affect to the fitness function.

Figure 9 illustrates the evolution chart of the fitness (expression without considering fuzzy goals) and performance values (C_i) obtained by the WMC-SSGA method when tuning the ATC summer FLC (PMVupp, PMVlow, and CO_2 improvements have been depicted with the same shade since they presents a very similar behavior with values near of 0%). The chart has been generated obtaining the values of the best individual (according to the fitness with fuzzy goals) in each generation. The improvement attained by the tuning process with respect to the On-Off controller solution is represented in vertical axis, where 0% stands for no improvement, negative value for a worsened result, and positive value for an improved result.

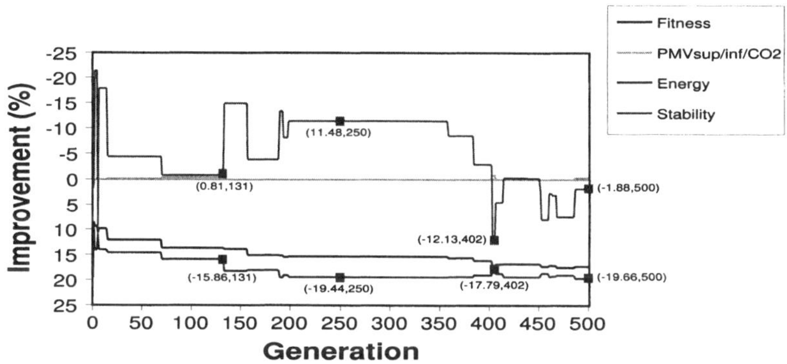

Fig. 9. Evolution of the WMC-SSGA in the ATC Summer model

Analyzing the chart, we can observe how, after some initial generations where the algorithm is being stabilized, the energy consumption is gradually decreased until the generation 131 where almost 16% of improvement is achieved. Stability and hence fitness are also improved during this period. After that, a significant improvement of the energy causes a worse stability to be obtained and the algorithm lies in a local optimum where an improvement of 19.4% for energy is obtained at the expense of stability, 11.5% worse than that of the On-Off controller. This is kept until the generation 402 where making the energy slightly worse involves finding a good stability result 12.1% better than the On-Off controller. This fact is derived from the restart action performed some generations before and it allows the algorithm to get away from the local optimum. From this generation to the end of the run, the energy is gradually improved with an acceptable stability that entails decreasing the fitness function value.

The obtained chart leads us to notice the restart influence and the convergence degree of this algorithm, and analyze the tuning process from the efficiency (time-consuming) point of view. From this angle, it is interesting to verify that a good solution where the energy consumption is improved in a 15.9% with the rest of performance values similar to the On-Off controller is obtained in less than 100 generations.

6.4 Experiments Developed on the Real Test Sites

Results are presented only for both CNRS ENTPE and ATC summer-season experiments. From now on, *experiments* is referred to the tests in the real sites. These experiments were performed using an FLC with the best DB selected by experts for each model.

At ATC (see Table 5), experimental results show that energy savings are interesting (12.5%). However, the stability criterion is far more important than initially expected. This could also be observed when new simulations have been performed with the same climatic conditions. The reason for this is partly due to the CO_2 concentration model. In the model, mixing is supposed perfect, which is not the case in the real test cells. Despite sensor being located close to extract fan, CO_2 concentration proved to be measured at much higher values than expected. Fan operation has therefore been more important and so did stability.

Table 5. ATC Summer model: Simulation results vs test results (two days period only)

SIMULATION	**Fitness**	**PMV>0.5**	**PMV<-0.5**	**CO_2**	**Energy**	**Stability**
On-Off	0.7189	0	0	10.13	344190	138
fuzzy	0.7135	0	0	0	304270	224.35
Difference (%)	**0.75**	**0**	**0**	**0**	**11.60**	**-62.57**
EXPERIMENT	**Fitness**	**PMV>0.5**	**PMV<-0.5**	**CO_2**	**Energy**	**Stability**
On-Off	0.7409	0	0	1015.5	350813	142
fuzzy	0.6881	0	0	1016.5	304031	188.62
Difference (%)	**7.12**	**0**	**0**	**-0.10**	**13.34**	**-32.83**

For CNRS ENTPE (Table 6), excellent results have been obtained with up to 30% energy savings. Experimental conditions created outdoor conditions from 21°C at night up to 31°C during the day. Outdoor air cooling potential in the morning is therefore quite important for these experiments, which explains these excellent results. On the other hand, stability proved to be very bad. A possible reason for this is a rounding problem within the controller. Actuator is operated with a small number of positions (4 for fan speed and 3 for mode) and rounding is required between fuzzy output and actuator signal, thus creating unstabilities.

Table 6. CNRS ENTPE Summer model: Test results (four days period only)

EXPERIMENT	Fitness	PMV>0.5	PMV<-0.5	CO_2	Energy	Stability
On-Off	8.07	0	0.21	4925	8493541	1280
fuzzy	6.97	0	0.11	4822	5951575	4330
Difference (%)	**13.63**	**0**	**47.62**	**2.09**	**29.93**	**-238.28**

Summarizing, it has been proved that energy consumption is greatly reduced during experimentation in real tests cells. Moreover, comparisons between simulations and experiments are in good agreement for the BEMSs designers. Therefore, the proposed technique has been demonstrated to be effective to solve this problem.

7 Concluding Remarks

In this contribution, the use of GAs to develop smartly tuned FLCs dedicated to the control of HVAC systems concerning energy performance and indoor comfort requirements is presented. Three efficient genetic tuning strategies considering different multicriteria approaches have been presented. Several FLCs have been produced and tested in laboratory experiments in order to check the adequacy of such control and tuning techniques. To run the proposed tuning techniques, accurate models of the controlled buildings (two real test cells) were provided by experts.

From this experimental study, we could say that probably the objective weighting method WMC-SSGA is the best option when trusted weights are available while in the case in which trusted weights can not be provided probably a most robust technique as MO-SSGA would be the best option. Moreover, the proposed techniques have yielded much better results than the classical On-Off controller showing the good behavior that FLCs can achieve on these kinds of complex multicriteria problems.

Regarding the experimentation in the real test cells, comparisons between simulations and experiments are in good agreement for the BEMSs designers, presenting significant energy savings in both cases. It shows the effectiveness of this automatic control technique to solve this problem.

The proposed tuning algorithms have an interesting advantage for industrial application: the consideration of goals to perform the multicriteria optimization. These goals significantly improve the tuning performance and it makes easier the expert's knowledge interpretation since the specification of goals, i.e., when each objective has been properly improved, seems to be easy to give. Furthermore, the use of fuzzy goals together with the penalization factor internally changes the initial proposed weights during the evolution of the WMC-SSGA algorithm, dynamically adapting the search direction in the space of solutions. It makes this method robust and more independent from the weight selection for the fitness function.

The results of this work should be ready for implementation in real buildings for the specific studied systems. An extended test of our prototypes will however be necessary before product marketing. Moreover, appropriate interfaces will have to be developed. First industrial applications of our results could therefore start approximately in two years.

This methodology could then be applied to other systems and progressively implemented at industrial level. However, the marketing potential should be particularly studied as well as the way by which they could be efficiently extended to other equipments and buildings.

Acknowledgements

This work has been carried out within the framework of the JOULE-THERMIE programme under the GENESYS [1] Project sponsored by the European Commission, Directorate-General XII for Energy.

We would like to thank the GENESYS [1] project partners for their careful implementation of the presented HVAC simulation models and for their invaluable assistance.

References

1. R. Alcalá, J.M. Benítez, J. Casillas, O. Cordón, R. Pérez, Fuzzy control of HVAC systems optimized by genetic algorithms, Technical Report #DECSAI-01-01-12, Department of Computer Science and Artificial Intelligence, University of Granada, Granada, Spain, 2001.
2. R. Alcalá, J. Casillas, J.L. Castro, A. González, F. Herrera, A multicriteria genetic tuning for fuzzy logic controllers, Mathware and Soft Computing (2001), to appear.
3. M. Arima, E.H. Hara, J.D. Katzberg, A fuzzy logic and rough sets controller for HVAC systems, Proc. of the IEEE WESCANEX'95 1 (NY, 1995) 133–138.
4. J.E. Baker, Reducing bias and inefficiency in the selection algorithm, in: J.J. Grefenstette (Ed.), Proc. of the 2nd International Conference on Genetic Algorithms, Lawrence Erlbaum Associates (Hillsdale, NJ, USA, 1987) 14–21.

5. P. Bonissone, Y. Chen, P. Khedkar, GA tuning of fuzzy logic controllers: A transportation application, Proc. of the 1996 IEEE Conference on Fuzzy Systems (FUZZ-IEEE'96), (New Orleans, LA, 1996) 674–680.
6. C.A. Coello, A comprehensive survey of evolutionary-based multiobjective optimization techniques, Knowledge and Information Systems 1:3 (1999) 269–308.
7. O. Cordón, F. Herrera, A three-stage evolutionary process for learning descriptive and approximative fuzzy logic controller knowledge bases from examples, International Journal of Approximate Reasoning 17:4 (1997) 369–407.
8. O. Cordón, F. Herrera, A. Peregrín, Applicability of the fuzzy operators in the design of fuzzy logic controllers, Fuzzy Sets and Systems 86 (1997) 15–41.
9. O. Cordón, F. Herrera, F. Hoffmann, L. Magdalena, Genetic fuzzy systems: Evolutionary tuning and learning of fuzzy knowledge bases (World Scientific, Singapore, 2001).
10. K. Deb, D.E. Goldberg, An investigation of niche and species formation in genetic function optimization, Proc. of the 3rd International Conference on Genetic Algorithms (1989) 42–50.
11. K. Deb, Multi-objective optimization using evolutionary algorithms, (John Wiley & Sons, 2001).
12. M. Delgado, M.A. Vila, W. Voxman, On a canonical representation of fuzzy numbers, Fuzzy Sets and Systems 93:1 (1998) 125–135.
13. D. Driankov, H. Hellendoorn, M. Reinfrank, An introduction to fuzzy control (Springer-Verlag, 1993).
14. L.J. Eshelman, The CHC adaptive search algorithm: How to have safe search when engaging in nontraditional genetic recombination, in: G.J.E. Rawlins (Ed.), Foundations of Genetic Algorithms (Morgan Kauffman, San Mateo, CA, 1990) 265–283.
15. C.M. Fonseca, P.J. Fleming, Genetic algorithms for multiobjective optimization: Formulation, discussion and generalization, in: S. Forrest (Ed.), Proc. of the 5th International Conference on Genetic Algorithms, Morgan Kaufmann (1993) 416–423.
16. C.M. Fonseca, P.J. Fleming, An overview of evolutionary algorithms in multiobjective optimization, Evolutionary Computation 3 (1995) 1–16.
17. M.P. Fourman, Compaction of symbolic layout using genetic algorithms, in: J.J. Grefenstette (Ed.), Genetic algorithms and their applications: Proc. of the 1st International Conference on Genetic Algorithms, Lawrence Erlbaum (1985) 141–153.
18. A.E. Gegov, P.M. Frank, Hierarchical fuzzy control of multivariable systems, Fuzzy Sets and Systems 72 (1995) 299–310.
19. P.Y. Glorennec, Application of fuzzy control for building energy management, in: Building Simulation: International Building Performance Simulation Association 1 (Sophia Antipolis, France, 1991) 197–201.
20. P.Y. Glorennec, Coordination between autonomous robots, International Journal of Approximate Reasoning 17:4 (1997) 433–446.
21. D.E. Goldberg, Genetic algorithms in search, optimization, and machine learning, (Addison-Wesley, 1989).
22. H.B. Gürocak, A genetic-algorithm-based method for tuning fuzzy-logic controllers, Fuzzy Sets and Systems 108:1 (1999) 39–47.
23. P. Hajela, C. Y. Lin, Genetic search strategies in multicriterion optimal design, Structural Optimization 4 (1992) 99–107.

24. F. Herrera, M. Lozano, J.L. Verdegay, Tuning fuzzy controllers by genetic algorithms, International Journal of Approximate Reasoning 12 (1995) 299–315.
25. F. Herrera, M. Lozano, J.L. Verdegay, Fuzzy connectives based crossover operators to model genetic algorithms population diversity, Fuzzy Sets and Systems 92:1 (1997) 21–30.
26. F. Herrera, M. Lozano, J.L. Verdegay, Tackling real-coded genetic algorithms: Operators and tools for the behaviour analysis, Artificial Intelligence Review 12 (1998) 265–319.
27. J.H. Holland, Adaptation in natural and artificial systems (Ann arbor: The University of Michigan Press, 1975) (The MIT Press, London, 1992).
28. S. Huang, R.M. Nelson, Rule development and adjustment strategies of a fuzzy logic controller for an HVAC system - Parts I and II (analysis and experiment), ASHRAE Transactions 100:1 (1994) 841–850, 851–856.
29. C. Karr, Genetic algorithms for fuzzy controllers, AI Expert (1991) 26–33.
30. J. Kiszka, M. Kochanska, D. Sliwinska, The influence of some fuzzy implication operators on the accuracy of a fuzzy model - Parts I and II, Fuzzy Sets and Systems 15 (1985) 111–128, 223–240.
31. F. Kursawe, A variant of evolution strategies for vector optimization, in: H.-P. Schwefel, R. Männer (Eds.), Proc. of the 1st workshop on Parallel Problem Solving from Nature, Springer-Verlag (1991) 193–197.
32. C.C. Lee, Fuzzy logic in control systems: Fuzzy logic controller - Parts I and II, IEEE Transactions on Systems, Man, and Cybernetics 20 (1990) 404–418, 419–435.
33. E.H. Mamdani, Applications of fuzzy algorithms for control a simple dynamic plant, Proc. of the IEEE 121:12 (1974) 1585–1588.
34. E.H. Mamdani, S.Assilian, An experiment in linguistic synthesis with a fuzzy logic controller, International Journal of Man-Machine Studies 7 (1975) 1–13.
35. Z. Michalewicz, Genetic algorithms + data structures = evolution programs (Springer-Verlag, 1996).
36. R. Palm, D. Driankov, H. Hellendoorn, Model based fuzzy control, (Springer-Verlag, 1997).
37. J.D. Schaffer, Multiple objective optimization with vector evaluated genetic algorithms, in: J.J. Grefenstette (Ed.), Genetic algorithms and their applications: Proc. of the 1st International Conference on Genetic Algorithms, Lawrence Erlbaum (1985) 93–100.
38. N. Srinivas, D. Kalyanmoy, Multiobjective optimization using nondominated sorting in genetic algorithms, Evolutionary Computation 2:3 (1994) 221–248.
39. G. Syswerda, J. Palmucci, The application of genetic algorithms to resource scheduling, in: R.K. Belew, L.B. Booker (Eds.), Proc. of the 4th International Conference on Genetic Algorithms (ICGA'91), Morgan Kaufmann (1991) 502–508.
40. P. Thrift, Fuzzy logic synthesis with genetic algorithms, in: R.K. Belew, L.B. Booker (Eds.), Proc. of 4th International Conference on Genetic Algorithms (ICGA'91), Morgan Kaufmann (San Mateo, CA, 1991) 509–513.
41. D.A. Van Veldhuizen, G.B. Lamont, Multiobjective evolutionary algorithms: analyzing the state-of-the-art, Evolutionary Computation 8:2 (2000) 125–147.
42. D. Whitley, J. Kauth, GENITOR: A different genetic algorithm, Proc. of the Rocky Mountain Conference on Artificial Intelligence, Denver (1988) 118–130.

43. D. Wienke, C. Lucasius, G. Kateman, Multicriteria target vector optimization of analytical procedures using a genetic algorithm — Part I. Theory, numerical simulations and application to atomic emission spectroscopy, Analytical Chimica Acta 265:2 (1992) 211–225.
44. R.R. Yager, On the construction of hierarchical fuzzy systems model, IEEE Transactions on Systems, Man, and Cybernetics 22 (1992) 1414–1427.
45. E. Zitzler, K. Deb, L. Thiele, Comparison of multiobjective evolutionary algorithms: empirical results, Evolutionary Computation 8:2 (2000) 173-195.

Intelligent Information Systems for Crime Analysis

K.P. Lam

Department of Systems Engineering & Engineering Management
The Chinese University of Hong Kong, Shatin, Hong Kong, China

Abstract. Decision support systems using intelligent techniques have found widespread usage in many disciplines, but the domain of crime analysis seems still widely unexplored. Apparently, one major reason is due to the complexity and multifaceted nature of crime that often defies a mechanistic, brute-force approach. Human ingenuity and intelligence in perceiving and identifying useful and unique crime-related features are of critical importance, albeit a black-box type computational approach is deemed highly inadequate. Through our earlier work in the AICAMS project, a computerized crime-map kiosk and facial identification kit system have been developed as basic crime analysis tools. This paper describes three specific applications as related to the human-machine integration aspects of spatial-temporal predictive modeling for the crime-map kiosk, of neural-network learning for mug shot matching for facial sketch, and of fuzzy expert systems for money laundering detection. The decision supporting roles of computational intelligence techniques are particularly noticeable, as evidenced in the evolution from their complete absence to increasing importance in the development of intelligent information systems for crime analysis.

1 The AICAMS Project

Nowadays many operations for effective law enforcement, such as crime investigation and detection, are information-intensive. The need for advanced information system has become increasingly critical for the police. The development of AICAMS (Artificial Intelligence Crime Analysis and Management System) [1] was initiated in late 1996, with a focus on applying knowledge-based system (KBS) technology to crime analysis. To facilitate the data and knowledge acquisition process, a prototype refinement approach was undertaken at Tuen Mun Police Station. An interactive map-based system integrated with a facial identification kit was successfully developed with user participation, and gained wide acceptance in subsequent implementation [3,4]. Coincidentally, research work and applications of similar nature have been initiated by other research groups [2,5,6], and attracted increased attention and funding which attest to the growing importance of the field.

During the initial development of AICAMS, we focused on constructing knowledge-based crime analysis tools for frontline officers. Two prerequisite components turned out to be essential: Relevant data sources and effective graphic user interfaces. Crime map pattern and suspect's facial composite have long been useful to police on beat patrol. In addition to conveying much graphical information with an intuitive appeal, these data sources provide means for further crime analysis. For instance, spatial-temporal analysis in the crime domain has been much emphasized by end users, especially in the management level. Image matching of facial sketches with mug shots are practically used by investigators in identifying potential suspects.

1.1 System architecture

The operational AICAMS version is an integration of three sub-systems: AICAMS-Fit, AICAMS-MapEntry and AICAMS-Map. AICAMS-Fit is a standalone system which can operate individually. AICAMS-Map and AICAMS-MapEntry are top-level systems running concurrently under ActiveX control. AICAMS-Map uses MapObject Version 1.2 and can access the AICAMS-Fit database. AICAMS-MapEntry provides means of data entry, database access, and facilitates the fine-tuning of the AICAMS-Map interface. Figure 1 depicts the system architecture of AICAMS. It shows the relationship between the three sub-systems of AICAMS. Normally the operator will input the suspect's facial description to the database through AICAMS-Fit, with subsequent addition of the details of the case to the AICAMS-Map database. This includes the case description, geographical information, voice message and linking the suspect information from AICAMS-Fit database to that case. The user can view individual cases under AICAMS-Map, with a backup-database operating when the main server is down.

1.2 User interfaces

Human/ machine interaction in AICAMS is mainly facilitated through the construction of various user interfaces. The most common use of AICAMS is for accessing stored crime information by beat officers who often prefer a simple touch-click and voice command interface. The retrieved information are mostly in map-based form (Fig. 2) or multimedia format (Fig. 3) for easy human interpretation, though specific crime and demographic details in text box can readily be prompted up if needed.

Other users, such as experienced crime investigators, prefer the use of map-based interface with access to more global information sources and analysis tools. In a research prototype of AICAMS, a district version with basic layer architecture for spatial analysis has been developed, where further crime case details, case-centered link analysis are also provided for human interpretation of similar cases. The basic concept for spatial analysis is to bind the crime case location to a grid layer or point layer and hence the long-term data distribution for supporting statistical trend analysis can be displayed or emphasized by these grid-point layers. Moreover, crime hot-spot discovery for proactive management can also be achieved by this grid-based or point-based analysis.

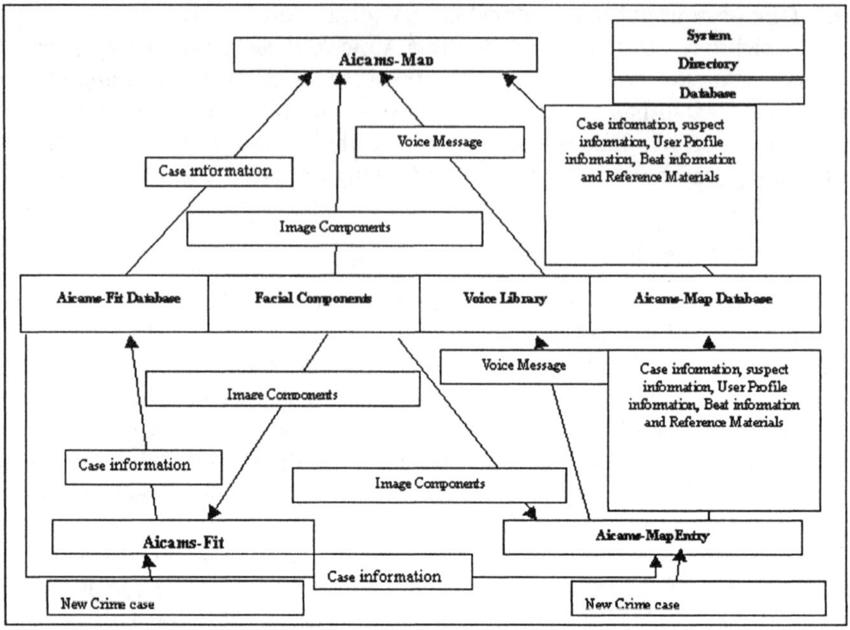

Fig. 1. System architecture of AICAMS

Using a layer architecture, we can assign grid-id to a crime case according to its co-ordinates. If the co-ordinates of a crime case are known, we can calculate the child cell contained that case in the next level of grid layer. Recursively we can get the grid-id for that case in terms of which level of grid-id we want to get. In our prototype, we assign level 20 grid-id for each crime case. As a result, we can bind the cases to any level of grid layer less than or equal to level 20. Fig. 4 shows the map-based interface developed for performing crime analysis on Tuen Mun, Tai Hing and Castle Peak, the three divisions of the Tuen Mun district.

1.3 Crime modeling and spatial-temporal analysis

Unlike many mechanical systems with well-defined physical equations, human behavior is often a very difficult measure to model and to predict. Crime is related with human behaviour, as well as with other cultural, social, political, and economic factors. To perform effective crime analysis, one needs to obtain good models on these different aspects based on our experience, knowledge, and understanding. In 1968 Nobel Laureate G. Becker's economic approach to crime [7] provided seminal insights on the use of models for crime analysis, especially in clarifying the roles of many vaguely defined societal and individual behaviors. Quantitative models obtained several years later [8] put the work under more practical context for statistical analysis and interpretation.

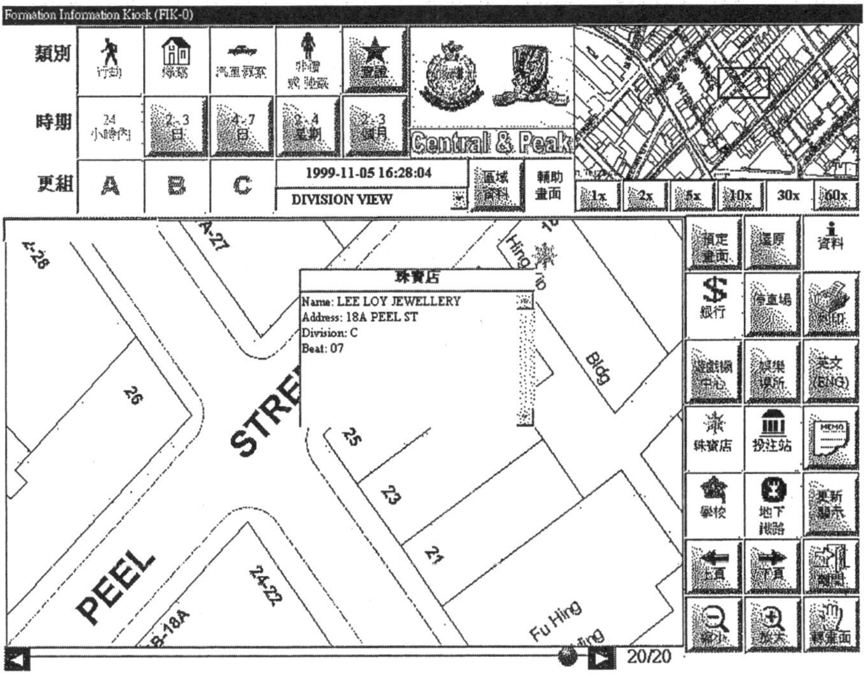

Fig. 2. Street map interface

Our previous description on AICAMS summarizes an information system infrastructure for effectively encoding data and disseminating crime information. However, without good models for analysis, the potential of such an infrastructure cannot be fully exploited. The multifaceted nature of crime demands for widely diversified models to handle different problems. During the early work of AICAMS [1], a simple rule-based system using decision matrix based on extracted parameters was developed for vehicle crime. Further work on rule induction and knowledge discovery, however, suffered from the lack of adequate data and good training cases.

The AICAMS pilot test in Tuen Mun provided map-related data for 1095 crime cases on burglary, robbery, sexual offences and vehicle crime. Following Ehrlich's work [8] and Anselin's extension [9] spatial-temporal linear autoregressive models for crime rate prediction can be readily formulated based on the Tuen Mun data:

$$C_t(r) = \alpha_1 C_{t-1}(r) + \cdots + \alpha_n C_{t-n}(r) + \sum_i^p \sum_j^m w_{i,j} P_{i,j}(r) \tag{1}$$

where $C(t,r)$ stands for the crime rate at time t with a spatial index r, and is affected by its past values from $t-1$ to $t-n$ and mp explanatory variables $P_{i,j}(r)$ (over p spatial indexes each with m past values). $w_{i,j}$ is the

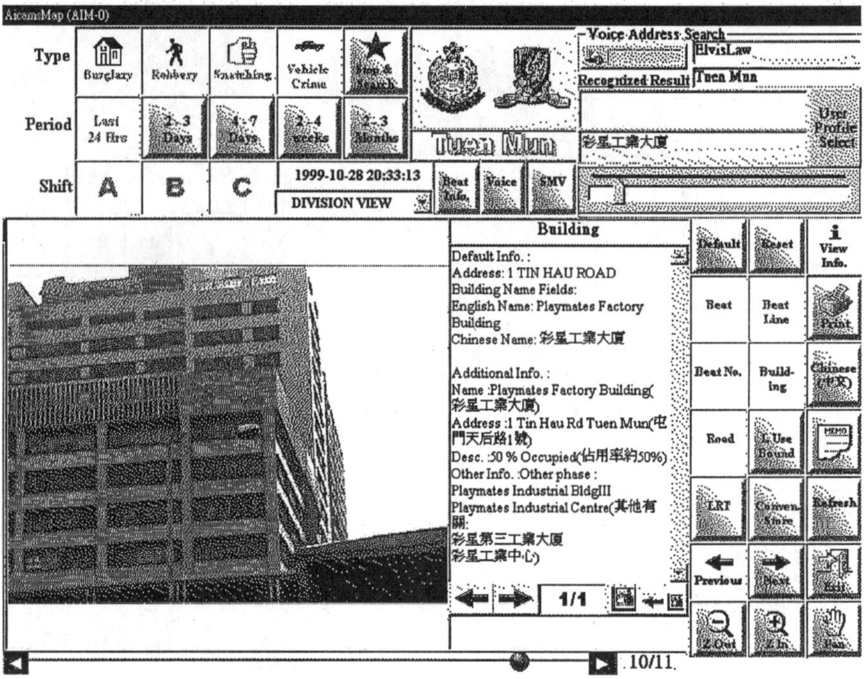

Fig. 3. Multimedia interface including voice command address search and image links

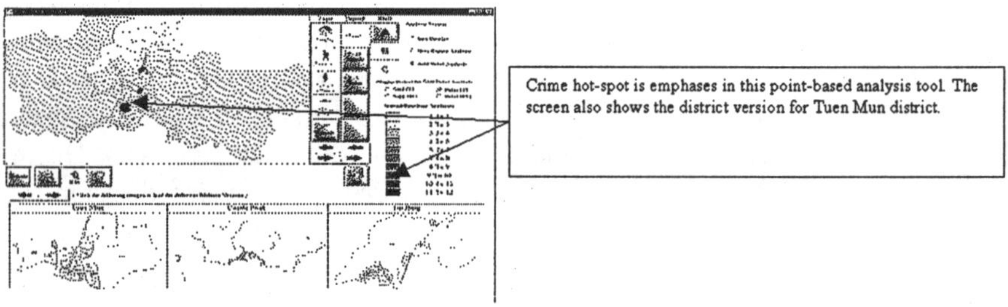

Fig. 4. Map-based interface for crime analysis

spatial weighting matrix and is usually time-invariant. A priori knowledge is used to define this matrix based on proximity consideration within the spatial neighborhood. The time index should be defined with respect to the sampling time chosen for the model in interest. For instance, for our Tuen Muen data over a year, we are interested in the weekly crime rate for a particular beat area (with spatial index r) and hence $C(t-1,r)$ would mean the weekly crime rate for that beat one week before.

The most difficult task of formulating Eqn. (1) is in identifying meaningful relationships between the predicted quantity (e.g. weekly crime rate) with the explanatory variables yet to be chosen. Usually if appropriate variables are chosen for the model, applying simple linear regression to Eqn. (1) would give reasonable results. For applications that require more accurate in-sample modeling, a nonlinear spatial-temporal model can be defined for a general nonlinear function $f(.)$ in line with similar work of Olligschlaeger [10].

$$C_t(r) = f(C_{t-1}(r), \cdots, C_{t-n}(r), w_{i,j}P_{i,j}(r)) \tag{2}$$

The effectiveness of applying neural networks for in-sample nonlinear spatial-temporal modeling is best illustrated for Anselin's Columbus crime data set [9] for the 49 neighborhoods. Fig. 5 shows a comparison on the modeling accuracy using linear spatial auto-regression (SAR) [11] based on Eqn. (1), backpropagation neural network [12], and radial basis function network [13] based on Eqn. (2). As observed, the radial basis function network almost gives a perfect fit for the data when compared with the acceptable results for linear regression. However, this usually comes at the expense of more sensitive out-sample performance for nonlinear models.

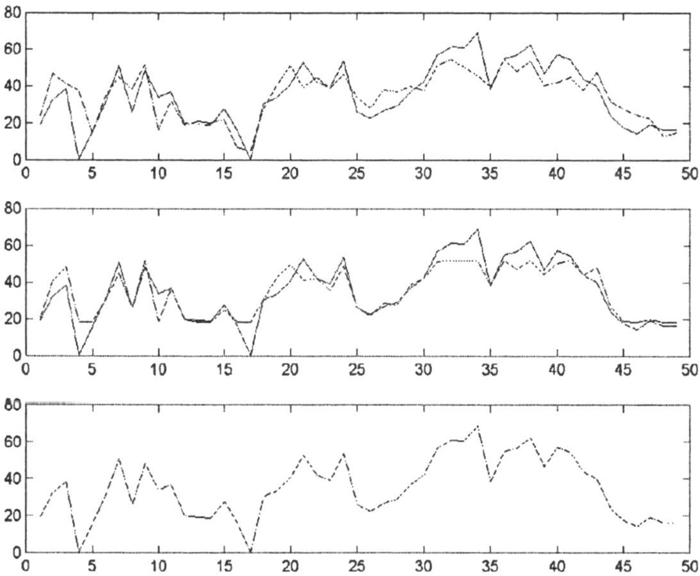

Fig. 5. Crime rate versus neighborhood (predicted and actual) (Top) Using spatial auto-regressive model; (Middle) Backpropagation neural network model; (Bottom) Radial basis function network model

2 Facial Recall and Matching

The forensic tasks of facial recall can be categorized into verbal description and generating hard-copy representations. In the latter, the witness is asked to select facial features from a library of transparent acetate sheets that match those of the culprit's face in his memory. The police artist or technician then combines them to form a composite face by superimposing them in a special frame. The manual process of facial recall is very time-consuming. However, time is of the essence in forensic face memory because it reduces memory. Computerization may help to speed up the generation of facial images and eliminate the time effect. AICAMS-Fit provides an efficient and effective tool for facial recall in order to replace the current manual process. The process of facial matching usually comes at a subsequent stage, after a facial composite has been generated by facial recall. It involves searching either manually or using automatic pattern recogntion software/ hardware for similar sketches and photographs (mug shots) of suspects.

2.1 AICAMS-Fit architecture

The general architecture of AICAMS-Fit is shown in Figure 6. Although the figure shows several components in the system, it is mainly made up of three components: the Facial Component Library, the Feature Selection Module and the Facial Construction Module.

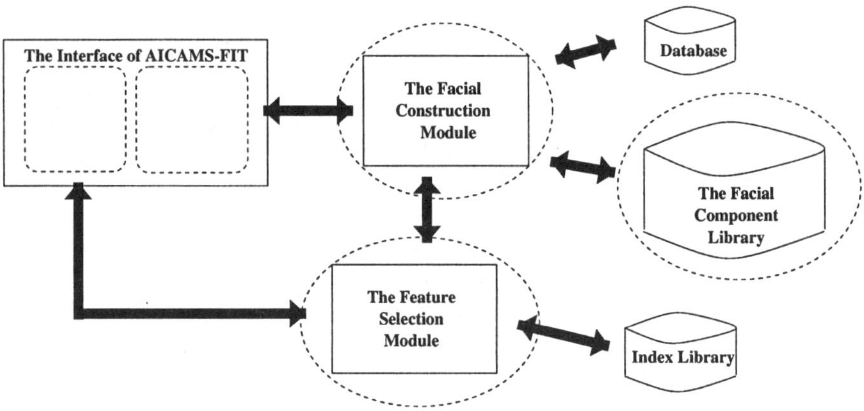

Fig. 6. The General Architecture of AICAMS-Fit

The Facial Component Library provides a library of facial feature templates of twelve feature classes. The twelve feature classes are shown in Table 1. The Feature Selection Module provides a friendly interface for the user to choose the desired facial features. The Facial Construction Module displays the selected facial features and combines them to form a composite face using a layer-by-layer representation and an overlay algorithm. Fig. 7 shows a user interface for AICAMS-Fit, which is also integrated within AICAMS in which suspect's facial composites are linked up with the the crime case database for efficient retrieval.

Major Features	Hair	Facial Outline	Eyes	Eyebrows	Nose	Mouth
Minor Features	Nose Sidelines	Moustaches	Beard	Wrinkles	Spectacles	Others (e.g scars)

Table 1. The twelve feature classes in AICAMS-Fit

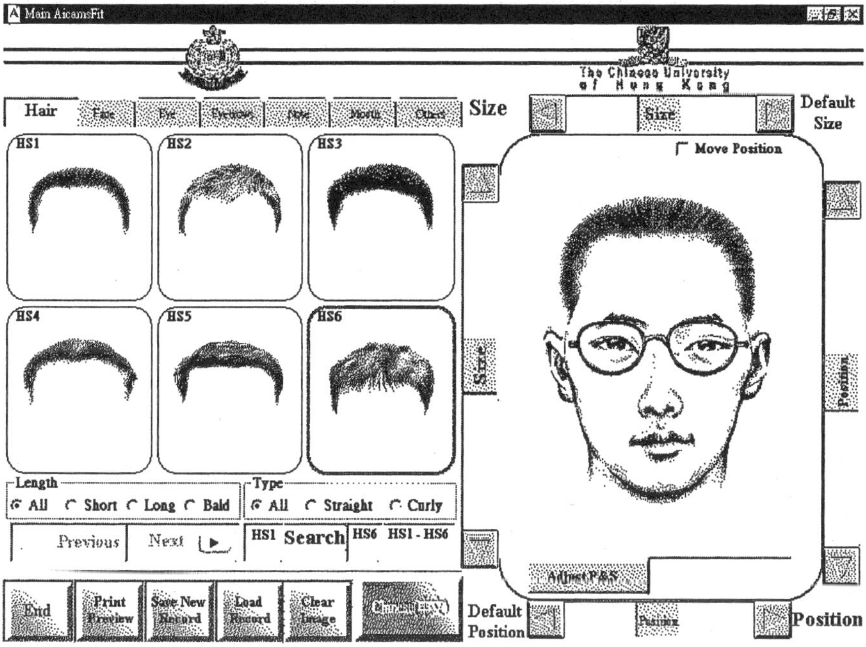

Fig. 7. AICAMS-Fit user interface

2.2 Sketch-to-sketch matching

In facial recall, a complete face is usually produced as a composite face which consists of at least the six major features as shown in Table 1. In terms of mathematical formulation, we can express this kind of composite face as $C = \{P_{i,k}, i = 1, 2, \ldots, m\}$, where m is the total number of feature classes, $k \in \{1, 2, \ldots, n_i\}$, n_i is the total number of facial features in feature class i and, $P_{i,k}$ indicates that the facial feature k of feature class i is selected.

Facial matching can be regarded as a matching of individual facial features. Traditional facial matching requires real time comparison of facial features. This is true either in feature-based matching or template-based matching because we cannot determine the similarity between facial features of two different faces prior to the matching process. However, if a face is made up of facial features from a facial feature library, such as the composite face constructed in the application of facial recall, AICAMS-Fit, batch mode comparison of facial features can be done. That is, we can pre-compare the facial features prior to the facial matching. When we want to compare two composite faces, we can just make use of the pre-computed values. Time can be greatly reduced for facial matching if the real-time comparison requires intensive computation.

A component-based encoding scheme is proposed to sketch matching. It encodes the similarity between each facial feature of the same feature class into a similarity matrix, so that the matching between two composite faces is made simpler. It can be as simplest as summing up the similarity value obtained from the similarity matrix of each feature class. Although it requires quite a lot of work in the preprocessing stage, it favors later process of facial matching. The idea of the encoding scheme is simple and intuitive. It is described as follows. Suppose there are two composite faces C_1 and C_2, where $C_1 = \{P_{i,k1}, i = 1, 2, \ldots, m\}$ and $C_2 = \{P_{i,k2}, i = 1, 2, \ldots, m\}$. They are nearly identical and differ only in the facial features of the feature class j, i.e. when $i = j$, $k1 \neq k2$. Hence, a measure of similarity between C_1 and C_2 will only take account of the similarity between $P_{j,k1}$ and $P_{j,k2}$.

$$Sim(P_{j,k1}, P_{j,k2}) = Sim(C_{j,k1}, C_{j,k2}) \tag{3}$$

where all facial features of $C_{j,k1}$ and $C_{j,k2}$ are identical except the one in feature class j.

2.3 Sketch-to-photo/photo-to-sketch matching

It is a common practice for law enforcement agencies to obtain facial photos (mug shots) of criminals. Usually, these photos are in frontal view and stored as grey-level images in a digital photo album. After the process of facial recall, investigators may be interested to know if the photo album contains a photographic image of face that is similar to the reconstructed facial composite. Thus, this involves a matching between facial photo and facial sketch and hence a more general type of facial matching is needed than that based on component similarity. The methods of PCA (Principal Component Analysis) [15] and LFA (Local Feature Analysis) [14] have been extensively pursued in research and applied in practical facial recognition work. In a research prototope of AICAMS-Fit with facial matching capability, a FaceIt[1] component based on LFA has been integrated to give promising matching results (Fig. 8).

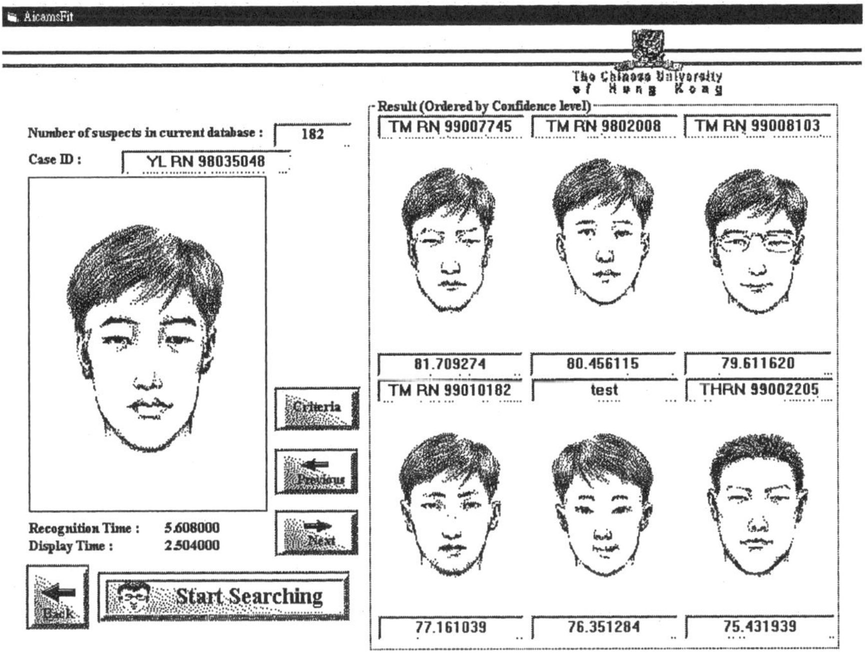

Fig. 8. Facial recognition using FaceIt within AICAMS-Fit (The numbers show the recognition accuracy in percentage)

2.4 A Hebbian network approach

Recently we studied a high-gain, post-annealing, generalized Hebbian algorithm (GHA) for facial image encoding, reconstruction, and matching. Using Sanger's extended work [16] on the GHA [17] we can formulate a learning equation for a single linear neuronal model.

$$\mathbf{y}(t) = W^T(t)\mathbf{x}(t) \tag{4}$$

$$\Delta W(t+1) = \eta(t)\{\mathbf{y}(t)\mathbf{x}^{\mathrm{T}}(t) - LT[\mathbf{y}(t)\mathbf{y}^{\mathrm{T}}(t)]W(t)\} \tag{5}$$

[1] FaceIt is a registered trademark of VISIONICS CORPORATION

where $\mathbf{x}(t)$ is the m-dimensional input vector and $\mathbf{w}(t)$ is the m-dimensional weight vector. $W(t) = [\mathbf{w_1}, \mathbf{w_2}, \ldots, \mathbf{w_m}]$ is the $m \times m$ weight matrix, and $LT[.]$ stands for the lower triangular matrix of [.].

Tne neural learning equation can further be generalized by extending the learning/ decay rate from a scalar $\eta(t)$ to diagonal matrices $A(t) = [\alpha_i(t)]$, $B(t) = [\beta_i(t)]$, for $i = 1, \ldots, m$, in order to take account of the different attraction of the eigenvectors.

$$\Delta W(t+1) = A(t)\mathbf{y}(t)\mathbf{x}^{\mathbf{T}}(t) - B(t)LT[\mathbf{y}(t)\mathbf{y}^{\mathbf{T}}(t)]W(t) \tag{6}$$

A high-gain approach is needed to approach the global attractors through bypassing most of the local minima. As the magnitude information of the eigenvector is not relevant, a normalization step is taken to constrain the weight vector in each learning step. This will retain the input variance information without an unnecessary growth of the magnitude of the weight vector. An annealing strategy is then required to get the precision. Various "cooling-off" annealing strategies can be imposed on the time-varying learning rate and decay rate, in order to reduce the magnitude of the input excitation and bring in the necessary accuracy to the estimated values.

Forensic application of facial sketch/ photo matching We consider a segmented principal component approach for facial recognition that uses low-dimensional eigenvectors and takes advantage of aggregating local feature similarities for global matching. Consider a typical facial image of 256×256 pixels with 8-bit pixel value (256 gray levels). Using 8×8 window blocks for the image there will be $32 \times 32 (= 1024)$ blocks. A window segment is referred to as a connected rectangular set of $p \times q$ (where $p, q \leq 32$) window blocks of the image. Hence, window segments of different sizes can be readily created on localized regions (e.g., eyes, nose, mouth) of the facial image.

For a given window segment, a segmented principal component analysis is to find the 64-dimensional eigenvectors for its $p \times q$ sample blocks. A Hebbian network with a single layer of 8 neurons is used for the evaluation. Each neuron receives 64 input signals, and the network as a whole generates 8 transformed outputs. The 8x64 weight matrix stores the eight learned eigenvectors, by applying the fast Hebbian learning algorithm. In many cases reasonably good reconstruction for the entire image (photo or sketch) can be obtained from the learned eigenvectors based on a single window segment of 256 blocks (the 8 middle row blocks). The images are obtained from a facial image database developed for ongoing research on facial memory recall and mugshots in law enforcement application.

Facial matching simulation Typical simulation runs under 10 epochs to give good results. The fast Hebbian network obtains almost identical image quality for the reconstructed image when compared with that using the exact first eight principal components (Fig. 9 for both facial sketh and photo). A simple facial matching scheme over the same window segment for different facial images is to measure the similarity between the transformed outputs (the so-called facial features space). Consider the ith transformed output (as mapped by the ith eigenvector), $y_i(u)$, $y_i(v)$, for Image u and Image v, respectively. Let $S_i(u, v)$ be the correlation coefficient between $y_i(u)$ and $y_i(v)$. The similarity between the two images is measured by the average of $S_i(u, v)$ for $i = 1, \ldots, n$, with n suitably chosen to give good discrimination. In the case of using more than one segments, a global measure of similarity can be defined as a weighted average of the calculated similarities over the individual segments. For simplicity, our evaluation test considers a single segment consisting of 8 middle row blocks (256 blocks), and n is chosen as 3. It is visually perceived that FaceA (Sketch) is close to FaceA (Photo) than to FaceB (Sketch and photo). The similarity measures using correlation testify to the effective discrimination of the recognition system, in matching both photo and sketch to the same person.

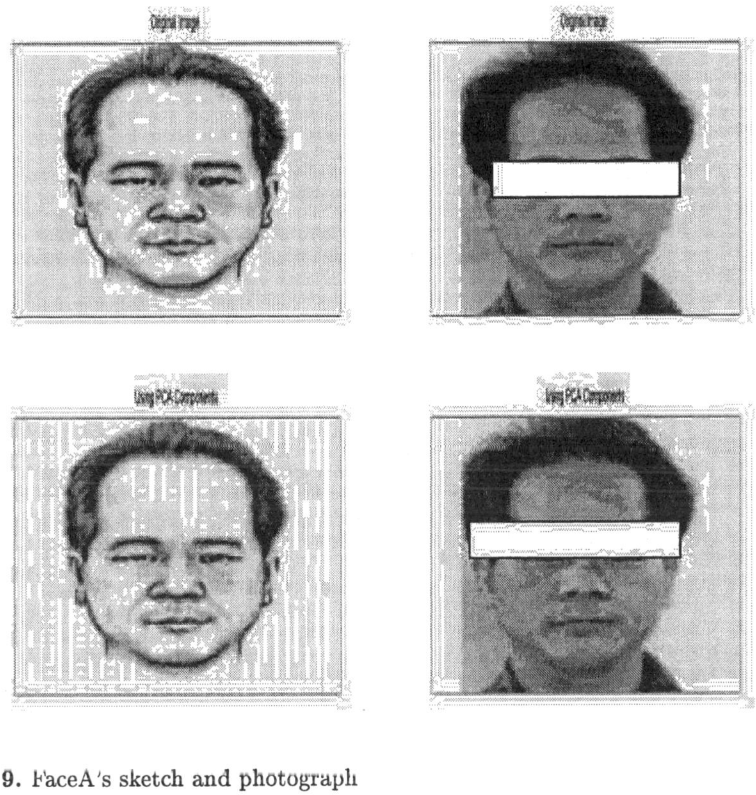

Fig. 9. FaceA's sketch and photograph

3 Money Laundering Detection

Money laundering refers to the act of disguising the origin or ownership of illegally gained funds to make them appear legitimate. It is a complex process of placing the profit from illicit activity into the legitimate financial system; and its detection has long been recognized as a very difficult task by law enforcement agencies and financial institutions worldwide. An annual amount of US$ 500 billion is believed to be obtained through money laundering from narcotics profits world-wide. Largely left undetected by law enforcement agencies, the money has become the primary fuel for propelling crime activities. The necessity of developing specialized software for detecting money laundering has been realized by leading countries, among which the 1995 OTA Report [18] and Financial Crimes Enforcement Network (FinCEN)'s AI targeting system [19] have been commonly cited.

Conventional methods of detecting money laundering are based on the principles of "Know Your Customer" (KYC) and "Know Your Business" (KYB). By knowing the type of business of an account and the established patterns, bank staff can detect abnormalities that should be investigated. Existing guidelines for reporting suspicious cases are established based on international standards that have been modified with specifics on local context. The forty recommendations of FATF [20] describe a general framework for international cooperations, the importance of customer identification and record-keeping rules, and guiding principles for increased deligence of financial institutions.The Hong Kong Monetary Authority (HKMA) have interpreted these recommendations and formulated more precise implementation [21] within a local context. This is then studied by individual bank to derive explicit instructions and rules, for training officers to become alert to abnormality from the normal customers' profiles (obtained basically through the KYC and KYB principles). Many of these valuable instructions and rules are available in the form of heuristics knowledge driven by key observation input parameters, behaviors, and patterns.

3.1 System architecture for a heuristics rule base

Although most of the bank rules are rather simple, they are also numerous and may become a memory drag for many inexperienced officers to use them effectively. In addition, the execution of some of those rules require accessibility of accounts information and transaction details in order to flag suspicious cases for further investigations. Our effort in developing a money laundering detecion system (MLDS) is to establish an information system to capture human expertise in a heuristics rule base based on fuzzy expert systems technology [22,23].

Our current version of MLDS prototype provides four main functional modules: Data Entry, ML Detection, ML Report Generation, and ML Case-based Analysis.The Data Entry Module is used for recording useful information needed for money laundering detection. Information, such as account information, suspicious company, suspicious person and transaction details, is required. The users can manage the information through the operations of addition, deletion, modification, and updating. The ML Detection module is used for detecting suspicious transactions that are related to the activities of money laundering. Before performing detection, a user should adjust the parameters that are needed in the detection process. These parameters are related to the criteria of capturing suspicious transactions. After parameter setting, user can select a suspicious person/company and his/its account, load the required knowledge base and run the inference engine. The result of detection is then displayed in the textbox, showing discovered suspicious activities and the corresponding transaction details.

The ML Report Generation module is used for generating a report that records the detail of a suspicious ML case. Users have to input the information of the reporter and choose the suspicious ML case to be reported. Finally, the ML Case-based Analysis module is used for analyzing the relationship between reported cases in order to find if a person/company is suspected as conducting money laundering. The relationship is presented as a link chart that can be viewed by different methods. The link chart is constructed based on a selected object that can be a person, a company, an account or a ML case.

The dataflow diagram for the MLDS prototype is shown in Fig. 10, illustrating the interactions among the four modules. Further details on the detection process dataflow are shown in Fig. 11. CLIPS [23] is used as the underlying experts system shell for knowledge encoding. Fuzzy rules are formulated for capturing the imprecise knowledge involved and can readily be implemented by some of the fuzzy system extensions to the basic CLIPS engine.

3.2 Detectable patterns

The rules (knowledge) used in the system are mainly extracted from the relevant guidelines. They include patterns classified under the following categories:

- Money Laundering Using Cash Transactions
  - Unusually large cash deposits made by an individual or company whose ostensible business activities would normally be generated by cheques and other instruments.
  - Substantial increases in cash deposits of any individual or business without apparent cause, especially if such deposits are subsequently transferred within a short period out of the account and/or to a destination not normally associated with the customer.

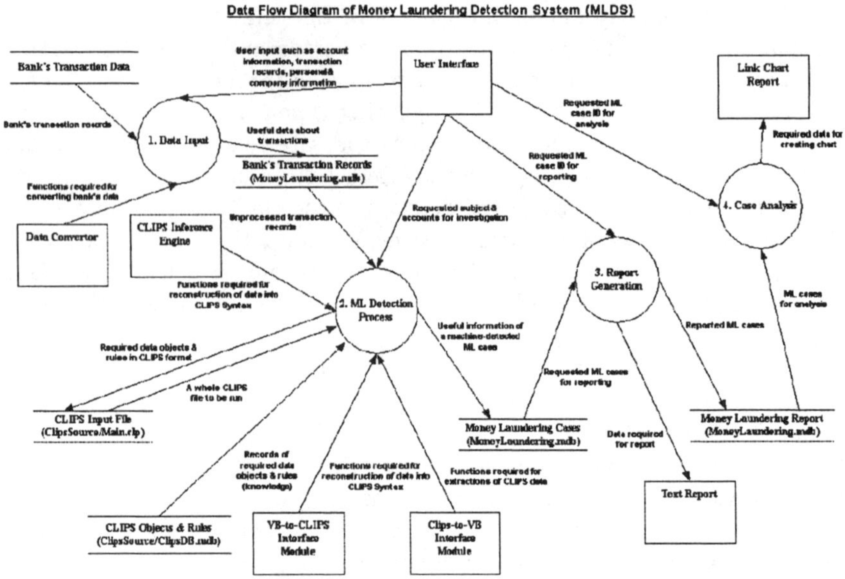

Fig. 10. MLDS system architecture

  - Company accounts whose transaction, both deposits and withdrawals, are denominated in cash rather than the forms of debit and credit normally associated with commercial operations (e.g. cheques, Letters of Credit, Bills of Exchange, etc.).
  - Frequent exchange of cash into other currencies.
  - Customers whose deposits contain counterfeit notes or forged instruments.
- Money Laundering Using Bank Accounts
  - An account opened in the name of a money changer that receives structured deposits.
  - The account activity is inconsistent with the customer's known background and previous account activity.
  - The account exhibits some of the recognized suspicious activity indicators. (Frequent Monday and Thursday Transactions)
  - Unusual Company Transaction (with unknown business nature, large amount transactions greater than HKD 500,000, and with no financial supporting documents e.g. LC)
  - Large Cash Transaction (greater than HKD 100,000 or Equivalent foreign currency)
  - Unusual Transaction Pattern - Matching of Deposits and Subsequent Withdrawals in Cash

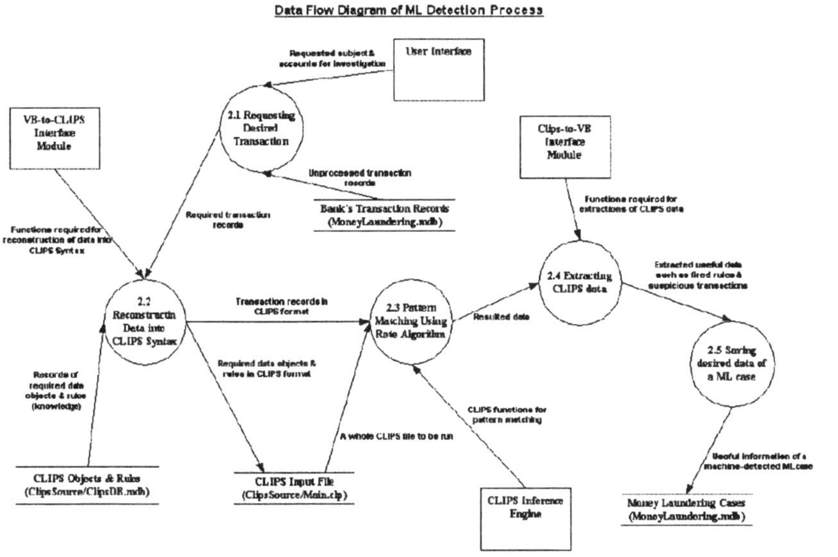

Fig. 11. Detection process in MLDS

- Unusual Transaction Pattern - Matching of Withdrawals and Subsequent Deposits in Cash
- Unusual Transaction Pattern - Temporary Depository: Matching of Deposits and Withdrawals within the same day
- Structuring / Smurfing of Cash Deposits

4 Conclusions

We have summarized our development work in AICAMS and three ongoing research directions currently pursued on crime analysis: the spatial-temporal predictive modeling for integrating a map-base with demographic and crime-related features, a Hebbian learning procedure for matching mug-shot and facial composite sketch, and a knowledge-based system capturing detection rules to uncover suspicious money laundering cases. All of the three applications require significant human input to define the precise role of machine-generated data and analysis for effective decision making. The development of intelligent information systems for crime analysis is not as straightforward a blind jacketing of AI software with crime data. On the contrary, it is a time-consuming and labor-intensive task that requires much human ingenuity and interpretation of voluminous data with the computational aid from machine.

5 Acknowledgement

The author would like to thank Jack Brahan, Hilton Chan, William Leung, Lewis Tse for their help on the AICAMS Project. C.S. Yip, M.Y. Law, and K.K. Lau provided excellent programming support. Financial supports from the Hong Kong Police and the Hong Kong Research Grants Council's Earmarked Grants are gratefully acknowledged.

References

1. Brahan, J.W., Lam, K.P., Chan, H., Leung, W. (1998) AICAMS: Artificial intelligence crime analysis and management system. Knowledge-Based Systems. **11**, 355–361
2. Hauck, R.V., Chen, H. (1999) Coplink: A case of intelligent analysis and knowledge management. Proc. Int. Conf. Information Systems (ICIS). 15–28
3. Information Technology Branch (1999) Training Manual for Facial Identikit Tool (FIT) & Formation Information Kiosk (FIK): Version 1.0, Hong Kong Police Force
4. Information Service Department (2000) Hong Kong Yearbook 1999: Public Order (Information Technology), Hong Kong SAR Government, 301
5. Charles, J. (1998) AI and Law. IEEE Intelligent Systems, January/February, 77–80
6. Brown, D.E. (1998) The Regional Crime Analysis Program (RECAP): A Framework for Mining Data to Catch Criminals. IEEE Int. Conf. Systems, Man, and Cybernetics, **3**, 2848– 2853
7. Becker, G.S. (1968) Crime and Punishment: An Economic Approach. The Journal of Political Economy, **76**, 169–217
8. Ehrlich, I. (1973) Participation in Illegitimate Activities: A Theoretical and Empirical Investigation. The Journal of Political Economy, **81**, 521–565
9. Anselin, L. (1988) Spatial Econometrics: Methods and Models. Kluwer Academic Publishers.
10. Olligschlaeger, A.M. (1997) Artificial Neural Networks and Crime Mapping. In Crime Mapping and Crime Prevention, D. Weisburd and T. McEwen (Editors), Criminal Justice Press, 313–347
11. LeSage, J.P. (1999) The Theory and Practice of Spatial Econometrics. Internal Report, Department of Economics, University of Toledo, 296 pages
12. Haykin, S. (1994) Neural Networks: A Comprehensive Foundation. Macmillan College Publishing Company, New York
13. Demuth, H. and Beale, M. (1998) Neural Network Toolbox: User's Guide Version 3.0. The MathWorks, Inc.
14. Penev, P.S. and Atick, J.J. (1996) Local Feature Analysis: A General Statistical Theory for Object Representation. Network: Computation in Neural Systems, **7**, 477–500
15. Turk, M. and Pentland, A. (1991) Eigenfaces for Recognition. Journal of Cognitive Neuroscience, **3**, 71–86
16. Sanger, T. (1989) Optimal Unsupervised Learning in a Single-Layer Linear Feedforward Neural Network. Neural Networks, **12**, 459–473

17. Oja, E. (1982) A Simplified Neuron Model as a Principal Component Analyzer. Journal of Mathematical Biology, **15**, 267–273
18. Office of Technology Assessment (1995) Information Technology for the Control of Money Laundering. Report OTA-ITC-630, U.S. Congress, 144 pages
19. Senator, T.E., et. al. (1995) The Financial Crimes Enforcement Network AI System (FAIS). AI Magazine, Winter, AAAI, 21–39.
20. Financial Action Task Force (FATF) (1996) The Forty Recommendations of the Financial Action Task Force on Money Laundering. FATF-VII, June
21. Hong Kong Monetary Authority (HKMA) (1997) Prevention of Money Laundering. Guideline No. 3.3, October
22. Giarratano, J. and Riley, G. (1994) Expert Systems: Principles and Programming. PWS Publishing Company
23. Giarratano, J. (1993) CLIPS User's Guide. Software Technology Branch, NASA Lyndon B. Johnson Space Center, May

Application of Fuzzy Decision Trees to Reservoir Recognition

X. Z. Wang*, D. S. Yeung**, E. C. C. Tsang**, and J. W. T. Lee**

*School of Mathematics and Computer Science
Hebei University
Baoding, Hebei, China
Email: wangxz@mail.hbu.edu.cn

**Department of Computing
Hong Kong Polytechnic University
Hong Kong
Email: {csdaniel,csetsang,csjlee} @comp.polyu.edu.hk

Abstract. This chapter reports a real application of fuzzy decision tree to a reservoir recognition in the logging area for oilfield exploration. Reservoir fluid recognition is an important but difficult task in providing a comprehensive explanation for logging. A good recognition method can provide reliable evidence for building a standard of explanation in a region. Since there is much vagueness in the reservoir fluid recognition and there are considerable differences of geological structure in different regions, it is very difficult to establish a uniform mathematical model to recognize the reservoir. The commonly used methods for reservoir recognition include empirical formula, synthetic evaluation, fuzzy clustering, etc. Unfortunately, these methods fail to meet many applications' requirements. For example, the empirical formula and synthetic evaluation methods could not handle fuzzy or vague data while the fuzzy clustering could not give a good recognition of oil-water layer. By applying the fuzzy decision tree induction method to the problem of reservoir recognition in an oilfield of northern China, we find the recognition results encouraging.

Keywords: Weighted fuzzy rules, learning, decision trees, approximate reasoning, reservoir recognition

1. Introduction

Knowledge acquisition, which has been universally considered as the bottleneck for developing expert systems, is a process of learning. Learning is the essential way to acquire wisdom for human beings as well as to acquire intelligence for any computer system. Learning from examples is one of the most important branches in machine learning. The major objective of learning from examples is to find a set of rules that provides a complete characterization of a given set of examples.

Knowledge could be acquired and represented in many forms such as decision trees, decision tables, neural networks, Petri nets, and especially Production Rules (PRs) which can be further extended to Fuzzy Production Rules (FPRs) and Weighted Fuzzy Production Rules (WFPRs). So far, FPRs have been used most widely for representing imprecise knowledge. With the increasing complexity of today's knowledge-based systems, the knowledge representation power of FPRs would be severely limited if important parameters such as weight and certainty factor are not considered [1-5]. In the process of knowledge interpretation, it is equally important to make use of FPRs with parameters to reduce undesirable effects in deriving a quality decision [2, 4]. However, automatically acquiring FPRs with parameters is a more important yet difficult problem than interpreting them. One method of acquiring FPRs with parameters is to map the FPRs to neural networks and then adjust the parameters repeatedly [6]. This chapter proposes another approach to generate FPRs together with their knowledge parameters.

There are many ways to automatically acquire imprecise knowledge. Induction of fuzzy decision trees is an important one. This method first generates a fuzzy decision tree and then extracts a set of fuzzy production rules from the tree. Generally speaking the fuzzy decision tree generation needs a heuristic method to select the expanded attributes. In our approach to fuzzy decision tree generation, a kind of new heuristic information is used. This chapter addresses a real application of fuzzy decision tree to a reservoir recognition problem in a logging area. Reservoir fluid recognition is an important but difficult task in providing a comprehensive explanation for logging. A good recognition method can provide reliable evidence for building a standard of explanation in a region. Since there is much vagueness in the reservoir fluid recognition and there are considerable differences in the geological structure in different regions, it is very difficult to establish a uniform mathematical model to recognize the reservoir. The commonly used methods for reservoir recognition include empirical formula, synthetic evaluation, fuzzy clustering, etc. Unfortunately, these methods do not satisfy the requirements of many practical applications. For example, the empirical formula and synthetic evaluation methods could not handle fuzzy or vague data while the fuzzy clustering could not give a good recognition of oil-water layer. By applying the fuzzy decision tree induction technique to the problem of reservoir recognition in an oilfield of northern China, we find the recognition results encouraging. This chapter is divided into six sections. In Section 1 an introduction is provided. In Section 2 a uniform fuzzy representation of mixed attributes with different value types is proposed to solve the problem of learning from examples. Section 3 introduces some concepts in FDT induction, a heuristics to generate a FDT and extract WFPRs are also provided. In Section 4 an inference algorithm for classifying an unknown object during the testing phase is proposed. Section 5 presents an application of using our FDT to solve a real problem in oilfield and the results are analyzed. Finally a conclusion is presented.

2. Fuzzy Representation of Attributes

2.1 Learning from fuzzy examples

Learning from examples is a type of inductive learning. The main task of learning from examples is to extract rules from given training examples.

Crisp learning problems assume that each attribute of the training example takes mutually exclusive values (crisp sets) and each example is classified into one and only one of the mutually exclusive classes [7]. However, when classes and attribute-values (linguistic terms) are not considered as crisp subsets but as fuzzy subsets instead, those classes and attribute-values will not be mutually exclusive. An example was given in [8] to explain this situation clearly. We call it learning from fuzzy examples. The main point here is to extract a set of fuzzy rules which would completely characterize the given set of examples.

2.2 Fuzzification of numerical numbers

The process of fuzzifying numerical numbers into linguistic term is often used to reduce information overload in human decision making process. Numerical values associated with the term "salary", for example, may be perceived in linguistic terms such as *high*, *average* and *low*. One way of determining the membership functions of these linguistic terms is by expert opinion or by people's perception. Fuzzy clustering based on self-organized learning can also be used to generate membership functions [9]. The following is an algorithm for generating certain type of membership functions.

Let X be a given data set which will be clustered into k linguistic terms T_j, $j = 1, 2, \ldots, k$. For simplicity, we assume the type of membership to be triangular.

Suppose that each pair of adjacent membership functions crosses at the membership value 0.5. The only parameters to be determined are the k centers $\{a_1, a_2, \ldots, a_k\}$. An effective method to determine these centers is Kohonen's feature map algorithm [9].

2.3 Handling of other attributes

The treatment of other types of attributes is briefly described as follows.

For symbolic attributes, the problem is trivial. This is because a Boolean vector could be considered as a special case of fuzzy vector. For example, if the universe of discourse of an attribute is {small, medium, big}, case 1 takes the value "big" and case 2 takes the value "small" for the attribute, then the two attribute values can be uniformly represented as (0, 0, 1) and (1, 0, 0) respectively.

For fuzzy number valued attributes, they can be treated by fuzzy clustering of fuzzy numbers. For example, let's consider triangular fuzzy numbers. It is well known that the set of all triangular fuzzy numbers is a complete metric space. This means that a distance measure can be defined on the space. According to this distance measure, a subset of triangular fuzzy numbers can be clustered into several linguistic terms. Consequently, each attribute-value can be regarded as a fuzzy vector on the space of linguistic terms. This is the uniform representation we expect. In addition, interval-valued attributes are regarded as a special case of fuzzy number valued attributes.

In the above discussions, it is implicitly assumed that an attribute has attribute-values of a single type. In fact, in many situations in which an attribute has attribute-values of mixed type can also be dealt with. For example, the Height of John is 2.00m, the Height of Tom is 1.80m, the Height of Bob is "short", and the Height of Mick is "tall". Such a situation has already been considered, for example, in [10]. In this chapter, the attribute Height takes two different types of values, namely, numerical and linguistic values. By computing the membership degrees of the numerical numbers belonging to the linguistic terms, one may obtain a uniform representation like Tom: 0.00/short + 0.15/medium + 0.85/tall and Bob: 1/short + 0/medium + 0/tall.

After applying the methods discussed in this section, we obtain a uniform representation of mixed attributes as shown in Table 1. From now on, all our discussions on fuzzy rule extraction are assumed to have the format as shown in Table 1.

Table 1. Result of uniform representation of mixed attributes Where A_j - attribute, $T_j^{(k)}$ - linguistic term (a fuzzy set on {1, 2, ---, N}), μ_{ij} - attribute-value (a fuzzy set on term space), $C^{(k)}$ - class, $\mu_{ij}^{(k)}$ and $\nu_i^{(k)}$ - membership degrees

No.	A_j $T_j^{(1)}\ T_j^{(2)} \cdots T_j^{(k_j)}$	$j = 1 \cdots n$	*Classification* $C^{(1)}\ C^{(2)} \cdots C^{(k_c)}$
1	$\mu_{1j} = \left(\mu_{1j}^{(1)}\ \mu_{1j}^{(2)} \cdots \mu_{1j}^{(k_j)}\right)$	$j = 1 \cdots n$	$\nu_1 = \left(\nu_1^{(1)}\ \nu_1^{(2)} \cdots \nu_1^{(k_c)}\right)$
2	$\mu_{2j} = \left(\mu_{2j}^{(1)}\ \mu_{2j}^{(2)} \cdots \mu_{2j}^{(k_j)}\right)$	$j = 1 \cdots n$	$\nu_2 = \left(\nu_2^{(1)}\ \nu_2^{(2)} \cdots \nu_2^{(k_c)}\right)$
	…… ……	……	…… ……
N	$\mu_{Nj} = \left(\mu_{Nj}^{(1)}\ \mu_{Njj}^{(2)} \cdots \mu_{Nj}^{(k_j)}\right)$	$j = 1 \cdots n$	$\nu_N = \left(\nu_N^{(1)}\ \nu_N^{(2)} \cdots \nu_N^{(k_c)}\right)$

3. Induction of Fuzzy Decision Trees with Parameters

In this section, the degree of importance of a fuzzy attribute-value contributing to the consequent will be introduced. Based on the degree of importance, a new heuristic for generating fuzzy decision trees with parameters will be given.

3.1 Preliminaries

We first recall some basic concepts here. The fuzzy set defined on a discrete universe of discourse X can be represented as a vector, called fuzzy vector. In this chapter, the membership function of a fuzzy set is denoted by itself. Let A be a fuzzy set, we define its α -cut, A_α, as $A_\alpha(x)=\begin{cases}A(x) & A(x)\geq\alpha \\ 0 & A(x)<\alpha\end{cases}$. It is worth noting that the α -cut defined here is different from the one given by common sense. The union and intersection of two fuzzy sets and the complement of a fuzzy set are defined as:

$$A\cup B(x)=Max(A(x),B(x)) \quad ,A\cap B(x)=Min(A(x),B(x)) \text{ and}$$

$$A^c(x)=1-A(x).$$

The cardinality measure (sigma count) of the fuzzy set A is defined by $M(A)=\sum_{x\in X}A(x)$. Obviously, it is a generalization of the number of elements of a crisp set. Throughout this chapter, the function $M(\cdot)$ always represents the cardinality of a fuzzy set.

Let A, B be two fuzzy subsets. We use $S(A, B)$ to represent the subsethood of A in B which is defined as $S(A,B)=M(A\cap B)/M(A)$. For a fuzzy rule taking the form {IF A THEN B}, S(A,B) usually denotes the degree of truth of this rule.

3.2 Degree of importance of attribute contributing to the consequent

We formulate our problem of learning from fuzzy examples as follows. Let us consider Table 1 in which there are N examples, n attributes $\{A_j\ (j=1,\cdots,n)\}$ and k_c classes $\{C^{(1)}\ C^{(2)}\cdots C^{(k_c)}\}$. The space of sample labels is X ={1, 2, ⋯ , N}, $T_j^{(k)}$ is the k-th linguistic term of the j-th attribute and is regarded as a fuzzy set on {1, 2, ⋯ , N} (corresponding to a column of the table $T_j^{(k)}=(\mu_{1j}^{(k)},\mu_{2j}^{(k)},\cdots,\mu_{Nj}^{(k)})^T$, $(1\leq j\leq n, 1\leq k\leq k_j)$, $\mu_{ij}=(\mu_{ij}^{(1)},\mu_{ij}^{(2)},\cdots,\mu_{ij}^{(k_j)})$ is the value of the i-th sample with respect to the j-th attribute and is considered as a fuzzy set on the term space $\{T_j^{(1)}\ T_j^{(2)}\cdots T_j^{(k_j)}\}$ $(1\leq i\leq N, 1\leq j\leq n)$ where $\mu_{ij}^{(k)}\in[0,1]$, and k_j is the dimension of a term space. The classification result of each sample is a fuzzy set on $\{C^{(1)}\ C^{(2)}\cdots C^{(k_c)}\}$ denoted $v_i=(v_i^{(1)},v_i^{(2)},\cdots,v_i^{(k_c)})$ $(1\leq i\leq N)$.

A fundamental problem we need to address is whether each attribute-value has the same degree of importance contributing to the classification result. To answer this question, we consider a simple crisp case (Table 2.) that is obtained by observing cars manufactured by two different factories.

Table 2. Car factory example

No.	Color	Max-speed	Size	Made in
1	White	Low	Full	Factory A
2	White	Mid	Full	Factory B
3	Black	Low	Compact	Factory A
4	Black	High	Compact	Factory B

Intuitively, some rules can be extracted from Table 2. For instance, it is observed that the *max speed* of cars made in Factory A is low. The second attribute *max speed* seems to be more important than the other attributes. This intuitive degree of importance of an attribute can be defined formally by the number of inconsistent examples increased due to removing an attribute from the table. An example x is called inconsistent with respect to a fixed example y if all attribute values of x are the same as the corresponding values of y but their classifications are different. It is easy to see that the number of inconsistent examples with respect to the first example in Table 2 is 2 when removing the *max speed* attribute, and is 0 when removing either *Color* or *Size*. Therefore, according to this viewpoint, the attribute *max speed* is considered to be indispensable with respect to classification and to be more important than the other attributes.

Based on the above intuitive idea, we introduce the following definition of degree of importance of the attribute value contributing to the consequent for the above formulated learning problem.

Definition 1. Let μ_{ij} be the value of the i-th example with respect to the j-th attribute $(1 \le i \le N, 1 \le j \le n)$, $\mu_{ij} = (\mu_{ij}^{(1)}, \mu_{ij}^{(2)}, \cdots, \mu_{ij}^{(k_j)})$, SM be a selected similarity measure. Define $\lambda_{ip} = Min_{q \neq j}\left(SM(\mu_{iq}, \mu_{pq})\right)$, $\sigma_{ip} = SM(v_i, v_p)$, and $\theta_i = \sum_p g^+(\lambda_{ip} - \sigma_{ip})$, where $g^+(x) = \begin{cases} x & x > 0 \\ 0 & x \le 0 \end{cases}$.

Then, the degree of importance of attribute value μ_{ij} contributing to the classification is defined as $(\theta_i \mu_{ij}^{(1)}, \theta_i \mu_{ij}^{(2)}, \cdots, \theta_i \mu_{ij}^{(k_j)})$ and the degree of importance of the j-th attribute A_j contributing to the classification result is defined as $(\sum_i \theta_i \mu_{ij}^{(1)}, \cdots, \sum_i \theta_i \mu_{ij}^{(k_j)})$. The key points included in definition 1 are that (1) the condition to check whether two attribute values or two clusters are the same is now replaced by the similarity measure between attribute-values or between classification-values; (2) the number of inconsistent examples becomes vague; and (3) the function g+ is used to express the degree of inconsistency.

It is noted that when all values of attributes and classification become either 0 or 1, the method of computing the degree of importance in definition 1 is identical to the intuitive idea in handling the attribute-values presented in Table 2. It is a direct generalization of the intuitive importance of attributes presented in Table 2.

3.3 Crisp and fuzzy decision trees

A fuzzy decision tree is a generalization of the crisp case. Before giving a rigorous definition, we introduce some symbols. For a given finite set X, let $F(X)$ denote the family of all fuzzy subsets defined on X, and the membership of a fuzzy subset is denoted by itself.

Definition 2. Let $\Omega_i \subset F(X)(1 \le i \le m)$ be m given groups of fuzzy subsets (m partitions), with the property $|\Omega_i| > 1$ ($|\bullet|$ denotes the cardinality of a crisp set). T is a directed tree satisfying:

(a) each node of the tree belongs to $F(X)$,
(b) for each non-leaf N whose son-nodes constitute a subset of $F(X)$ denoted by Γ, there exists $i(1 \le i \le m)$ such that $\Gamma = \Omega_i \cap N$, and
(c) each leaf corresponds to one or more values of classification decision.

Then, T is called a fuzzy decision tree. Each group of fuzzy subsets, Ω_i, corresponds to an attribute and each fuzzy subset corresponds to a value of the attribute.

The fuzzy decision tree, regarded as a generalization of the crisp case, is more robust in tolerating imprecise information. There are many methods of representing fuzzy decision trees. Essentially, the representation used in this chapter can be regarded as fuzzy partitioning of fuzzy sets.

3.4 A new heuristic algorithm for generating fuzzy decision trees with parameters

There are two key points in the process of constructing fuzzy decision trees. One is the selection of the expanded attribute. Tree is expanded at a node according to the values of its attributes (which are fuzzy subsets), *i.e.,* child-nodes of the node are generated. The other is the determination of leave nodes. A node is usually regarded as a leaf if the relative frequency of one class at the node is greater than or equal to a given threshold value. Usually if the cardinality of the node is less than a given threshold, then it is regarded as an empty leaf.

Noting that every node is a fuzzy set defined on the universe of discourse $\{1, 2, \cdots, N\}$, we consider $(1, 1, \cdots, 1)_N$ as the root, namely, the first candidate node. A general learning algorithm for generating fuzzy decision trees can be described as follows:

> **WHILE** there exist candidate nodes
> **DO** selcct one using the given search strategy; if the selected one is not a leaf, then generate its son-nodes by selecting the expanded attribute using a heuristic algorithm. These son-nodes are regarded as new candidate nodes.

From existing references, two powerful heuristic algorithms guiding the selection of expanded attributes in the fuzzy decision tree generation can be found. One is Fuzzy ID3 that is based on minimum fuzzy entropy [11], and the other is based on the reduction of classification ambiguity [8]. Mainly, these two heuristics are based on the uncertainty reduction. In the following, we introduce a new attribute-importance-based heuristic for generating fuzzy decision trees. According to the discussion in section 3.2, we know that different attributes have different degrees of importance contributing to the fuzzy classification, thus we hope the most important attribute is selected as the expanded attribute at each non-leaf. This is one of the purposes of introducing this heuristic. Another purpose is to acquire weight parameters of fuzzy production rules.

In the process of expanding a non-leaf node on the fuzzy decision tree, our proposed algorithm selects the expanded attribute by using the maximum degree of importance and the relative frequency of a class at a node as the leaf criterion. That is, the node will be regarded as a leaf if the frequency exceeds a given threshold. The relative frequency of a class F at a node B is defined as $f(B|F) = M(F \cap B)/M(B)$ where both the class F and the node B are fuzzy sets on the universe of discourse $\{1, 2, \cdots, N\}$. It is worth noting that

$f(B \mid F)$ is just the degree of subsethood of B in F. When the node B is a leaf, the value $f(B \mid F)$ will be regarded as the classification truth level of the node B. Systematically, we formulate our heuristic algorithm for generating fuzzy decision trees as follows:

Step 1. Given the cut-standard $\alpha(\alpha \in (0,1])$ and the leaf-standard $\beta(\beta \in (0,1])$.

Step 2. Use α to cut the Table 1 for reducing the classification ambiguity.

Step 3. Consider the root node $(1, 1, \cdots, 1)_N$ as the first candidate node.

Step 4. Randomly select a candidate non-leaf node, for any attribute which has not been used in forefather-nodes of this non-leaf node, compute the degree of importance of the attribute contributing to classification.

Step 5. Select an attribute with maximum degree of importance to classification as the expanded attribute. According to the expanded attribute, generate son-nodes of the candidate non-leaf node. These son-nodes are considered as new candidate nodes.

Step 6. For each of these son-nodes, if the relative frequency of a certain class exceeds β or the membership sum of the considered son-node is less than a small positive number, then this son-node is labeled as leaf.

Step 7. If all nodes are leaves then stop else go to Step 4.

3.5 Extracting weighted fuzzy rules from the generated fuzzy decision tree

A fuzzy production rule is, in short, a rule which can have linguistic terms (fuzzy sets) appearing in its antecedent and consequent. Furthermore, several knowledge parameters such as certainty factors and local weights can be assigned to the entire rule or to each proposition for the purpose of enhancing the representation power of rules [4]. We call these rules with several knowledge parameters Weighted-Fuzzy-Production-Rules, in short, WFPRs. WFPRs have the following form: **IF** antecedent **THEN** consequent (*CF*; *Lw*), where the antecedent comprises one or more proposition(s) connected by AND; the consequent is a single fuzzy set; *CF* is the certainty factor of the rule; *Lw* is a set of local weights specified for the propositions in the antecedent. The purpose of introducing the certainty factor is to show how certain the relationship between the antecedent and the consequent is. The local weight, which is assigned to each proposition in the antecedent of a rule, is to indicate the relative degree of importance of the proposition contributing to its consequent.

Before we proceed any further, an understanding of the following three issues would help us gain insight concerning the importance of WFPRs.

(1) Is it necessary to introduce knowledge parameters in fuzzy production rules? One critical assumption being made by this chapter is the fact that WFPRs (i.e. those knowledge parameters introduced in fuzzy rules) are necessary and useful. One may ask why do we prefer WFPRs instead of the common fuzzy production rules (CFPRs)? Or equivalently, what is the advantage of WFPRs over CFPRs?. The answer to this question can be found in [3, 4, 5]. Here we will only give an illustrative explanation for the need of the knowledge parameter *L*ocal *W*eight.

In medical diagnosis systems it is common to observe that a particular symptom combining with other symptoms may lead to a possible disease. Doctors often assign a weight to each symptom in order to show the relative degree (weight) of importance of each symptom leading to the consequent (a disease). This means that for a disease one symptom is possibly more important than the other symptoms. Therefore, local weights assigned to propositions are necessary.

(2) In what situation the WFPRs have a better performance than CFPRs? Generally speaking, if we believe that propositions in the antecedent part have different degrees of importance, a number of rules executed in an inference path leading to a specific goal, or the same rule employed in various inference paths leading to distinct final goals have relative degrees of importance, it may be expected that WFPRs have a better performance than CFPRs.

A good learning algorithm should strike a balance among complexity, comprehensiveness and accuracy. In addition to being more accurate, one should note that the method of production rules offers a better explanation capability than other methods such as neural network or discriminant analysis.

(3) How to determine the knowledge parameters? Is it possible to roughly determine the necessary knowledge parameters while generating the set of fuzzy rules? The present chapter provides an affirmative answer to this question.

Usually, the knowledge parameters are acquired by consulting with domain experts. Of course, this is an iterative and time-consuming process. To avoid or reduce this consultation workload, we would like to present a method that can generate not only a set of fuzzy rules but also a rough estimation of the necessary parameters. This is one of the main purposes of this chapter. Since the estimation of these parameters is rough, it is very likely that they need to be put into a connectionist system for further refinement. However, this refinement problem will not be dealt with in this chapter.

Let us now consider the generated fuzzy decision tree with parameters. The transformation from the tree to a set of WFPRs is described as follows:

(a) Each path of branches from root to leaf can be converted into a rule. The antecedent of the rule represents the attributes on the traversed branches from the root to the leaf; and the consequent of the rule represents the fuzzy cluster at the leaf.
(b) The parameters representing the degree of importance of the attributes (attribute-values) along the branches (which have been used to find the expanded attribute) are regarded as the local weights of the rule.
(c) The certainty factor of a weighted fuzzy production rule having the form

$$IF \quad < A_1 \text{ and } A_2 \text{ and } \cdots \text{ and } A_n >$$
$$THEN\ B\ (CF,\ Lw_1, Lw_2 \cdots, Lw_n)$$

is computed by the following equation

$$CF = \sum_{i=1}^{N}\left(\wedge_{j=1}^{n} A_j(i) \wedge B(i)\right) \Big/ \sum_{i=1}^{N}\left(\wedge_{j=1}^{n} A_j(i)\right)$$

where A_j and B are fuzzy sets on $\{1,2,\cdots,N\}$ and $\wedge$ denotes the minimum.

4. Weighted Fuzzy Rule Evaluation Method

In this section, a weighted fuzzy rule evaluation method (i.e. an inference algorithm) is proposed. This similarity-based method is a variant of the one presented in [8]. The weights are incorporated into the process of computing the similarity degree between the given fact and the antecedent part.

Assuming that a group of weighted fuzzy rules has been generated and a fuzzy object with unknown classification is given. If the given fuzzy object for the antecedent in a fuzzy rule cannot match exactly with the antecedent of the rule, the consequent should be drawn via an approximate reasoning algorithm. There have been two types of well known approximate matching methods. One is based on Zadeh's Compositional Rule of Inference, and the other is by means of similarity between two fuzzy sets. The main advantages of similarity-based methods are their computational simplicity and their ability to assign parameters like certainty factors and weights. A detailed comparison among the similarity-based fuzzy reasoning methods can be found in [3].

In the non-fuzzy matching process, only one rule can be applied for each object. But in the fuzzy case, many rules can be used at the same time. More specifically, for a given object and a set of fuzzy rules, the object can be matched with any one of the rules in the set and the consequent for the rule being matched is drawn with a degree of truth. Consequently, an object with unknown classification may be classified into several classes with different degrees of truth. The following is our proposed algorithm for approximate reasoning in which the inference result is a fuzzy set on the space of all clusters (i.e. a possibility distribution $(p_1, p_2, \cdots, p_n)$ in which the j-th component indicates the possibility of the object belonging to the j-th cluster.)

An approximate reasoning algorithm. Let $B = (\mu_1, \mu_2, \cdots, \mu_n)$ be the object to be classified, F be a group of weighted fuzzy production rules extracted from a fuzzy decision tree, and let there be k clusters in total. The initial state of B with respect to the classification is set to be $(p_1, p_2, \cdots, p_k) =$ (0, 0, $\cdots$, 0) in which the j-th component indicates the possibility (the degree of truth) of the object belonging to the j-th cluster. SM is a given similarity metric.

Step 1. From the rule set F select a fuzzy rule R that has not been applied to this object B:

$$IF \ < A_1 \ and \ A_2 \ and \cdots and \ A_n >$$
$$THEN \ C \ (CF, \ Lw_1, Lw_2 \cdots, Lw_n)$$

where A_j and C are fuzzy sets defined on the universe of discourse.

Step 2. Compute SM_j, the degree of similarity between μ_j and A_j. If μ_j is a numerical number then SM_j is regarded as the membership degree of $\mu_j \in A_j$. If μ_j is a fuzzy set (linguistic term) then $SM_j = SM(\mu_j, A_j)$.

Step 3. Compute SM_W, the overall degree of similarity between the given object $B = (\mu_1, \mu_2, \cdots, \mu_n)$ and the antecedent of the rule, by $SM_W = \sum_{j=1}^{n} \frac{SM_j \cdot Lw_j}{Lw_1 + Lw_2 + \cdots + Lw_n}$.

Step 4. Suppose that the consequent C is the i-th cluster $(1 \le i \le k)$, compute α_i, the certainty degree of the given object B belonging to the i-th cluster, by $\alpha_i = CF \cdot SM_W$.

Step 5. Modify the possibility vector $(p_1, p_2, \cdots, p_k)$ by $p_i = Max(\alpha_i, p_i)$ and label that the rule R has been applied to the object B.

Step 6. If in the set F there exists at least one rule which has not been applied to the given object B, then goto Step 1.

The evaluation result of the proposed algorithm, $(p_1, p_2, \cdots, p_k)$, is a fuzzy set defined on the space of all clusters. It is noted that when two or more rules are applied to classify the object into the same cluster with different degrees, the algorithm selects the maximum as the membership value to the cluster. When an object is classified into several clusters with different degrees, but only one classification result is required, one can select the cluster with the highest membership degree. That is, if a crisp classification result must be generated from $(p_1, p_2, \cdots, p_k)$, one can choose the i_0-cluster where i_0 satisfies $p_{i_0} = \max_{1 \le i \le k} p_i$. Moreover, if there is more than one component reaching the maximum simultaneously, our method will not give the crisp result.

An approximate reasoning algorithm for a set of fuzzy rules has also been suggested by Yuan and Shaw in [8]. By comparing our proposed algorithm with the one by Yuan and Shaw, one can find the following main differences. (1) Our algorithm incorporates the weights in computing the overall degree of similarity between the given object and the antecedent part whereas Yuan and Shaw's algorithm does not. (2) Our algorithm uses the weighted average to compute the overall degree but Yuan and Shaw's algorithm uses the minimum operation to compute it. Moreover, both algorithms use the maximum operation to aggregate the degrees of truth of classification for different rules.

Generally speaking, the training accuracy of fuzzy learning algorithms cannot reach 100%. But for this particular application by making use of the principle of maximum degree of classification of truth, both matching algorithms can fully yield correct classification results. Although the matching algorithm proposed in [8] can also yield correct classification, our proposed method has the smaller overall ambiguity of classification with respect to the training set. Here, for a fuzzy vector $(p_1, p_2, \cdots, p_k)$ (normalized with descending order) defined on the space of all clusters, the ambiguity of classification [8] is defined as

$$Ambig = \sum_{i=1}^{k}(p_i - p_{i+1})\ln i, \quad \text{where} \quad p_{k+1} = 0.$$

At the end of this section, we investigate the sensitivity of classification accuracy to missing attributes (attribute-values). For a given set of fuzzy rules and a group of objects, one problem we are interested in is, if an attribute (or most of its values) is (are) missing due to problems encountered in recording or observing, how does the missing attribute affect the classification accuracy of the set of fuzzy rules to the group of objects? In other words, how sensitive is the classification accuracy to the missing attribute? In this chapter, the effect of a missing attribute to the classification accuracy is called robustness.

It is clear that the robustness depends heavily on the degree of importance of attributes contributing to classification. Common heuristic algorithm provides no information on the degree of importance of attributes in the process of tree generation. However, since our heuristic algorithm has employed the degree of importance of attributes to select the expanded attributes, a formula is needed to compute the overall degree of importance of attributes contributing to classification.

Let's consider a fuzzy decision tree generated by our heuristic algorithm and an attribute *A*. According to **Definition 1**, for each node *N*, *A*'s degree of importance to classification at the node *N* (including its child-nodes) can be defined as $I_N(A) = \Sigma_i \theta_i$. The overall degree of importance of the attribute *A* to classification is defined as

$$OI(A) = \Sigma_{N \in S}\left(\left(M(N)/\Sigma_{N \in S} M(N)\right) \cdot I_N(A)\right),$$

where *M* is the cardinality function and *S* is the set of all nodes of the tree.

Attributes used in the tree generation (the set of weighted fuzzy production rules) can be arranged according to their overall degree of importance. We expect that missing attributes with small overall degree of importance will not be sensitive to the classification accuracy. In other words, the result that the deficiency of attributes with small overall degree of importance has less influence on classification accuracy is expected. Clearly, this is a significant result since it provides us with, in view of the classification accuracy, how much influence the loss of an attribute has.
One may argue that the root attribute is always the most important attribute for any decision tree. But this is not necessarily the case. [12,13]

5 Application to Reservoir Recognition

5.1 Preliminaries of logging explanation

Reservoir recognition is the main task of logging explanation. According to logging data and earthquake data, the underground structure of the explored area can be divided into a number of layers such as water layer, dry layer, oil layer, sand layer, gas layer, and so on. Those layers are usually called reservoir. How to obtain the reservoir distribution is the first crucial issue in logging explanation. There have been a number of techniques such as earthquake speed inverse and earthquake feature parameter to get the reservoir distribution in a specified explored area. Usually determining reservoir distribution is considered as phase 1 of logging explanation.

Another crucial issue for logging explanation is how to recognize the layers (reservoir) obtained in phase 1. In other words, phase 2 aims to determine whether the considered layer is oil layer or other kind of layers. This chapter will focus on phase 2. To recognize the reservoir, a number of techniques such as Bayesian estimation, linear regression, fuzzy comprehensive evaluation have been developed. Owing to the complexity of underground structure, it is hard to say which specified technique is generally better than the others. It depends on the specified explored area. In this chapter, we will apply the fuzzy decision tree induction to the reservoir recognition.

The comprehensive explanation of logging-data refers to a series of synthetic analysis and calculation based on a large volume of observed curves and data. The aim of the comprehensive explanation is to investigate the partition of oil-layers and the correlation among osmosis stratum, oil-gas-water, and other geological problems encountered during the process of oil-layer exploration. Comprehensive explanation of logging-data is developed with the help of computer technology. Each logging method has its own special usage and advantages in a specific application range, so it reflects some quality of stratum medium under specified conditions. Geological situation is so complex that different objects being measured can end up with the same curve. Therefore it is necessary to synthetically analyze logging-data by using different curves and data.

A group of logging-items, which are combined together based on some specific purpose, is called a logging-series. Generally speaking, the selection of a logging-series should be based on how well the following activities are performed:

(1) Dividing the drilling well section plane in detail into precisely measuring stratum;
(2) Examining the rock quality as precisely as possible and partitioning the osmosis stratum;
(3) Providing framework to judge and investigate oil-gas-water;
(4) Computing the oil (water) saturation degree, the porosity, the osmosis ratio and the mud content of oil layer.

The common logging section plane is mainly classified into two basic types: the sand-mud-rock and the carbonate. However, some particular region may have special type of logging section such as ointment rock section. These rock section planes may possess a wide range of greatly distinct characteristics of physical geography. It implies that the logging-series being used for each type should be also different. After obtaining the logging-curves and data based on the selected logging-series, the following information must be available in order to conduct the logging explanation precisely:

5.1.1The characteristics of physical geography in the oilfield

It includes the following:

(1) Structural characteristics and oil-gas depository types of oil field;
(2) Distributed rules and the main standard layer of oil-layers in each period;
(3) The rock characteristic, the physical characteristic and the oil-bearing characteristic.

5.1.2 The first materials that directly reflect stratum situation

It includes:

(1) All kinds of logging materials;
(2) The materials of drilling well samples;
(3) The materials of gas-logging;
(4) Stratum data and experimental measure data.

5.1.3 The changing trend of content of oil layer

This information can help reduce the difficulty of logging explanation and make it easy to contrast the changing trend of content of oil layer. We generally divide objective layers into: the oil layer, the stripped oil layer, the dried layer, the oil-water layer, and the water layer. These layers can have two kinds of changing trend:

(1) Oil layer →stripped oil layer →dried layer;
(2) Oil layer →oil-water layer→water layer.

5.2 The common logging explanation methods

One of the main tasks of the logging explanation is to determine the parameters of oil-gas reservoir by using the logging materials. These parameters include the degree of porosity, the osmosis ratio, the oil(water) saturation degree, the mud content, and etc. They are the main parameters for crosswise prediction of reservoirs.

There are many methods to determine the parameters of a reservoir. The main method includes the congregation graph, the experience formula, and the statistical optimization. A brief introduction is given below:

5.2.1 Congregation graph method

In the process of artificial logging explanation, congregation graph method is a commonly used one. It is a kind of 2-D geometric graph that is established according to the relation between the logging response curves and the reservoir's parameters. There are many kinds of congregation graph due to different design principles and usages. For example, the resistivity-porosity congregation graph, the density-neutron congregation graph, and the sound wave-neutron congregation graph are all well known ones. The first kind of congregation graph is mainly for determining the geometric factor's constant of shape of hole. The latter two kinds are mainly used to calculate the degree of porosity, determine rock quality, and compute the mud content, the water-bearing saturation degree, etc. Since the congregation graph mostly adopts empirical formula to determine the geometric shape on graph, the precisions of the parameters obtained by using congregation graph is usually not good.

5.2.2 Empirical formula method

Every logging method always corresponds to a set of theoretical formula that describes the relation between the logging curves and the geological factors. However, in actual logging practice, since there are many factors effecting the logging curves and it is difficult for some factors to be measured precisely, the theoretical formula is rarely applied to the practical explanation. The empirical formula, which is obtained by conducting a considerable number of experiments, is regularly used. The following empirical is a typical one.
$Sw = Fr \cdot Rw / Rt$, where Sw is water-bearing saturation degree; Fr is stratum resistivity factor, Rw is stratum water resistivity; and Rt is stratum resistivity.

5.2.3 Statistical optimization method

There exist a number of geology models in terms of the geophysical characteristics. These models include: the resistivity model, the W-S model, the double-water model, and the uniform geometric model. In accordance with every model's characteristic and the reflected geophysical characteristic, the corresponding logging response equation for every logging method can be established. Then statistical optimization method is applied to compute the parameters of the reservoir, and the Global method is a most typical one.

Global method is a kind of method that deals with the logging explanation. This method aims at:

(1) using all probable information, including logging measure, geology constraints and local information;
(2) seeking the best way appropriate for the complicated information system;
(3) improving the potentiality, i.e., ease of introducing new logging curves and explanation model;
(4) providing a powerful artificial control of explanation results.

The above mentioned goals require:

(1) using N-dimensions rather than 2-D congregation graph technology;
(2) observing and collecting physical data based on sound statistical techniques so as to get the most possible answer;
(3) establishing a flexible modular software system, in which logging method is a separate, individual component.

In order to explain the basic principle of this method, we consider the double-mineral model. The input to the Global program is the set of all samples.

(1) All obtained logging curves which are environmentally rectified can be defined as follows

$$a = (b, N, t, \mathrm{Rx}, \mathrm{Rt}, GR, SP)$$

(2) The set of geology parameters is Rw, clay parameter, mud quality, etc. In such stratum, the output is

$$x = (\varphi, V_{et}, S_w, S_{xo}, p)$$

The relationship between the input and the output can be represented by a set of response equations. For example, the density relationship can be represented as

$$\rho_b = \varphi S_{xo}\rho_{mf} + \varphi(1 - S_{xo})\rho_n + V_{et}p_{et} + (1 - \varphi - V_{et})p_{ma} \, .$$

All equations can be expressed as $a_i = F_i(x) \quad i = 1,2,\cdots$.

The approximation solution of the above-mentioned equation makes the residual difference $(a_i - F_i(x))$ of every logging equation as small as possible.

The main idea of the Global method can be described as follows. For a given set of logging values, the aim is to determine the unknown vector x such that the non-correlation function tends to the minimum. Compared with some existing conventional methods, the Global method is relatively advanced. However the method has inherent defects that cannot be overcome by itself. When the dimension of the unknown vector is larger than the number of logging equations, the solution is not satisfactory. In addition, the hypothesis that requires the geology parameters be kept constant on some region is not correct, and this will greatly limit the practical applications of this method.

5.3 Calculation of logging reservoir parameters

In the Minning14 -20 regions (in an oilfield of northern China), the extension appearance of the well radius is serious, which leads to a situation where the nature potential and gamma curves cannot effectively reflect the stratum characteristics. In light of this gravel rock stratum which is strongly non-uniform, several methods had been attempted to solve this problem. We also made use of the HL (Hodges-Lehmann) method to optimize the reservoir parameters in order to obtain precise results of the logging explanation. HL method is a kind of statistical technique. In logging explanation, we often adopt different methods to compute the same parameter such as porosity degree, mud content, etc. Because of the existence of many stochastic and fuzzy factors, for a fixed parameter, its values which are obtained by different methods are often different to some extent. One problem is how to maximally reduce the deviation from the true value. HL method offers a solution to this problem.

5.3.1 Determining mud content by using HL method

Natural gamma and potential nature logging curves cannot effectively reflect the mud content in the oil region. Using the porosity logging curve and the deep side to respectively calculate the mud content, the result is not good. Based on the analysis of the logging materials and the core data, we select three different methods to determine the mud content. By applying the HL method to the synthetic processing, the best selection of the mud content is obtained.

Three methods of calculating the mud content are given as follows:

(1) $SH_1 = \frac{R_{SH}}{R_t}$, $SH_2 = [\frac{R_{t\min}}{R_t} \cdot \frac{R_{t\max} - R_t}{R_{t\max} - R_{t\min}}]^{1/B}$,

where $B = \begin{cases} 1 = R_{t\min} / R_1 > 0.5 \\ 0.5 / R = R_{t\min} / R_1 = 0.5 \\ 1 - R_{t\min} / R_t = R_{t\min} / R_1 < 0.5 \end{cases}$, R_{SH} is near mud-stone resistivity, R_{tmin} and R_{tmax} are the minimum and maximum of resistivity of processing layer. Generally speaking, the aforementioned SH_1 and SH_2 reflect the minimum and maximum of the mud content, so $SH(1) = min(SH_1, SH_2)$

(2) Combined neutron sound wave method

$$SH_1 = \frac{\varphi_N - \varphi_{N\min}}{\varphi_{NSH} - \varphi_{N\min}} \qquad SH_2 = \frac{\Delta t - \Delta t_{N\min}}{\Delta t_{SH} - \Delta t_{\min}},$$

where Δt_{SH}, φ_{NSH} are the time difference and the neutron value of near mud-stone sound wave respectively, $\Delta t_{\min}$, $\varphi_{N\min}$ are the minimum of the time difference of sound wave and neutron values. We set $SH(2) = min(SH_1, SH_2)$

(3) Combined neutron density congregating method

$$SH(3) = \frac{\frac{\varphi_N - \varphi_{Nma}}{\varphi_{Nmf} - \varphi_{Nma}} - \frac{\rho_{ma} - \rho_b}{\rho_{ma} - \rho_{mf}}}{\frac{\varphi_{NSH} - \varphi_{Nma}}{\varphi_{Nmf} - \varphi_{Nma}} - \frac{\rho_{ma} - \rho_{SH}}{\rho_{ma} - \rho_{mf}}},$$

In which φ_{Nmf}, ρ_{mf} are the neutron value and the density value of mud filtrate, φ_{NSH}, ρ_{SH} are the neutron value and the density value of mud-stone; and φ_{Nma}, ρ_{ma} are the neutron value and density value of framework.

We can calculate the mud content V_{SH} by using the HL method to process the above mentioned *SH(1), SH(2)* and *SH(3).*

5.3.2 Computing porosity φ and water-bearing saturation *Sw*

Because of the serious uniformity in this region, the use of a single method to calculate the porosity degree would produce much defects. By analyzing the materials of the logging and the core, we select three combined porosity methods by using the HL method. The combinatorial parameters are:

$$\Phi(1) = \frac{\Delta t - \Delta t_{ma}}{\Delta t_f - \Delta t_{ma}} - V_{SH} \frac{\Delta t_{SH} - t_{ma}}{\Delta t_f - \Delta t_{ma}}$$

$$\Phi(2) = \frac{\rho_{ma} - \rho_b}{\rho_{ma} - \rho_{mf}} - V_{SH} \frac{\rho_{ma} - \rho_{SH}}{\rho_{ma} - \rho_{mf}}$$

$$\Phi(3) = \frac{\varphi_N - \varphi_{Nma}}{\varphi_{Nmfma} - \varphi_{Nma}} - V_{SH} \frac{\varphi_{NSH} - \varphi_{Nma}}{\varphi_{Nmf} - \varphi_{NmA}}$$

Note that the double water model formula is used to calculate the water-bearing saturation S_w.

5.3.3 Calculating the averaged value of granularity

We find that the gravel content is relative to the averaged value of granularity according to the core granularity materials. Using regression technique, we obtain the following correlation formula

$$M_d = (V_d + 9.6547)/75.1679,$$

where the formula for computing the gravel content V_d is given by

$$V_d = \frac{134.5 - H \cdot (23.776 \cdot \Delta t - 217.761 \cdot \rho_b + 43.2 \cdot \varphi_N)}{26.1 \cdot H + 3.2},$$

in which H is the real deep measured in meter, m; Δt is the real measured difference measured in second per minute, μ s/m; ρ_b is the real density measured in gram per cubic centimeter, g/cm^3; and φ_N is the real neutron value in percentage, %.

5.4 Fuzzy learning process of reservoir fluid recognition

The problem of determining the quality of a reservoir fluid is inherently vague in nature. ISDDATA fuzzy clustering method, also called the *k*-mean fuzzy clustering, is a rather popular technique in handling the reservoir recognition problem. It firstly pre-classifies the samples and then adjusts the membership function until reaching a relatively steady situation. After the number of classes is determined, the classes are integrated. Finally the system outputs a new file of cluster center parameters. The recognition precision of this method is not stable and particularly depends on the geology structure and the refined explanation of this region. Furthermore, the recognition of oil-water layer is not very good.

Generally speaking, the combination of reservoir parameters can reflect the quality of reservoir fluid rather than the separate reservoir parameters for the gravel rock layer which has much non-uniformity and heavy mud materials. Based on the special geology structure of this local region and the previous work of the block explanation in this region, we selected 5 sets of parameters as the evaluation indexes to reflect the reservoir fluid quality. After properly selecting a set of samples, we obtain a set of weighted fuzzy rules of recognizing fluid by using fuzzy decision tree learning and apply them to recognize the unknown layer.

5.4.1 Selecting 5 sets of parameters representing the quality of reservoir fluid as evaluating indexes

We have:

(1) Stratum cementation index $m_F = a + b\dfrac{\log S_w}{\log \varphi}$,in which a and b are stratum cementation coefficients, S_w is water-bearing saturation degree, and ϕ is the efficient porosity.

When the efficient porosity degree is definite, m_F is increasing with the reduction of the water-bearing saturation degree. As the efficient porosity degree is increasing, m_F is correspondingly increasing.

(2) The ratio of the efficient porosity degree to the resistivity porosity $RPER = \varphi_e / [\varphi_R (1 - V_{SH})]$, in which φ_e is the total porosity of stratum, $\varphi_e = \varphi \cdot (1 - V_{SH})$; φ_R is the resistivity porosity, $\varphi_R = (a \cdot R_w / R_o)^{1/m}$; φ is the efficient porosity; V_{SH} is the mud content; R_o is the real resistivity value; Rw is the stratum water resistivity; a and m are the stratum cementation

coefficient and coefficient respectively. When there is oil gas in the pore, $\varphi > \varphi_R$, and $RPER$>1; when the stratum is full of water, $\varphi = \varphi_R$, $RPER \approx 1$.

(3) M-power of the visual stratum water resistivity $R_{wa} = R_o^{1/m} \cdot \varphi_e \cdot (1 - V_{SH})$, in which R_{wa} is the visual stratum water resistivity, R_o is the real stratum water resistivity. Due to the impact of oil gas, R_{wa} of oil layer will be larger than that of the water layer apparently.

(4) The logarithm value of the degree of the hole-path camber

$$TAA = 0.5\{Log[\varphi_e(1-V_{SH})]+LogRo+LogRw\}$$

Stratum hole-path camber reflects the situation of the oil-bearing of stratum. If the stratum contains oil gas, the hole-path camber will hoist.

(5) The ratio of the reconstruction waterline resistivity R_0^* to real resistivity $SRTO = (b \cdot R_0^* / R_t)^{1/N}$ in which $R_0^* = a \cdot R_w / \varphi^{m1}$, $m1 = f(V_{SH}, m) = f(M_d, m)$. If oil gas is contained, $SRTO$ will be reduced.

5.4.2 Selecting and fuzzifying training data

Because of the existence of vagueness in reservoir recognition, we will adopt the following principles for the selection of training data in this block:

(1) Its original data must be relatively precise and the environmental impact is small;
(2) The characteristics of oil layers, gas layers, water layers and dried layers are clear, and their memberships are larger than 75%.
(3) The impact of the surrounding rock is small and its thickness is larger than 2 *m*.

We selected 7 attributes which are the 5 sets of parameters specified in section 5.4.1, *STRO*, m_F, *PRER*, R_{wa}, *TAA*; and the layer thickness *H* and the hydrocarbon content *Q*. The first 6 attributes are continuous-valued, but the last one is a fuzzy-valued vector defined on the universe of {the first class hydrocarbon content *Q1*, the second class hydrocarbon *Q2*, the third class hydrocarbon *Q3*, the forth class hydrocarbon *Q4*}. The classification value of every sample is a fuzzy vector on the universe of {water layer, oil layer, gas layer, dried layer}. Based on the standard of sample selection, the membership of some layer is larger than 0.75 at least.

We select 142 layers as training samples from five wells. The formats for the values of every sample are shown in Table 3.

Table 3. Sample values

No	STRO	mF	PRER	Rwa	TAA	H	Q				CLASS			
							Q1	Q2	Q3	Q4	water layer	oil layer	gas layer	dried layer
1	0.569	2.21	2.29	2.11	0.47	11	0.7	0.3	0.0	0.0	0.2	0.8	0.0	0.0
2	0.688	1.98	2.26	1.35	0.34	9.5	0.5	0.4	0.1	0.0	0.7	0.3	0.0	0.0
	...	...	...	...	...	...	...	...	...	...	...	...	...	...

The calculation of all kinds of hydrocarbon content can be obtained by using conventional methods in terms of logging curves. The vagueness of classification is the characteristics of the fluid reservoir. It can be described by linguistic terms. For example, the first sample in Table 3 can be described as water-bearing oil layer and the second one can be regarded as oil-bearing water layer. The fuzzification of the first 6 continuous-value attributes is based on the algorithm mentioned in the previous section. It first determines the number of linguistic terms and then generates the cluster centers. Consequently the number of linguistic terms for these attributes, $STRO$, m_F, $PRER$, R_{wa}, TAA and H is 3, 3, 2, 4, 3 and 2 respectively.

5.4.3 Training and explanation

Using the algorithm proposed in this chapter, we trained the data set in Table 3 and obtained 28 weighted fuzzy production rules. The training accuracy is 99%. We applied the training results to process the real materials of 56 wells involving 602 layers in this region. The comparison with the original explanation is shown in Table 4.

Table 4. Comparative result

	<1500m	1500~2500m	2500~3500m	total
Number of processed layers	217	261	124	602
Number of real-testing layers	63	84	62	209
Number of the original correct explanation layers	52	63	44	159
Correctness rate	83%	75%	71%	
Correct number of explanation layers by using our technique	62	76	52	190
Correctness rate	98%	90%	84%	

As an example, we only selected two wells (denoted by No.1 and No.2) for demonstration. The detailed explanation for some layers on these two wells is given below.

No.1 well: the processed well section is 2513-2613m including a patched layer 2516.5-2514.5m, whose thickness is 2 m. The original explanation is a gas layer, but our processed result is oil-bearing water layer. The membership matched to gas layer is smaller than 0.15, so the layer cannot be regarded as a gas layer obviously. The result of real-testing verifies that our explanation result is correct. The 56th layer is 2563.3-2560.6m whose thickness is 2.7 m and the original explanation is a water layer, but our explanation result is water-bearing oil layer. The result of real-testing is that the daily yield of oil is 61.47T and that of water is 12.71m^3. In addition, the 58th, 59th, 60th and 61th layers of this well were considered as gas layers in their second explanation, but their first explanation results are water layers. Our processed result shows that the membership of the water layer exceeds 0.75, so we suggest to keep the first explanation result.

For the No. 2 well, the processed segment is 3333-3420m. The original explanation is that the 5th layer whose thickness is 23.6 m is a gas layer, but our processed result is that this layer is oil/water layer. (The membership of the oil layer is 0.50; the membership of the water layer is 0.40, and the membership of the gas layer is 0.10). The result of real-testing is that the daily yield of oil is 83 T and that of water is 49 m^3 and there is no gas.

The original explanation of this explored area is based on the fuzzy comprehensive evaluation model. The real drilling data has showed that the explanation has poor performance. By using the fuzzy decision tree induction technique, the correct recognition rate has about 15% increase (compared with their original explanation). The result shows that, in this specified explored area, the fuzzy decision tree induction application to reservoir recognition is very successful. Due to the 15% increase of correct recognition rate, the artesian well company can reduce 3 drilling wells for making clear the underground structure of the specified area. Since reducing one drilling well can save about HK$10 million, the artesian-well company highly assesses the achievements of the project.

6 Conclusion

In this chapter we apply a new heuristic algorithm for generating fuzzy decision trees with parameters to the reservoir recognition problem in a specified logging area. Using the learned fuzzy rules to explain 602 layers of 58 wells in an oilfield of northern China, the explanation correctness rate reach 93% which has been verified by testing with real data. Owing to about 15% increase of correct recognition rate, the artesian-well company can reduce 3 drilling wells for making clear the underground structure of the specified area. The results show that our suggested method is quite effective in the presence of much vagueness that exist in the sample data.

References

[1] S. M. Chen, "A weighted fuzzy reasoning algorithm for medical diagnosis," Support Systems, vol. 11, pp.37-43, 1994

[2] H. C. Tseng and D. W. Teo, "Medical expert system with elastic fuzzy logic," 3rd IEEE International Conference on Fuzzy Systems, vol. 3, pp.2067-2071, 1994

[3] D. S. Yeung and E. C. C. Tsang, "A comparative study on similarity-based fuzzy reasoning methods," IEEE Trans. System, Man, Cybernetics, vol. 27, no. 2, pp.216-226, 1997

[4] D. S. Yeung and E. C. C. Tsang, "Weighted fuzzy production rules," Fuzzy Sets and Systems, vol. 88, pp.299-313, 1997

[5] D. S. Yeung and E. C. C. Tsang, "A Multilevel weighted fuzzy reasoning algorithm for expert systems", IEEE Transactions on System, Man and Cybernetics, vol. 28, no. 2, pp.149-158, 1998

[6] N. K. Kasabov, "Learning fuzzy rules and approximate reasoning in fuzzy networks and hybrid systems," Fuzzy Sets and Systems, vol. 82, pp.135-149, 1996

[7] J. R. Quinlan, "Induction of decision trees," Machine Learning, vol. 1, pp.81-106,1986

[8] Y. Yuan and M. J. Shaw, "Induction of fuzzy decision trees," Fuzzy Sets and Systems vol. 69, pp.125-139, 1995

[9] T. Kohonen, "Self-Oganization and Associate Memory," Springer, Berlin, 1988

[10] S. Mitra and S. K. Pal, "Self-organizing neural networks as a fuzzy classifier," IEEE Trans. on System, Man and Cybernetics, vol. 24, no. 3, pp. 385-399, 194

[11] K. J. Cios and L. M. Sztandera, "Continuous ID3 algorithm with fuzzy entropy measures," Proc. IEEE Internat. Conf. On Fuzzy Systems (San Diego, CA, 8-12, Match 1992) pp.469-476

[12] D. S. Yeung, X. Z. Wang, E. C. C. Tsang, Learning Weighted Fuzzy Rules from Example with Mixed Attributes by Fuzzy Decision Trees, in Proceedings of IEEE International Conference on System, Man, and Cybernetics (October 12-15, 1999, Tokyo, Japan) pp. 349-354

[13] X.Z. Wang, E.C.C. Tsang, D.S. Yeung, A Comparative Study on Heuristic Algorithms for Generating Fuzzy Decision Trees, IEEE Transactions on Systems, Man, and Cybernetics, Part B, vol.31(Issue.2), April 2001, pp. 215-226

Introducing *SACRA*: A Decision Support System for the Construction of Cattle Diets

David A. Pelta[1], José L. Verdegay[1], and José M. Cadenas[2]

[1] Depto. de Ciencias de la Computación e I.A.
E.T.S. Ing. Informática
University of Granada
Periodista Daniel Saucedo Aranda s/n
18071 Granada, Spain

[2] Depto. de Ing. de la Información y las Comunicaciones.
Fac. de Informática
University of Murcia
Campus de Espinardo
30071 Murcia, Spain

Abstract. Diet problems are usually modelled as linear programming problems and then solved by any adequate method. Although this seems a valid approach, is almost impossible to have a perfect knowledge of all the data taking part into a particular diet problem. Such uncertainty must be managed in some way and here is where the theory of fuzzy sets and systems can show all its power. We address a problem arising in Argentinean farms where the cattle is not confined. As a consequence, it is impossible to assure the amount of food each animal will eat and then, it has no sense to design diets with exact requirements. In order to solve the problem, we formulate it as a fuzzy diet problem and provide a computational tool to help cattlemen in the diet design process.

1 Introduction

A particular key problem that presents features of vagueness, economic relevance and human interaction is the so called Diet Problem (*DP*), which is of upmost importance in a variety of practical situations. The goal of this problem is to find the cheapest combination of foods that will satisfy all the daily nutritional requirements of either a person or animal. The problem may be formulated as a linear programming one where the objective is to minimize the cost and satisfy the constraints representing the nutritional needs.

In practical applications is almost impossible to have a perfect knowledge of some or all the data taking part into the *DP* at hand, and then it is usual to approximate those values, data and/or models in different ways. Fuzzy Sets and Systems based methodology is an appropriate tool for dealing with this kind of lack of precision or uncertainty. In particular, Fuzzy Linear Programming (*FLP*) can show all its power in the framework of Decision Support Systems (*DSS*) because it is useful for modelling the inherent vagueness in the data that the user often has available, and it may help to find solutions for problems in which finding an optimal solution is not easy.

Hence, it makes sense to consider Fuzzy Diet Problems (*FDP*) that have to be understand as *DP*'s in which either a concept or part or all of the data defining them may be considered as fuzzy.

To deal with general *FLP* problems the authors developed *PROBO*, an interactive *DSS* for solving general *FLP* problems which collect all the relevant solution techniques [1,2]. Now, in order to practically solve an specific *FDP*, we present a new *DSS* named *SACRA* (a Spanish acronym for Decision Support System for Diets Construction) that may be viewed as an extension of *PROBO* in the sense that no knowledge about *FLP* nor solution techniques are required to operate with it. The appropriated modules of *PROBO* are used as black boxes, and *SACRA* interacts with them. From this point of view *SACRA* is highly friendship and very interactive. The initial tests carried out with *SACRA*, showed a high level of satisfaction from the side of the decision makers.

In order to present *SACRA* and to show its main features, this chapter is organized as follows: first we will introduce the *DP*, and hence the *FDP*. Then, the essence of classical *FLP* models will be refreshed and the basis of *PROBO* system will be described. After that, the statement of the problem to solve will be presented and then *SACRA* architecture and operation will be fully described. At the end, some final comments are pointed out[1].

2 Fuzzy Diet Problems

As it has been said some lines above, the so called Diet Problem presents well known features of both vagueness, economic relevance and human interaction and therefore it is of upmost importance in a variety of practical situations. The goal of the *DP* problem is to find the cheapest combination of foods that will satisfy all the daily nutritional requirements of either a person or animal.

The conventional *DP* can be addressed, at least, in two ways according to either a linguistic or a mathematical formulation. When a basic linguistic formulation is used one tries to minimize the "cost of the menu" subject to the nutrition requirements, as for instance: "Eat enough but not too much Nutrient A", and this may be stated for each of the nutrients taking part into the problem. Also additional constraints can be added to the *DP*; for instance, one can put constraints on the amount of which foods are consumed: "Eat at least a certain minimum number of rations of fish but not more than the maximum number of rations of beef you want".

With regard to address a mathematical version of this model, we need to define some variables. Let consider a set of foods $\mathcal{F} = \{x_1, \ldots, x_n\}$, a set of nutrients $\mathcal{N} = \{A_1, \ldots, A_m\}$, and the following parameters:

[1] Part of the material of this chapter, extended with application examples, was accepted for publication in the European Journal of Operational Research

c_j = cost of the food j, $j = 1, \ldots, n$
x_j = amount of food j to eat, $j = 1, \ldots, n$
a_{ij} = amount of nutrient i in food j, $j = 1, \ldots, n, i = 1, \ldots, m$
p_i = min. amount of nutrient i required, $i = 1, \ldots, m$
P_i = max. amount of nutrient i allowed per day, $i = 1, \ldots, m$
m_j = min. amount of food j desired per day, $j = 1, \ldots, n$
M_j = max. amount of food j desired per day, $j = 1, \ldots, n$

Hence, in a first approximation the model can be mathematically formulated as a linear programming problem where the objective is to minimize the cost while the nutritional needs (the constraints) are satisfied:

$$\begin{array}{ll} Min \sum_{j=1}^{n} c_j x_j & \\ s.t: & \\ \quad p_i \leq \sum_{j=1}^{n} a_{ij} x_j \leq P_i, & i = 1, \ldots, m \\ \quad m_j \leq x_j \leq M_j, & j = 1, \ldots, n \end{array} \tag{1}$$

As said before, it is absolutely clear that in real applications the perfect knowledge of the exact data taking part into the possible *DP* models is almost impossible, and then it is usual to approximate those values, data and/or models in different ways: for instance by rounding off the data. However it would be much more appropriate to model those vague data by means of pertinent formalisms and in particular, if it is the case, by means of fuzzy sets. As this is the case for the *DP*, it makes perfect sense to assume Fuzzy Diet Problems (*FDP*) which will be managed as *DP* in a *FLP* context.

Roughly speaking a general mathematical programming problem assumes an objective and a constraint set. Conventionally, the objective is modelled by a n-variable function $z(x_1, x_2, \ldots, x_n)$, where the optimal value of x_j, $j = 1, \ldots, n$, is to be determined in a constrained set defined by means of restrictions given by functions $g_i(x_1, x_2, \ldots, x_n) \stackrel{\geq}{\underset{\leq}{=}} b_i$, $i = 1, \ldots, m$.

Fuzzy optimization problems generalize this definition by assuming that the decision maker has a vague knowledge about some elements of the problem. From an operational point of view this vagueness is modelled by membership functions which are associated to the presumed vague element defining the problem. When the objective or the constraints or both elements are defined by linear functions, *FLP* problems arise. Fuzzy Linear Programming is one of the most studied topics in the fuzzy area and many models and methods dealing with these problems can be found in the literature [4]. The reader may also refer to [5] for a recent and updated review on the topic.

As the main aim of this paper is not to focus on *FLP* problems and methods, but using them as tools to be implemented in a *DSS*, let us only recall that *FLP* models can be assumed as associated to any of the following prototype problems:

a) Problems with fuzzy constraints, in which the decision maker may permit a flexible satisfaction of the restrictions, that is, s(h)e allows small violations in the accomplishment of the constraints. This problems are to be formulated as:

$$\begin{array}{ll} Min \sum_{j=1}^{n} c_j x_j & \\ s.t: & \\ \quad \sum_{j=1}^{n} a_{ij} x_j \lesssim b_i & ,i = 1,\ldots,m \\ \quad x_j \geq 0 & ,j = 1,\ldots,n \end{array} \tag{2}$$

where for each row, constraint, $\exists\, \mu_i \in F(R)$ such that $\mu_i : R \to [0,1]$, $i = 1,\ldots,m$ which describes the accomplishment of the i-th constraint for any $x \in R^n$ ($F(R)$ is the set of fuzzy real numbers).

b) Problems with a fuzzy objective, in which the information on the coefficients of the objective function is vague, and then they may be modelled by fuzzy numbers:

$$\begin{array}{ll} Min \sum_{j=1}^{n} \tilde{c}_j x_j & \\ s.t: & \\ \quad \sum_{j=1}^{n} a_{ij} x_j \,\square\, b_i & ,i = 1,\ldots,m \\ \quad x_j \geq 0 & ,j = 1,\ldots,n \end{array} \tag{3}$$

where for each $j = 1,\ldots,n$, $\exists\, \mu_j \in F(R)$ such that $\mu_j : R \to [0,1]$, defining the fuzzy numbers in the objective function and,

c) Problems with fuzzy coefficients, in which one can consider fuzzy numbers in both the coefficients of the technological matrix and the coefficients of the right hand side of the constraint set:

$$\begin{array}{ll} Min \sum_{j=1}^{n} c_j x_j & \\ s.t: & \\ \quad \sum_{j=1}^{n} \tilde{a}_{ij} x_j \,\square^f\, \tilde{b}_i & ,i = 1,\ldots,m \\ \quad x_j \geq 0 & ,j = 1,\ldots,n \end{array} \tag{4}$$

where each $\tilde{a}_{ij}$ and $\tilde{b}_i$ are fuzzy numbers defined by respective membership functions and the symbol $\square^f$ stands for a ranking function between fuzzy numbers which must determined for each problem.

Every combination of the three situations is allowed, so it makes sense to think of the following different *FDP*:

- *FDP* with fuzzy constraints which, coherently from the above described general *DP* model, now will be addressed as:

$$\begin{array}{ll} Min \sum_{j=1}^{n} c_j x_j & \\ s.t: & \\ \quad p_i \lesssim \sum_{j=1}^{n} a_{ij} x_j \lesssim P_i, & i = 1,\ldots,m \\ \quad m_j \lesssim x_j \lesssim M_j, & j = 1,\ldots,n \end{array} \tag{5}$$

This reflects a situation where the decision maker allows some violations in the accomplishment of the constraints on both nutrients and foods that will determine the diet under consideration. In this way, the decision maker will propose the margins to define all of the membership functions, $\mu_j : R \to [0,1]$, $j = 1, \ldots, n$, and $\mu_i : R \to [0,1]$, $i = 1, \ldots, m$, corresponding to fuzzy constraints on foods and nutrients respectively.

- *FDP* with a fuzzy objective in which, for each food $j = 1, \ldots, n$, there is some vagueness on its corresponding cost which is modelled by means of a fuzzy number defined by a membership function $\mu_j \in F(R)$ such that $\mu_j : R \to [0,1]$. The problem can be formulated as follows:

$$\begin{array}{lll} Min \sum_{j=1}^{n} \widetilde{c}_j x_j & & \\ s.t: & & \\ \quad p_i \square \sum_{j=1}^{n} a_{ij} x_j \square P_i, & i = 1, \ldots, m & \\ \quad m_j \square x_j \square M_j, & j = 1, \ldots, n & \end{array} \tag{6}$$

- *FDP* with fuzzy numbers in the technological matrix, that is, *FDP* in which the amount of nutrient i in the food j, $i = 1, \ldots, m$ and $j = 1, \ldots, n$, is defined (perhaps linguistically) by a vague data, and hence it is modelled by means of a fuzzy number given by a membership functions μ_{ij} such that $\mu_{ij} : R \to [0,1]$. The minimum and maximum amount of nutrient i required, can be also modelled by means of fuzzy numbers. Then this kind of problems can be formulated as follows:

$$\begin{array}{lll} Min \sum_{j=1}^{n} c_j x_j & & \\ s.t: & & \\ \quad \widetilde{p}_i \square^f \sum_{j=1}^{n} \widetilde{a}_{ij} x_j \square^f \widetilde{P}_i & i = 1, \ldots, m & \\ \quad m_j \square x_j \square M_j, & j = 1, \ldots, n & \end{array} \tag{7}$$

Besides, it is also possible to assume, as in the above model, that the constraints on the amount of foods desired are also fuzzy. In this last case one has the following *FDP* model:

$$\begin{array}{lll} Min \sum_{j=1}^{n} c_j x_j & & \\ s.t: & & \\ \quad \widetilde{p}_i \square^f \sum_{j=1}^{n} \widetilde{a}_{ij} x_j \square^f \widetilde{P}_i & i = 1, \ldots, m & \\ \quad \widetilde{m}_j \square^f x_j \square^f \widetilde{M}_j, & j = 1, \ldots, n & \end{array} \tag{8}$$

Now, fuzzy numbers are used in the technological matrix and in the margins of the constraints. It must be noted again that the ranking function between fuzzy numbers $\square^f$ has to be determined in each case [3].

It is clear that these *FDP* are just particular models of that *FLP* in (1), and therefore they can be solved with the help of any oriented *DSS*. In this framework, Cadenas et.al [1,2] presented *PROBO*, a *DSS* devoted to manage this kind of problems.

PROBO is a model-oriented, user-dependent, interactive system providing solution techniques for solving classical *LP* problems, that is *LP* problems with no fuzziness at all, as well as *FLP* problems with either fuzzy constraints or fuzzy costs or fuzzy coefficients in the technological matrix. Obviously, every combination of these three last models can be also managed.

Given the problem to solve, *PROBO* carries out a syntactic and semantic analysis in order to detect possible problem data errors first, and second to recognize the features of the model (fuzzy restrictions, fuzzy cost, fuzzy coefficients or crisp). In *PROBO* these four models are respectively associated to four modules **Res-Fuz, Cos-Fuz, Gen-Fuz** and **Classic**.

PROBO interacts with the decision maker, who selects several methods and evaluates their solutions until (s)he is satisfied. The system, through consecutive questions to the user, shows her(him) the different solutions which can be obtained, regarding the particular model (s)he is working with, and the solution method chosen. Among these solutions, the decision maker will select the one (s)he is interested in according to her(his) preferences (criteria).

In order to obtain the maximum profit using *PROBO*, it is essential to have some background on the foundations of *FLP*.

3 Description of the Problem

The problem we address here is the determination of diets for cattle in a field located at Buenos Aires (Argentina) which is fully dedicated to livestock farming. The livestock is not confined. Animals are changed daily between plots and the area for each animal varies as a function of the fodder available. This productive system is the most usual in the referenced region.

The feeding of both fattening categories (male and female calf in process of weaning) is based on winter fresh fodder (wild rye-grass and oat), summer fresh fodder (sorghum) and perennial pastures (Alfalfa and Australian barley).

In order to increment the fattening paces, the previous categories are supplemented with grains and hay which are available at the establishment. The quantities of the supplements are determined in terms of the requirements of the animals (taken from tables published by the National Research Council Board on Agriculture and Natural Resources, USA) and of some costs. The consumption and requirements of the livestock varies as a function of the weight, age, sex and weather conditions.

The main aim is to design a diet which enable an animal of certain characteristics, to achieve a certain level of daily weight gain referred as *ADPV*.

So, in principle, it is necessary to first determine a set of foods and second their corresponding amounts (in kg.)

Each food provides certain amount of nutrients and the diet must verify some requirements on them. The nutrients are determined by the expert and we will consider here the *dry matter intake, energy, crude protein* and *rumen degradable Protein* (*PDR*).

As it was previously said, the animals are living in a open field, so it is impossible to determine how much food each one will eat. As a consequence, the nutritional requirements are hard to achieve for each animal, so it has no sense to design diets verifying the requirements exactly but allowing certain tolerances.

Under this scenario, it seems natural to model the situation as a fuzzy diet problem with fuzzy restrictions which corresponds to the formulation defined in Eq. 5. In principle, we consider $x_j \geq 0, j = 1, \ldots, n$ and we allow the definition of fuzzy goals $(z \lesssim z_0)$ [6,7].

In order to solve the given problems, we use the module **RES-FUZ** from *PROBO*, which implements all the available methodologies to manage the corresponding models.

4 Description of *SACRA*

Having in mind both, the characteristics of the problem and the functionalities provided by *PROBO*, we present here *SACRA*: a decision support system to help in the construction of diets for cattle.

SACRA is based on *PROBO* and three essential principles were in mind at its design stage:

- to be user friendly
- to serve as a guide to the expert, enabling him to design its own diets
- to be extensible, in order to support the addition of other functionalities

SACRA is organized in four modules, each one responsible for a set of actions clearly defined. In this way, extensions and/or modifications are quite easy to implement.

The first module, *INIT*, captures all the procedures related with the initialization of the environment, menus, database, etc. The so called module, *PROBO*, is responsible of solving the problem with the most convenient model. Then the system provides two modules improving and extending the interactivity of *PROBO*, which are mainly oriented to users who do not know (and they do not want to) the theoretical aspects related to solving mechanisms about fuzzy linear programming.

The first of such modules, *INPUT-DATA*, enables the user the input of the problem data (foods and requirements) in an easy and friendly fashion. The second one, *ANALYSIS-RESULTS*, has the responsibility of showing the results provided by *PROBO* in a convenient way. Also, it allows the user to inspect the characteristics of the proposed solutions, or to evaluate other ones he wish to test.

In the following sections, we will describe each module with detail. We will not describe here *PROBO* because its main characteristics were shown previously.

4.1 Module *INIT*

When *SACRA* is executed, the initial screen shown in Fig. 1 appears and the module *INIT* is called and executed. The variables, database and application indexes are initialized in a transparent way to the user.

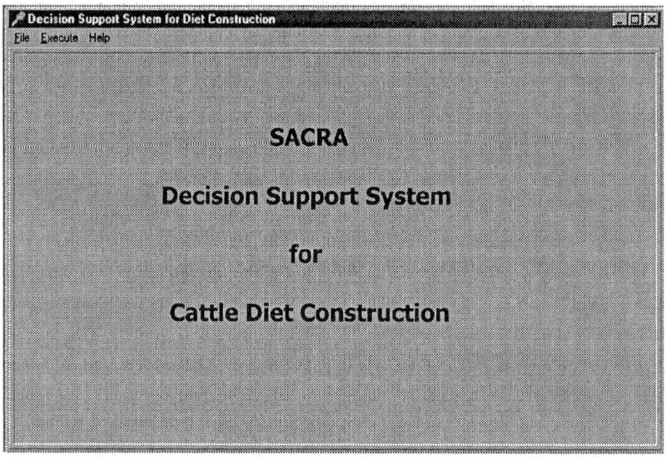

Fig. 1. Initial Screen

The initial screen presents a menu with two main options: *File* and *Execute*. From them, the rest of the modules of *SACRA* can be called. Let's describe the *INPUT-DATA* and *ANALYSIS-RESULTS* modules.

4.2 Module *INPUT-DATA*

For each problem to be solved, the user has to input the nutritional requirements of the animals under consideration and then, he should select the foods to compose the diet. The system provides a database of foods of about 100 records, each one with a large set of characteristics.

Fig. 2 shows the data screen of *SACRA*. The management of the database is comprised in the command bar *Database Administration Commands*, where buttons are provided for the addition, deletion, modification and sequential

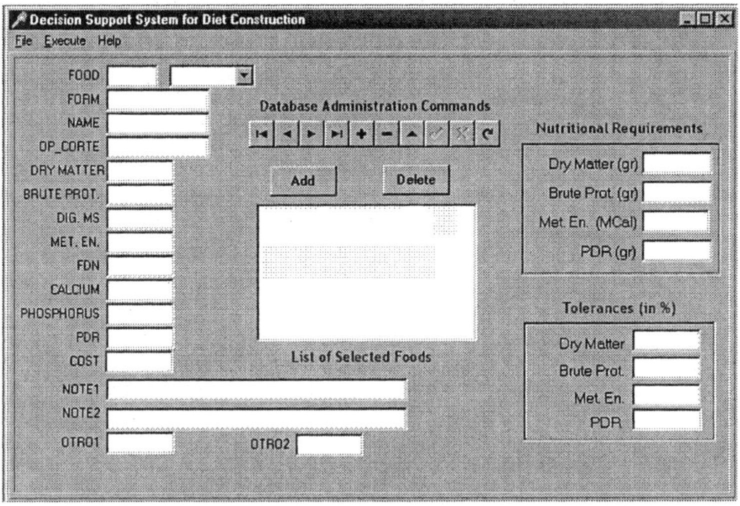

Fig. 2. Data Screen of *SACRA*

browsing of records. The List Box *Available* enable the user to get the first record of a given category.

The left side of the data screen shows the current record of the food database with the whole set of nutritional characteristics.

Two buttons are available to manage the list of selected foods. The *Add* button, adds the current record to the list of current foods while the *Delete* button eliminates the selected item from the list.

Two panels are used to input the nutritional requirements to achieve and the maximum variations tolerated for each one. The variations are expressed as a percentage of the corresponding requirement.

Once the foods are selected and the requirements and their tolerances are established, the user has to execute the procedure *Solve Problem* trough the *Execute* menu. Then the module *PROBO* is started. A dialog box appears and the user has to indicate if he wish to set up an economic goal for the diet or not. If the first option is chosen, then he has to choose from a classical goal (where just the goal should be given) or a fuzzy one (where to goal and a tolerance should be given). After that, the *OK* button must be pressed.

4.3 Module *ANALYSIS-RESULTS*

After *PROBO* is executed in background, the screen of results appears showing the set of founded solutions. This screen is shown in Fig. 3.

If the user provides no goal to achieve, at most 11 solutions will be shown, each one corresponding to a degree of constraint satisfaction. Each solution will give the amount of each selected food in the diet. The obtained solutions are displayed in cost increasing order.

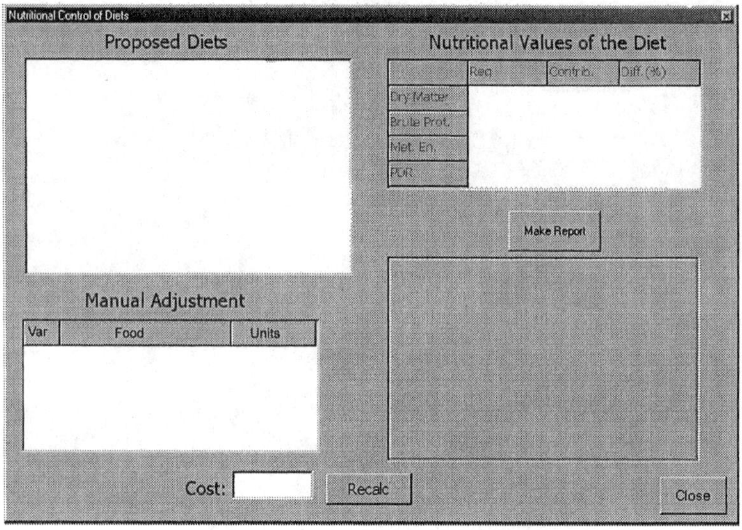

Fig. 3. Results Screen of *SACRA*

Each time a solution from the list is selected, the system shows the nutritional characteristics of such diet. In this way, the user can evaluate the solution not just from an economic point of view, but also from a nutritional one checking where a lack or excess of nutrient arise.

Usually, the diets were constructed by hand and after several talks with the expert we decide to provide the option to the user to design its own diets.

In order to do this, the user must type in the cells of the grid located below *Manual Adjustment*, the desired quantities for each food of the diet. Then pressing the *Recalc* button, the cost and nutritional characteristics of this new solution are calculated and then shown.

The system also uses a pie chart to display the percentage of each food in the diet. The chart is updated each time a new solution is selected or the *Recalc* button is pressed.

Finally, the *Make Report* button enable the user to obtain a printed version of the characteristics of the whole set of solutions or just the characteristics of a selected one.

5 Discussion

The design of diets for cattle it is far from a trivial task. When the livestock is confined, the problem could be well modelled as a classical linear programming problem. But for our case, where the livestock is not confined, such option is not realistic.

So, in order to address the problem, we proposed to model it as a fuzzy linear programming problem with fuzzy constraints and we developed a novel decision support system called *SACRA*.

The tool is now under testing by cattlemen at Buenos Aires, Argentina and the initial results obtained show a high level of satisfaction from the potential users.

Acknowledgments

This research is supported in part by Projects TIC2002-04242-C03-02 and TIC2002-04021-C02-01.
David A. Pelta is a grant holder from the National Council of Research (CONICET) of Argentina.

The authors are indebted to Héctor Pelta, from the National Institute of Farming and Livestock Technology (INTA, Argentina), for his help with the formulation of the problem and his support in the design of *SACRA*.

References

1. J.M. Cadenas, Design and implementation of an Interactive System to solve Fuzzy Optimization Problems (in Spanish), *Phd Thesis*, University of Granada (1993).
2. J.M. Cadenas,J.L. Verdegay, PROBO: an interactive system in fuzzy linear programming. *Fuzzy Sets and Systems 76 (1995), 319-332.*
3. M. Delgado, J.L. Verdegay and M.A. Vila, A General Model for Fuzzy Linear Programming, *Fuzzy Sets and Systems* 29 (1989) 21-29.
4. M. Fedrizzi, J. Kacprzyk and J.L. Verdegay, A Survey of Fuzzy Optimization and Fuzzy Mathematical Programming, in M. Fedrizzi, J. Kacprzyk, M., Roubens, Eds., *Interactive Fuzzy Optimization and Mathematical Programming* (Springer Verlag, Berlin, 1991).
5. H. Rommelfanger and R.Slowinski. Fuzzy linear programming with single or multiple objective functions. In R. Slowinski, editor, *Decision Analysis, Operations Research and Statistics*, volume 5 of *Handbooks of Fuzzy Sets Series*, pages 179–213. Kluwer Academic Publishers, Boston, 1998.
6. J.L. Verdegay, Fuzzy Mathematical Programming, in M.M. Gupta and E. Sanchez, Eds., *Fuzzy Information and Decision Processes* (North-Holland, Amsterdam, 1982) 231-237.
7. H.J. Zimmermann, Description and Optimization of Fuzzy Systems, *International Journal of General Systems* 2 (1976) 209-215.

Prediction of Parthenium Weed Dispersal Using Fuzzy Logic on GIS Spatial Image

Andrew Chiou[*] and Xinghuo Yu[**]

[*]Faculty of Informatics and Communication
Central Queensland University
Rockhampton, 4702 Queensland, Australia
E-mail: a.chiou@cqu.edu.au

[**]School of Electrical and Computer Engineering
RMIT University
Melbourne, 3001 Victoria, Australia
E-mail: x.yu@rmit.edu.au

Abstract. This paper demonstrates the framework and methodology how weed population dynamics can be predicted using rule-base fuzzy logic as applied to GIS spatial image. Parthenium weed *(parthenium hysterophorus* L.*)* infestation in the Central Queensland region poses a serious threat to the environment and economic viability of the infested areas. Government agencies have taken steps to control and manage existing infestation and to curb future spread of this noxious weed. One of the tools used in these strategies is the prediction of parthenium weed population. Conventional weed forecasting methods utilises discrete values in exponential models and linear algorithms extensively. Attempts at predicting weed dispersal relied heavily on accuracy of the original charts or images to yield reasonable results. Using these methods, results of weed population forecasting are only as reliable as the data originally provided. This paper demonstrates that by using GIS spatial image categorised into themes, a fuzzy logic based forecasting methodology can be performed. Fuzzy logic is best suited to this type of problem as its requisite is in handling approximate data. The methodology would allow us to: 1) predict weed population reasonably well based on approximate data, 2) take into consideration additional parameters without re-writing the algorithm, 3) refine large-scale forecasts to suit localised situations, 4) allow users to determine and inspect individual infestation factors, and 5) adapt the methodology to other weed or plant species. This paper also briefly introduces a new fuzzy If-Then operand, Case-Of, specifically used in this methodology.

1. Introduction

In Central Queensland, parthenium weed *(parthenium hysterophorus* L.*)* has demonstrated the ability to cause significant environmental, health, and financial impacts on the pasture and grazing industries if not controlled or managed properly (Adamson, 1996; Chippendale and Panetta, 1994). However, due to the limited resources and capabilities of government agencies and individuals, there is a need of an alternative mean in disseminating expert knowledge. One of the alternative is in the development of an expert advisory system, *P-Expert*, that can be deployed to end users in providing expert knowledge and recommendation in the control and management strategies of parthenium weed. P-Expert's usefulness to the community lies in its different number of advisory outputs that is presented to the user to cater to different requirements. These are, management strategies, control strategies, herbicide use, biological control strategies, explanatory function, and parthenium weed forecasting (Chiou, Yu and Adsett, 2002; Chiou, Yu and Adsett, 2001; and Chiou, Yu and Lowry, 2000). In this paper, a detailed description is presented on the methodology employed in using fuzzy rules to perform forecasting in the prediction of parthenium weed population dynamics in Central Queensland.

The forecasting mechanism is carried out on spatial images of known infested sites and growth influence factors (eg. rivers, roads, soil type). These spatial images is separated and categorised into individual layers, where each layer is further sub-classified into themes. By applying a fuzzy membership function onto each of this themes, it is possible to aggregate the individual values into a final weed infestation factor. This infestation factor is then further applied to the actual spatial image of known infested sites to yield a weed population forecast (Chiou and Yu, 2001).

By using fuzzy rules in this methodology, it overcomes some of the weaknesses encountered with more conventional methods of predicting parthenium weed infestation. Currently, the process of charting infested areas are still being carried out manually and hence, available data are only estimates. By applying conventional methods to these estimates, results obtained would only be as reliable as the accuracy of the original data. In this case, fuzzy logic is best suited to this type of problem as its original requisite is in handling approximate data (Saint-Joan and Desachy, 1995). Furthermore, by applying the methodology presented in this paper, it not only allows weed prediction at a large-scale level, but also at a localised level. By allowing qualitative data (eg. anecdotal references, local information) to be included in the forecasting process, the fuzzy logic mechanism will be able to utilise it to refine large-scale forecasts to obtain localised forecasts.

2. Forecasting Parthenium Weed Population

In parthenium weed control and management strategies, it is of utmost importance that preventive measures must be taken to hinder weed growth in un-infested areas (Rejmanek, 2000). In past and present practices, this is achieved by destroying isolated outbreaks and the edges of saturated areas to maintain economic thresholds (Auld, Menz and Monaghan, 1978; Norris, 1999). That is, the cost of destroying known infestation should not be greater than the cost of actual damages caused by the infestation itself. This strategy is only as successful or optimal as far as the weed forecast results permits. However, only a few studies and attempts have been made to study weed dispersal (Auld and Coote, 1990; Ghersa and Roush, 1993; Rejmanek, 2000).

Established work in predicting weed population dynamics utilises discrete values in linear algorithms and exponential models extensively (Auld and Coote, 1980; Auld, Hosking and McFadyen, 1982; Cousens, 1985; Mortimer, McMahon, Manlove et al., 1980). Such attempts relied heavily on accuracy in the original charts or images for the forecasting mechanism to be able to yield reasonably acceptable or useful results (Rozenfelds and Mackenzie, 1999). However, due to coverage of existing infested sites (approximately 170,000km^2 or 10-percent of Queensland), this can only be achieved by proper and efficient mapping tools where such accuracy cannot be practically attained in the original spatial image. Data gathering were normally done manually via visualisation, that is, an official would travel in person to the site and give an approximate measurement of that infested site. And in many instances, data collection is accomplished by referring to anecdotal references given by landowners themselves. Majority of spatial data collected in such a manner are the only existing data available used for forecasting parthenium weed population in this region.

In addition, conventional forecasting algorithms normally do not have provisions to take into account other influencing factors (eg. gas exchange, photoperiod). At best, other factors are only taken into account under special circumstances. Conventional methodology requires that the algorithm be re-written each time an additional factor is taken into consideration as a supplementary variable. Given that results may be affected by these factors, it is nevertheless difficult to determine which factor or factors played a major role in influencing the weed dispersal. Moreover, conventional forecasting algorithms normally predict future infestation on a global scale (ie. large-scale), ignoring local conditions. Landowners whose property is smaller than the scale of the given global resolution would find local forecast results of minimal use (Daehler, 1998).

The parthenium weed prediction methodology in P-Expert was developed using fuzzy logic techniques to: 1) predict weed population reasonably well based on approximate data, 2) take into consideration additional parameters without re-writing the algorithm, 3) refine large-scale forecasts to suit localised situations, 4) allow users to determine and inspect individual infestation factors, and 5) adapt the methodology to other weed or plant species.

3. Methodology

This section describes the methodology used in the prediction of parthenium weed population using spatial images from GIS datasets. The overall methodology is in four phases: pre-processing, thematic forecast, global forecast and local forecast. All references made to examples refer to the diagram in Figure 1.

3.1 Phase 1: Pre-processing

This step involves preparing the proper data structures to facilitate analysis of the spatial image datasets. The spatial image representation of GIS datasets is first separated into different layers. Each layer is sectioned into a matrix, where each cell represents a resolution of 10^2m in this demonstration. However in practice, this resolution is determined by the level of detail in the original charts and the resolution required by the fuzzy membership functions. Adhering to standard GIS practices (Demers, 2000, p98-109; Tomlin, 1990, p46-54), each layer comprise of themes, $\tau_1, \tau_2 \ldots \tau_n$. Themes are made up of features, $Fs_1, Fs_2 \ldots Fs_n$, indicating natural or man-made phenomenon found in maps (eg. rivers, roads, buildings). In Figure 1., each theme will contain only one feature. Each of the themes are represented as raster images in the matrices in each layer. In practice, this method of structuring spatial image helps accommodate fuzzy variables and membership function (Dragicevic and Marceau, 1999).

In Figure 1., the fuzzy representation is: 1) all layers represent a universe of discourse, and 2) each theme represents a fuzzy variable. There are two special themes, T_1 and τ_{null}. T_1, is a theme reserved for spatial images representing known locations of infested sites. While, τ_{null} is a reserved theme for regions labelled as *null-zones*. Null-zones areas are *non-negotiable* regions, where no known plant species will propagate at all. In actual weed control and management practices, these areas are occupied by man-made structures, such as buildings, washdown and quarantine areas. Null-zones are significant in the forecasting process as explained in the following sections. Hence in Figure 1, the following themes are as follows:

$$
\begin{aligned}
&T_1 : \text{known infestation sites} \\
&\tau_1 : \text{river feature} \\
&\tau_2 : \text{road feature} \\
&\tau_3 : \text{soiltype feature} \\
&\tau_{null} : \text{washdown areas}
\end{aligned} \tag{1}
$$

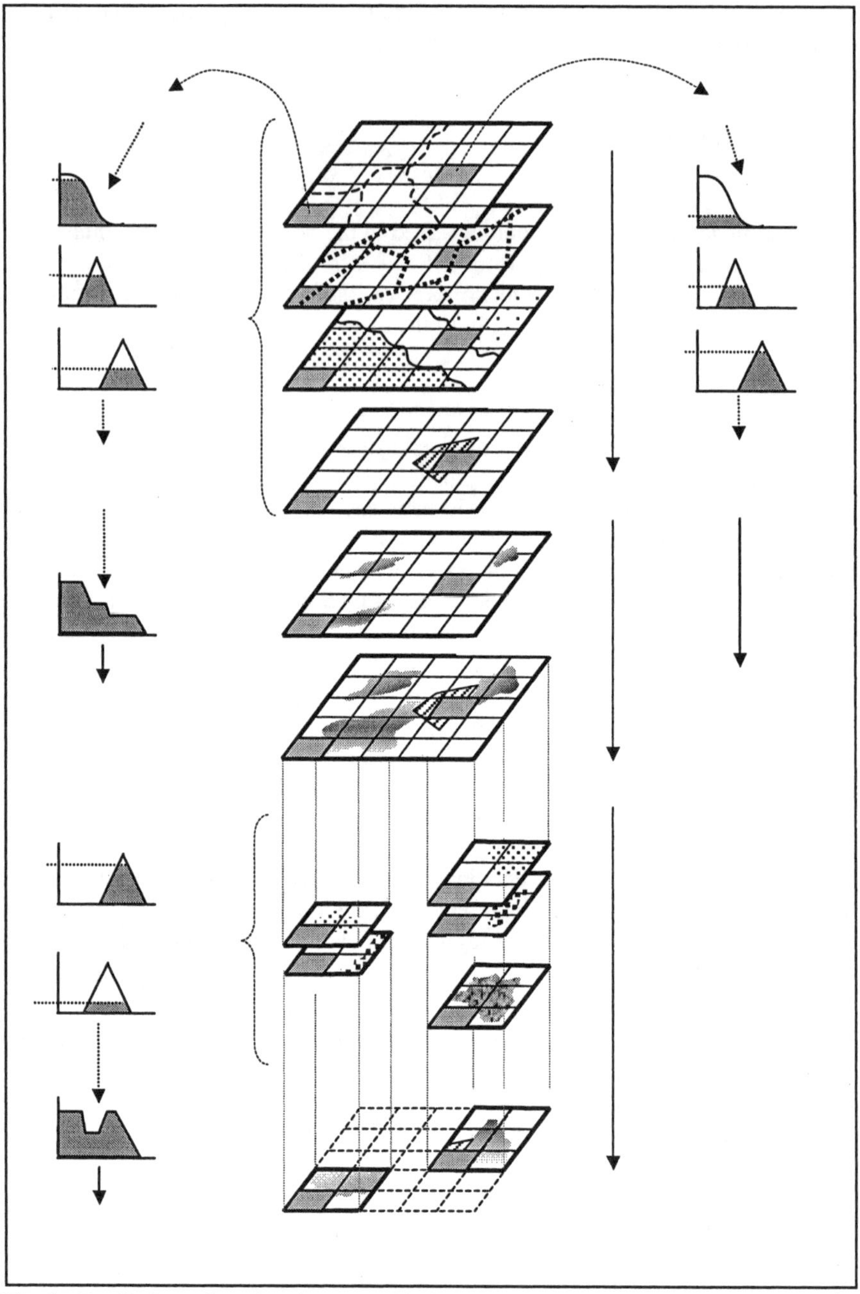

Fig 1. Prediction of parthenium weed infestation using fuzzy logic applied to GIS spatial image. The methodology requires each feature (eg. rivers, roads, soil type) to be layered into themes to facilitate the assignment of appropriate fuzzy membership functions.

3.2 Phase 2: Thematic Forecast

This phase requires that a thematic forecast be carried out individually on each τ_1, τ_2 and τ_3. In this way, each factor that influences the propagation of parthenium weed can be segregated from other factors. Forecasting is performed layer by layer on each corresponding cell to determine its infestation factor, that is, a fuzzy consequent membership value. Membership parameters can be derived from the dispersal rates of weeds given by Cousens (1995, p57-85). In each theme, a thematic forecast function is assigned to all cells, such that the function is a fuzzy If-Then rule to determine the infestation factor. The function for each current cell, C_m, in τ_n, its predicted consequent infestation factor, I_n is,

$$\mathbf{f}_{\mathbf{layer}}(\tau_n\!: C_m) = \textbf{If } (C_m \text{ is } \mathit{proximity\_Fs_n}) \textbf{ Then } (\mathit{infestation\_factor} \text{ is } I_n) \tag{2}$$

Using rule (2) based on the example in (1), the actual fuzzy If-Then rule implementation block for thematic forecast, excluding special themes, τ_{null} and T_1 is,

begin
  If (*current_cell* is *close_to_river*) **Then** (*infestation_factor* is *likely*$_1$)
  If (*current_cell* is *close_to_road*) **Then** (*infestation_factor* is *likely*$_2$)
  If (*current_cell* is *close_to_soiltype*) **Then** (*infestation_factor* is *likely*$_3$)
end (3)

Hence, the thematic forecast phase has the function,

$$\mathbf{f}_{\mathbf{theme}}(T_1\!: \tau_1, \tau_2, \tau_3) = (T_1\!: \tau_{1f}, \tau_{2f}, \tau_{3f}) \tag{4}$$

Where, τ_{1f}, τ_{2f} and τ_{3f} are the resulting themes after $\mathbf{f}_{\mathbf{layer}}$ function has been performed on all cells at each layer of the respective themes.

Note that the predicate I_n (ie. *likely*$_n$) of the linguistic variable, *infestation_factor* from (2) and (3) are not identical in each theme. Since each theme has only one membership function assigned, linguistic predicates such as *low*, *medium* or *high* do not apply. Instead I_n is used to indicate the fuzzy sets of different membership functions for each of the level. For example, I_1, indicates the *infestation_factor* for τ_1 and I_2, indicates the *infestation_factor* for τ_2.

3.3 Phase 3: Global (Large-Scale) Forecast

In this phase, all the layers in the previous thematic forecast, τ_{1f}, τ_{2f}, τ_{3f}, is mapped onto the known infestation theme, T_1, to form a composite, T_2, to result in the forecast of infestation sites on a global scale. It must be pointed out that the term global does not refer to a worldwide scale, but rather to the whole given area represented on a spatial image (eg. state, city, county). Based on parthenium weed dispersal and life-cycle parameters, the length of time between T_1 and T_2 is approximately 28 days. Inserting the themes T_{1f}, τ_{1f}, τ_{2f} and τ_{3f} from (1) into the following global forecast function we have,

$$\mathbf{f}_{\mathbf{global}}(T_1\text{: } \tau_{1f}, \tau_{2f}, \tau_{3f}, \tau_{null}) = T_2 \tag{5}$$

Therefore, by recursively iterating (5), we can obtain a forecast in multiples of approximately 28 days. The actual implementation for this process would be the aggregating of all the consequent infestation factor, I_n, of each corresponding cell on each layer to form a new membership function. Hence, this process would result in a DOF (degree of fulfilment). The DOF is a straightforward implementation of a number of standard defuzzification methods (Berkan and Trubatch, 1997, p70-73). The DOF for every cell is mapped onto every cell in T_1, to result in T_2. Thus, for every cell in τ_{1f}, τ_{2f}, and τ_{3f}, we have an implication that maps onto a corresponding cell in T_1.

$$\tau_{1f}(C_{m,\,1f}) \;\Phi\; \tau_{2f}(C_{m,\,2f}) \;\Phi\; \tau_{3f}(C_{m,\,3f}) \rightarrow T_1(C_m) \tag{6}$$

where Φ is an aggregation operator.

Recall from (1) we have a special theme τ_{null}, that requires special treatment. As this theme represents null-zones with cells of null values, the standard aggregation process for τ_{null} cannot be carried out. Instead, a cell from τ_{null} that has a null value shall nullify all previously aggregated values from τ_{1f}, τ_{2f}, τ_{3f} for each corresponding cell. Thus, from (6) if the aggregation process encounters a null-zone, then the resulting forecast for that cell in T_1 is also null,

$$\tau_{1f}(C_{m,\,1f}) \;\Phi\; \tau_{2f}(C_{m,\,2f}) \;\Phi\; \tau_{3f}(C_{m,\,3f}) \;\Phi\; \tau_{null}(\text{null}) \rightarrow T_1(\text{null}) \tag{7}$$

However, not all regions in a τ_{null} theme is a null-zone. In the case where the aggregation process encounters a cell that is not a null-zone in τ_{null}, the cell in question is simply ignored and the process reverts to (6).

Note that null-zones in parthenium weed prediction is of great significance in best management practices, as by law, it is required that 'quarantine' areas such as washdown locations are made available in heavily infested sites to allow vehicles or persons to be sanitised to prevent spread of seeds to un-infested areas (Parthenium Action Group, 2000). Hence, the emphasis placed on this issue.

3.4 Phase 4: Local Forecast

In a global forecast, all factors influencing parthenium weed growth can mostly be categorised as *static themes*. Static themes normally do not and are not expected to change in the time frame of the forecasting process. However, the results from such a prediction are more useful on a global scale than for localised condition. As most spatial images only indicate infested sites on larger scales, it does not cover sites that may be considered as insignificant in the original data collection process. Landowners on these properties can normally give useful input (ie. anecdotal references, local knowledge) to the forecasting process that are not available elsewhere. Coincidentally, most of these inputs coincide with factors termed as *dynamic themes*. These factors can be conveniently layered into individual themes as in Phase 1.

In Figure 1., the themes are,

$$\tau_4 : \text{local weather}$$
$$\tau_5 : \text{local flood areas} \quad (8)$$

In this case, to obtain a local forecast of parthenium infestation, these dynamic factors are mapped against the original global forecast obtained from (5). Hence, the function is,

$$\mathbf{f}_{\mathbf{local}}(\mathbf{f}_{\mathbf{global}}(T_1 \colon \tau_{1f}, \tau_{2f}, \tau_{3f}, \tau_{null}), \tau_4, \tau_5) = T_3 \quad (9)$$

Observe from (9), that since anecdotal references are the source of data and where the membership values are arbitrarily adjusted, the dynamic themes need not undergo the thematic forecasting phase as their static theme counterparts, therefore is represented by τ_4, τ_5 without the subscript *f*. T_3 is the resulting local forecast. Since the size of the spatial image that T_3 represents is less than or equal to T_2 (ie. $T_3 \leq T_2$), there can be more than one single instance of T_3. That is, as many T_3 as needed can be generated, each independently from other occurrences of T_3. Each instance of T_3 can be generated from additional sets of dynamic themes other than the ones shown in (8). Note that dynamic themes can also contain previously unchartered null-zones.

4. Meta Consequent Function

In the case of local forecasting that has different consequence depending on different localised condition, an operator has been defined specifically as part of P-Expert's meta consequent function (Chiou, Yu and Lowry, 2000). The operator *Case-Of*, is a passive operator. That is, it does not operate directly on the operands, but rather perform as a navigational function. In this way, in the instance of a one-to-many mapping consequent, the multi consequent function in P-Expert will allow branching to different membership function within the same fuzzy variable. This will allow the same variable to have ownership over different sub-domains. In practice, this would have the form,

(**If** *x* is *a* **Then Case-Of** C_m)
{**Case-Of** C_{m1}:y_1 is b_1;

.

.

Case-Of C_{mn}:y_n is b_n } (10)

where, C_m is a cell from a dynamic theme. For example,

(**If** *weather* is *wet* **Then Case-Of** *location*)
{ **Case-Of** *Rockhampton:* *infestation_factor* is $likely_1$;
Case-Of *Mackay:* *infestation_factor* is $likely_2$;
Case-Of *Gladstone:* *infestation_factor* is $likely_3$ }

5. Example

5.1 Progressive Linear Dispersal: A Pedagogical Example

The basis of the range of generic weed dispersal parameters is based on Cousens and Mortimer (1995) dispersal vector, where dispersal is likely related to plant height, distances are given as multiple of height (H). Even though Cousens and Mortimer stated that this dispersal vector data was based on very limited available data, conforming it to fuzzy membership in this project was uncomplicated. There are many other factors for parthenium weed dispersal, such as, wind direction, season and soil type. In this example, these factors are omitted as not to detract readers from the underlaying methodology. However, even with some of these factors omitted, it yields a totally acceptable prediction of weed infestation as we shall see. This is to the credit of using fuzzy logic that inherently retain its robustness towards incomplete or limited knowledge of other influencing factors.

Using known fact that a mature parthenium weed has the potential of growing up to an average height of 1 metre, we have H=1. As parthenium weed is categorised as the 'shakers' type, where propagule dispersal is by vigorous shaking of the branches before the seedlings fall off the parent plant and is carried further by the wind factor, we shall use the wind-assisted/'shaker' factor in the dispersal vector (Table 1).

From the table, we can determine that the maximum distance (d_{max}) parthenium weed can disperse is over 5 metres (with wind assistance) from the parent plant. And approximate 50-percent (d_{50}) of the seedlings can be found dispersed unaided within 2 to 3 metres of the parent plant. In this example, only wind assisted dispersal factors for the 'shakers' type is developed into fuzzy membership functions to simplify explanation.

Hence, the antecedent (LHS) linguistic variable, *area*, is concerned with how *far*, *midpoint* or *near* the proximity of an infested cell. The consequent (RHS) is the *infestation factor* with the predicates *low*, *medium* or *hight*.

The If-Then fuzzy rule block is,

If *area* is *near* **Then** *infestation factor* is *high*
If *area* is *midpoint* **Then** *infestation factor* is *medium*
If *area* is *far* **Then** *infestation factor* is *low*

Table 1. Cousens and Mortimer (1995) weed dispersal vector. The distance of dispersal is a multiple of the plant height (H). The maximum dispersal distance is d_{max} and d_{50} is the dispersal distance of where 50-percent of propagules are found.

Vector	d_{max}	d_{50}
Unaided	1H	0.5H
Explosive	3H	05-1.0H
Wind:		
'Gliders/parachuters'	100 m	2H
'Shakers'	5H	2H
'Tumblers'	Several km	?
Water	>100 m	?
Animals:		
Farm & birds	>100 m	>100 m
Ants	20 m (?)	5 m (?)
Cultivation	2-5 m	0.5 m
Harvesting	20-100 m	5-50 m

Hence, the fuzzy sets for LHS and RHS are in Figure 2.

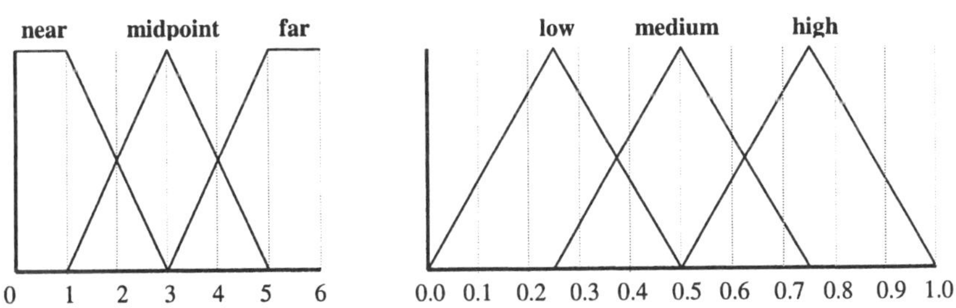

Fig 2. The membership for the LHS predicates is the distance of predicted area from a known infestation and the RHS memberships are the level of infestation factor.

In this purely pedagogical example, if weed dispersal is linear (ie. propagule spreads toward a single direction with wind assistance in that same direction), it is possible to obtain a prediction showing a progressive level of weed infestation. As parthenium weed requires approximately 28 days from seed to full maturity capable of flowering, prediction for this particular weed's dispersal dynamics is set at a cycle of every 28 days, $Cycle_{28.}$

Therefore, to predict the uni-direction of weed dispersal for three parthenium weed cycle can be implemented in the following program block,

```
For Cycle28 = 1 To 3
   Predict_Dispersal(Cycle28)
Next Cycle28
```

Figure 3 shows this progression using a vector of linear cells. Each cell represents a $1m^2$ area and the level of infestation through time is uni-directional (towards North). The originating cell at $Cycle_{28}$ *0* has an infestation point of 60-percent saturation. After a cycle of 28 days, the predicted dispersal is Northwards covering five more cells (ie. within 5 to 6 metres of the original infested site). However, the saturation of each cell gets lighter as it departs further from the original cell.

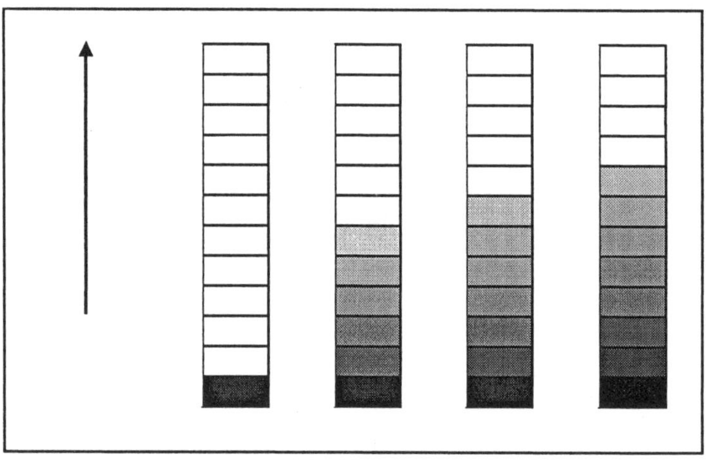

Fig 3. Progressive linear dispersal in a uni-direction showing the original infestation through a period of three cycles. Heavy infestation is shaded with a darker colour band (in this chapter, colour bands have been converted to a grey-scale). In practice, infestation of 10-precent is normally ignored.

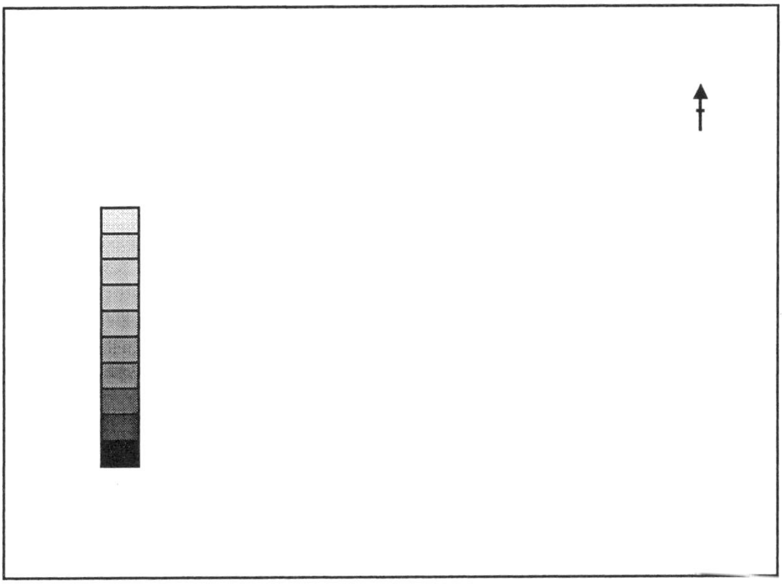

Fig 4. The original known parthenium weed infestation as at $Cycle_{28}$ 0 recorded on a 50 by 50 grid. Each cell is an area of $1m^2$. For readability, all other features (eg. contours, landmarks) of the area have been removed.

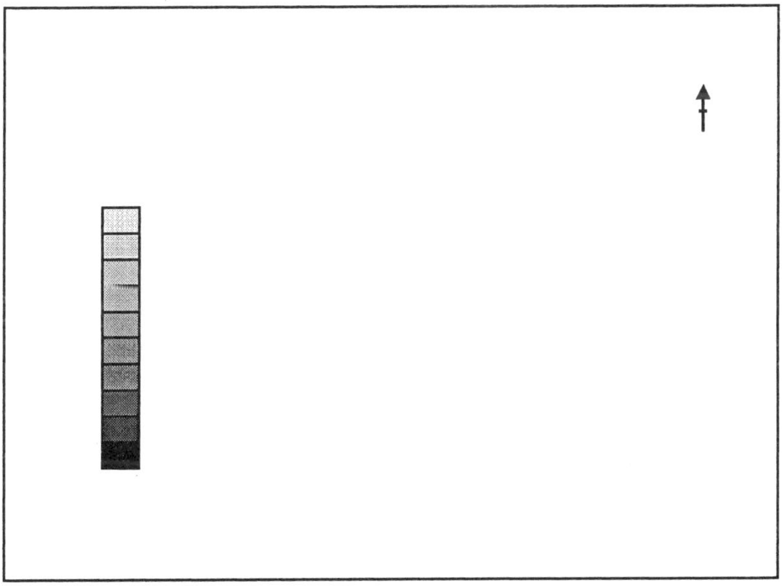

Fig 5. Actual known infestation of the same 50 by 50 grid local area after six cycles (approximately six months).

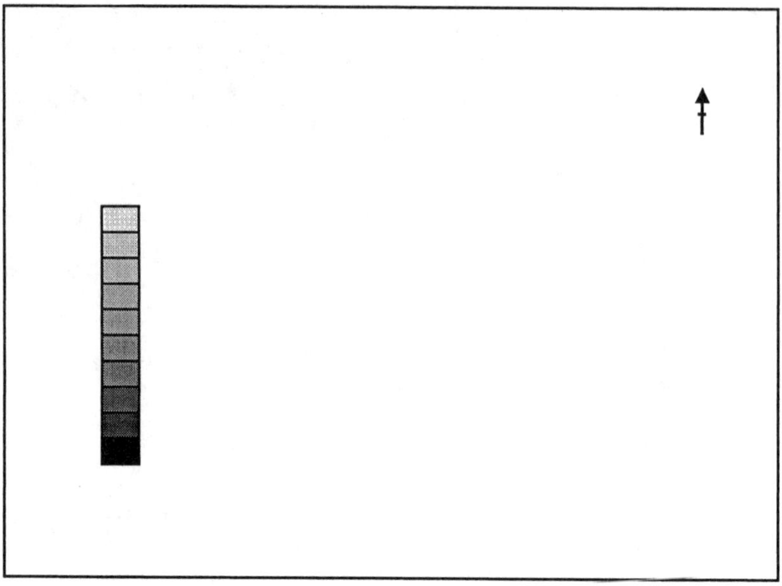

Fig 6. This is the predicted dispersal of parthenium weed based on the original known infestation from Figure 4 using fuzzy algorithm.

Fig 7. The prediction of parthenium weed dispersal from Figure 6 re-plotted in three-dimensional image in *P-Expert* to help identify high infestation areas.

As the prediction is repeated for two more cycles (*Cycle*$_{28}$ *2* to *Cycle*$_{28}$ *3*), the dispersal gets further away from the original infested site. Note that the level of saturation does not refer to the height or maturity of the plant, but refers to the approximate number of physical weed found within each 1m^2 area. Also, an area that has been infested can only disperse to un-infested areas, it cannot reinfest itself. In this case, a previously infested area increases its saturation by 10-percent for every cycle. Therefore, a previously infested area can increase the number of plants in its current area through several cycles until it attains 100-percent saturation. That is, the number of plants cannot increase anymore within that given area. In this case, all subsequent seedlings borne after the 100-percent saturation point has been attained will not germinate due to competition for resources within the same infested area. In this case, the parthenium weed completely dominates the pasture, resulting in a monoculture where no other plants can survive within this area.

5.2 Scale-Colour Bands

In the *P-Expert* expert system, the level of predicted infestation for each cell is indicated by different scale-colour bands (ie. dark colour = heavy infestation, light colour = light infestation). These colours are not arbitrarily assigned, but is part of the requirements for GIS spatial image representation using scale-colour bands as meaningful legends. In practice, it is extremely difficult for users to read actual printed numbers on spatial images. Hence, users have to rely on scale-colour for efficient visual cues. Even though it is common for current hardware display units to be capable of displaying and outputting spatial images with greater than 1,024 scale-colours, this level of accuracy is not fully utilised here as the required scale-colour goes up to a maximum 100 (ie. 100-percent saturation).

6. Case Study

The parthenium weed infestation prediction methodology presented in this paper is part of a three-year long on-going project, P-Expert Project, jointly funded by Australian Research Council (ARC), Queensland Department of Natural Resources (QDNR) and Central Queensland University (CQU). Printed spatial images (ie. maps) of known location of parthenium weed infestation were provided by QDNR for the years, 1981, 1985, 1989, 1992 and 2000 for the state of Queensland. A specific region of known parthenium weed location, covering approximately 300,000km^2 of Central Queensland has been selected for the case study. The themes and features of the selected region were entered into a specialised GIS editor specially developed for this project with provision to include fuzzy If-Then rules and associated variables and membership functions. The membership functions were adjusted and fine-tuned from one known period to another known period until predicted infested sites corresponded within the known sites. In this way, the adjusted membership functions can be applied to future predictions.

7. Results

A 50 by 50 metre (shown as 50 by 50 grid on the spatial image) local area was selected from the property of a volunteer stakeholder. This area has not been significantly infested by parthenium weed except for several isolated known patches. Throughout six parthenium weed cycles (approximately 6 months), records and manual field survey was completed by the stakeholder. For each of the cycle, the known infestation was drafted onto a grid map of that selected area.

Based on the first original grid map at $Cycle_{28}$ 0, the fuzzy prediction algorithm performed a predicted dispersal of parthenium weed of the selected area for six cycles. As it is impractical to compare the actual physical number of weeds in this sort of survey, comparison of results is performed solely on spatial images with scale-colour banding. In this chapter, the colour bands has been reproduced as a grey-scale. Figure 4 shows the original area, with known isolated patches. Figure 5 and 6 respectively shows the previously known and predicted dispersal for the parthenium weed after six cycles. As can be seen from the two images, the predicted dispersal closely matches the known infestation. In actual practice, the level of accuracy shown by the predicted dispersal in Figure 6 is totally adequate and acceptable to help decide the next course of action to carry out.
In addition, P-Expert helps user to expediently visualise the predicted percentage saturation of the area by providing real-time three-dimensional imaging (Figure 7). The level of dispersal is represented by the height of the charted area. The height of the charts do not represent the height of the actual plants. In this way, the three-dimensional image help users quickly and efficiently identify potential high infestation areas.

8. Conclusion

Parthenium weed infestation in the Central Queensland region poses a serious threat to the environment and also economic viability of areas occupied by agricultural industries. Government agencies have taken steps to control and manage existing infestation and also to curb future spread of the noxious weed. One of the tools used in these strategies is the prediction of parthenium weed population. With the fore knowledge of where and what location the next outbreak would occur, preventive steps can be taken to hinder the weed dispersal from actually taking place. Conventional weed forecasting methods utilises discrete values in exponential models and linear algorithms extensively. Attempts at predicting weed dispersal relied heavily on accuracy in the original charts or images to yield reasonable results. However, since early data were poorly collected, the results of weed population forecasting are only as reliable as the data originally provided. In this case, fuzzy logic is best suited to this type of challenge as its original requisite is in handling approximate data.

This chapter shows, how by using GIS spatial image datasets categorised into themes, a fuzzy logic based forecasting methodology can be performed. This methodology would allow us to: 1) predict weed population reasonably well based on approximate data, 2) take into consideration additional parameters without re-writing the algorithm, 3) refine large-scale forecasts to suit localised situations, 4) allow user to determine and inspect individual infestation factors, and 5) adapt the methodology to other weed or plant species.

References

Adamson, D. C. (1996). *Determining the Economic Impact of Parthenium on the Australian Beef Industry: A Comparison of Static and Dynamic Approaches. MSc Thesis.* Brisbane, University of Queensland.

Auld, B. A. and Coote, B. G. (1980). "A Model of a Spreading Plant Population." *OIKOS* 34: 287-92.

Auld, B. A. and Coote, B. G. (1990). "Invade: Towards the Simulation of Plant Spread." *Agricultural Ecosystem & Environment* 30: 121-128.

Auld, B. A., Hosking, J. and McFadyen, R. E. (1982). "Analysis of the Spread of Tiger Pear and Parthenium Weed in Australia." *Australian Weeds* 2: 56-60.

Auld, B. A., Menz, K. M. and Monaghan, N. M. (1978). "Dynamics of Weed Spread: Implications for Policies of Public Control." *Protection Ecology* 1: 141-148.

Berkan, R. C. and Trubatch, S. L. (1997). *Fuzzy Systems Design Principles: Building Fuzzy IF-THEN Rule Bases*, IEEE Press.

Chiou, A., Yu, X. and Adsett, K. (2002). P-Expert Project: Development of a Prototype Expert System for Control and Management of Parthenium in Central Queensland. *Transforming Regional Economies and Communities with Information Technology.* Green Greenwood Publishing. in press.

Chiou, A. and Yu, X. (2001). Prediction of Parthenium Weed Infestation Using Fuzzy Logic Applied to Geographic Information System (GIS) Spatial Image. *Proceedings of 10th IEEE Inernational Conference on Fuzzy Systems 2001,* Vol. 3. University of Melbourne.

Chiou, A., Yu, X. and Adsett, K. (2001). Parthenium Weed Problem in Regional Queensland: The Role of an Expert Advisory System. *Proceedings of the Third International Information Technology in Regional Areas Conference.* Central Queensland University, CQU Press.

Chiou, A., Yu, X. and Lowry, J. (2000) P-Expert: A Prototype Expert Advisory System in the Management and Control of Parthenium Weed in Central

Queensland. *Second International Discourse With Fuzzy Logic In The New Millennium*. Physica-Verlag.

Chippendale, J. F. and Panetta, F. D. (1994). "The Cost of Parthenium Weed to the Queensland Cattle Industry." *Plant Protection* 9: 73-76.

Cousens, R. (1985). "A Simple Model Relating Yield Loss to Weed Density." *Annals of Applied Biology* 107: 239-252.

Cousens, R. and Mortimer, M. (1995). *Dynamics of Weed Population*. Cambridge, Cambridge University Press.

Daehler, C. (1998). Comparison of Region-Specific Models for Predicting Plant Invaders: Prospects for a General Model. *VIIth International Congress of Ecology Proceedings*: 103.

Demers, M. N. (2000). *Fundamentals of Geographic Information Systems*. New York, John Wiley & Sons.

Dragicevic, S. and Marceau, D. J. (1999). "Spatio-Temporal Interpolation and Fuzzy Logic for GIS Simulation of Rural-to-Urban Transition." *Cartography and Geographic Information Science* 26(2): 125-137.

Ghersa, C. M. and Roush, M. L. (1993). "Searching for Solutions to Weed Problems." *BioScience* 43(2): 104-109.

Mortimer, A. M., McMahon, D. J., Manlove, R. J. and Putwain, P. D. (1980). The Prediction of Weed Infestations and Cost of Differing Control Strategies. *Proceedings 1980 British Crop Protection Conference - Weeds*. Brighton, England. 2: 415-422.

Norris, R. F. (1999). "Ecological Implications of Using Thresholds for Weed Management." *Journal of Crop Production* 2(1): 31-58.

Parthenium Action Group (2000). *Parthenium Weed: Best Management Practice*. Brisbane, DPI Publications.

Rejmanek, M. (2000). "Invasive Plants: Approaches and Predictions." *Austral Ecology* 25(5): 497-506.

Rozenfelds, A. C. F. and Mackenzie, R. (1999). The Weed Invasion in Tasmania in the 1870s: Knowing the Past to Predict the Future. *12th Australian Weed Conference Papers and Proceedings*: 581-583.

Saint-Joan, D. and Desachy, J. (1995). A Fuzzy Expert System for Geographical Problems: an Agricultural Application. *Proceedings of IEEE International Conference on Fuzzy Systems*. 2: 469-476.

Tomlin, C. D. (1990). *Geographic Information Systems and Cartographic Modeling*. Englewood Cliffs, NJ, Prentice-Hall Inc.

Druck: Strauss Offsetdruck, Mörlenbach
Verarbeitung: Schäffer, Grünstadt